Sonar to Quartz Clock

Technology and Physics in War, Academy and Industry

SHAUL KATZIR

Tel Aviv University

OXFORD
UNIVERSITY PRESS

OXFORD
UNIVERSITY PRESS

Great Clarendon Street, Oxford, OX2 6DP,
United Kingdom

Oxford University Press is a department of the University of Oxford.
It furthers the University's objective of excellence in research, scholarship,
and education by publishing worldwide. Oxford is a registered trade mark of
Oxford University Press in the UK and in certain other countries

Published in the United States of America by Oxford University Press
198 Madison Avenue, New York, NY 10016, United States of America

British Library Cataloguing in Publication Data

Data available

Library of Congress Control Number: 2023935109

ISBN 978-0-19-887873-5

DOI: 10.1093/oso/9780198878735.001.0001

Printed and bound by
CPI Group (UK) Ltd, Croydon, CR0 4YY

Cover image: Marison frequency standard clock, 1928,
courtesy of AT&T Archives and History Center,
Walter Cady, Methods of Maintaining Electric Currents,
and Cady NB 25, p. 48, courtesy of Walter Guyton Cady Papers,
Archives Center, National Museum of American History, Smithsonian Institution

CONTENTS

ACKNOWLEDGEMENTS

The origins of this book go back to 2004, when I embarked on an archival study of Walter Cady's research on piezoelectricity. Since then the history of the technological application of piezoelectricity has been one of my central research projects. I could not undertake such a study without the support that I have received during such a long period: from the Alexander von Humboldt Foundation for the research taking place in Department I (structural changes in systems of knowledge) of the Max Plank Institute for the History of Science in Berlin (2008–10), where I stayed for an additional year supported by the Fritz Haber Institute of the Max Plank Society; and from the Gerda Henkel Foundation with the European Research Council (M4HUMAN Programme 16/EU/11) as a Marie Curie senior research fellow at the Cohn Institute for the History and Philosophy of Science and Ideas at Tel Aviv University (2012–14), where I consequently became a faculty member. The Humboldt Foundation and the Max Planck Institute continued to assist me with summer visits in the years that followed. I am grateful for the stimulating and warm environment provided by the staff and students at the Cohn and the Max Planck Institutes, which have been my intellectual home. In writing the book I was furthered helped by a visit to the Deutsches Museum, Munich and a personal grant of the Israeli Science Foundation (1255/18).

During the many years of working on this project, I have stayed at a few intuitions and presented parts of my research at many formal and informal occasions, enjoying comments and suggestions from more people than I can now remember; many of their ideas and criticism found their place in the work. I am grateful to all who commented on some parts of the work. In particular, I would like to thank Jaume Navarro, Jed Buchwald, Kijan Espahangizi, Jeremiah James, Ed Jurkowitz, Maria Rentetzi, and Ido Yavetz.

Collecting the sources for this multi-national history required the assistance of many institutions and individuals. Shigehisa Hirose, Kenji Ito, Jan Kotůlek, James Nye, and Chris S. McGahey were generous enough to provide me with archival sources and to translate texts from Japanese and Czech. Massimiliano Badino, Oksana Kuruts, Alexei Kojevnikov, and my late colleague Giuseppe Castagnetti helped me understand Italian texts and translated and summarized ones from Russian. It is a pleasure to thank them again. I would like express my appreciation of many archives' staff: Alison Oswald and her colleagues at the Archives Center of the National Museum of American History; Catherine Kounelis at La Centre de resources historiques de l'École Supérieure de Physique et de Chimie Industrielles de la Ville de Paris; George Kupczack of the AT&T Archives and History Center, Warren, NJ; and the staff of the National Archives of the UK and US (College Park) for their help, and of the Department of Manuscripts & University Archives, University Library, Cambridge.

Parts of this book have been published previously as articles, which are partly reproduced here in a modified form with the permission of their publishers: 'Who Knew Piezoelectricity? Rutherford and Langevin on Submarine Detection and the Invention of Sonar', *Notes and Records of the Royal Society*, 66 (2012): 141–57 (in Ch. 2); 'From Ultrasonic to Frequency Standards: Walter Cady's Discovery of the Sharp Resonance of Crystals', *Archive for History of Exact Sciences*, 62 (2008): 469–87, and 'War and Peacetime Research in the Road to Crystal Frequency Control', *Technology and Culture*, 51 (2010): 99-125 (together in Chs. 3 and 4); 'Frequency and Time Standards from Acoustic to Radio: The First Electronic Clock', in Lara Huber and Oliver Schlaudt (eds), *Standardization in Measurement: Philosophical, Historical and Sociological Issues* (London: Pickering and Chatto, 2015), 111–24 (in Ch. 5); 'Variations and Combinations: Invention and Development of Quartz Clock Technologies at AT&T', *ICON,* 22 (2016): 78–114 (in Ch. 6); 'Pursuing Frequency Standards and Control: The Invention of Quartz Clock Technologies', *Annals of Science,* 73 (2016): 1–39 (in Ch. 7); 'The Shaping of Interwar Physics by Technology: The Case of Piezoelectricity', *Science in Context,* 31 (2018): 321–50 (in Ch. 9). I am indebted to the editors and referees of these papers and of this book for their thoughtful suggestions.

LIST OF ABBREVIATIONS

Journals and publications

AIEE Transactions	Transactions of the American Institute of Electrical Engineers
Comptes rendus	Comptes rendus hebdomadaires des séances de l'Académie des sciences *(France)*
IRE Proceedings	Proceedings of the Institute of Radio Engineers
NPL report [for particular year]	Annual Report of the National Physical Laboratory for the Year . . . (Teddington: National Physical Laboratory, published in the year following the report)

Archives

ACNMAH	Walter Guyton Cady Papers, 1903–1974, Archives Center, National Museum of American History
ADM	Records of Admiralty, Naval Forces, Royal Marines, Coastguard, and related bodies at the National Archives of United Kingdom (UKNA)
AIP	Niels Bohr Library, American Institute of Physics, Cady's dossier
AT&T Archives	AT&T Archives and History Center, Warren, NJ
Cady, Diaries	Walter Cady's diaries kept by the Rhode Island Historical Society MSS 326—Cady Family Papers
Cady's papers	see ACNMAH
CUL	Department of Manuscripts & University Archives, University Library, Cambridge, (papers of Ernest Rutherford, Joseph J. Thomson and John William Strutt, Lord Rayleigh)
L (with numbers)	papers of Paul Langevin at La Centre de resources historiques de l'École Supérieure de Physique et de Chimie Industrielles de la Ville de Paris. (ESPCI)
USNA	National Archives (United States of America), Records of the national academy of science, London, Paris and Washington offices.

Others

AT&T	American Telephone and Telegraph Company
BoS	Bureau of Standards
EPCI	École municipale de physique et de chimie industrielles de la ville de Paris
GE	General Electric
GR	General Radio
LFTI	Leningrad Physico-Technical Institute
NB	Notebook
NG	Notgemeinschaft der Deutschen Wissenschaft
NPL	National Physical Laboratory
NRC	National Research Council
PTR	Physikalisch-Technische Reichsanstalt
RCA	Radio Cooperation of America
RLC circuit	resistor, inductor, and capacitor
UCL	University College London

INTRODUCTION

During the First World War and in its aftermath scientists, engineers, and inventors employed piezoelectricity for an innovative system of submarine detection, known today as sonar, for a method for determining frequencies of electromagnetic oscillations, and for the quartz clock, to mention only the most important inventions. Sonar became the basis for all ultrasound scanners. Crystal frequency control was, and still is, crucial for electromagnetic telecommunication, and is ubiquitous in the contemporary world, among others, as the basis for quartz clocks and watches, which alone sell in billions annually. Military and government organizations and commercial companies recognized the importance and potential of these methods, from detecting enemy submarines to lowering costs of telephone calls, through improving commercial and military telecommunications; thus, they invested in the field's research and development.

Yet, before the First World War none of these institutions showed any interest in piezoelectricity. The phenomenon, which is the generation of electric charge by changes in pressure and the converse generation of mechanical strain by electric tension in asymmetric crystals, had not been employed beyond the scientific laboratory. Its study and usage were confined to physics. The practical application of piezoelectricity, thus, required transforming knowledge from physics to technology. The current study, however, defies any purported linear model, according to which knowledge flows unidirectionally from basic ('pure') science to applied science and onto technology.[1] It is rather a story of bidirectional transmission and transformation of knowledge between scientific research and technical research and development, whose findings and devices fed back into physics. The technical employment of piezoelectricity shaped the field not only through technological research and development but also by directing the questions investigated within the scientific study. Industrial companies and government and military organizations not only adopted and applied knowledge arriving from physics, but also

---

[1] Benoit Godin, 'The Linear Model of Innovation: The Historical Construction of an Analytical Framework', *Science, Technology & Human Values* 31 (2006): 639–67.

Sonar to Quartz Clock. Shaul Katzir, Oxford University Press. © Oxford University Press (2023).
DOI: 10.1093/oso/9780198878735.003.0001

expanded knowledge of the phenomenon and steered (directly and indirectly) academic research into topics of their interest.

This book suggests an integrative history of science and technology that explores how researchers applied scientific knowledge and expertise to invent and improve practical devices and methods, how they studied the new technics, ways to improve them, and connected phenomena, and how the scientific field was consequently changed. It thus analyses different kinds of research on the spectrum between 'pure' research conducted without any connection to potential applications and the study of a particular device or method with the hopes of improving it. These studies were carried out in various settings, from universities through national research institutes and temporary groups formed to develop war technics, to military and industrial laboratories. Looking at the various research practices and settings, without assuming a fixed correspondence between the kind of research and the location of the researcher, this examination offers a non-dichotomous, nuanced, and historically based analysis of the important and much-discussed relationships between science and technology. Assuming a continuity between different research practices, I do not neglect differences between science and technology. This integrative examination fits the close and complex connections between physics and technology in the research of piezoelectricity and in the modern world in general. It is through such close studies of concrete historical cases that we can better understand the dynamics and complexity of the mutual development of technology and science.

Although they mutually influence each other, I regard physics, technology, and society as separate entities that sometimes overlap. As historical entities they evolved through a specific process whose borders and properties continued to blur and change also during the period discussed here while also maintaining distinct cores. For example, Ernest Rutherford's study into the structure of the atom was clearly in physics, while his work on designing underwater microphones during World War II was clearly a kind of engineering (Ch. 2). The differences, I claim, are real, i.e. in historical events, whereas the distinctions separating the entities are analytical. As such the distinctions are not necessary, but they are useful for representing and understanding the historical events, i.e. the research and choices taken in the different fields. As with the dividing lines between colours, different observers may disagree where exactly to draw the lines, but we should agree that blue is distinct from green.

At the beginning of the twentieth century piezoelectricity was studied within physics, which was a well-established discipline, with a strong tradition, self-identity, and quite clear borders. In science sociologists Anne Marcovich and Terry Shinn's terms, research in physics followed a disciplinary regime:

> The disciplinary regime is strongly defined by its self-referencing orienta-
> tion. As regards research topics, they are drawn from within the discipline,
> and relate both to disciplinary history and inertia and to where disci-
> plinary practitioners perceive the future of their discipline to lie. The
> discipline similarly sets its internal criteria for the evaluation of its research
> results. . . . The disciplinary regime itself constitutes its own market. Prac-
> titioners are the consumers of their own productions. Research output is
> directed to peer disciplinary colleagues.[2]

Scientists choose research topics that extend active empirical and theoretical
realms of research, explore related phenomena and variations on fruitful experiments (as judged within the discipline), raise the precision or the range
of known empirical data, elucidate apparent anomalies and discoveries, and
empirically test theories and conjectures.

Most of the research examined in this book, however, addresses problems
set by external groups due to their deemed usefulness, rather than questions
set by the discipline itself. A central goal for these studies, set by various forces
and agencies in society, was to create and improve practical devices and methods. Some of these studies engaged directly with the development of practical
instruments, from sonar to quartz clocks; others rather extended knowledge
about natural phenomena deemed relevant for improving such instruments.
To assist in the analysis of such activities I distinguish between 'technics' (in
plural)—as the tools and methods of their use (including 'know-how')—and
'technology'—propositional knowledge concerning technics. The sources of
technology are wide ranging: from systematic data collected about the performance of the devices and methods used in practice, through their study
and that of possible alternatives, to research into natural phenomena related
to the technics used and scientific knowledge about them. In the twentieth
century, technology was often very close to and interacted with the natural
sciences. The distinction between technics and technology is similar to that
which exists in continental European languages, for example, between *Technik*
and *Technologie* in German, which virtually disappeared in English, although
dictionaries still mention 'technics'.[3]

---

[2] Anne Marcovich and Terry Shinn, 'Regimes of Science Production and Diffusion: Towards
a Transverse Organization of Knowledge', *Scientiae Studia* 10 (2012): 38–9.

[3] On the terms in English and other European languages see Eric Schatzberg, *Technology:
Critical History of a Concept* (Chicago: University of Chicago Press, 2018), especially 7–9. The
German term was also a source for Lewis Mumford's use of 'technics' in his famous *Technics
and Civilization* of 1934. Similar to its use here, he preferred 'technics' as a more concrete term
than 'technology'. Although it helped Mumford to stress the crucial role of tools already in

Technology, like science, includes theory and empirical findings. Research engineers, who carry out much of its development, conduct their own original research.[4] Despite some differences in attitude, scientists and engineers use mainly the same research methods. The main difference between their endeavours, claim historians of technology Edwin Layton and Walter Vincenti, is found in their goals. Technological (or engineering) research aims at providing useful information for practical technical design.[5] Its goals are, thus, set by considerations external to the structure of knowledge of the field. For example, during the First World War, researchers studied piezoelectric vibrators, due to their possible use for sonar rather than because of internal developments in the field (Ch. 3). The distinction by goals rather than by methods fits quite well Marcovich and Shinn's analysis of research regimes. Disciplinary logic is characterized by topics and questions of research rather than by ways of studying them. The study of piezoelectric vibrators followed 'utilitarian' logic of research, aimed at solving particular problems set by

use before the Industrial Revolution, he applied the term also to the 'modern technics' of his time, as done here. Mumford, however, did not employ the term to distinguish between systematic knowledge and tools as I suggest. Lewis Mumford, *Technics and Civilization* (London: Routledge, 1946 [1934]); Rosalind H. Williams, 'Lewis Mumford's Technics and Civilization', *Technology and Culture* 43 (2002): 139–49. Donald Cardwell advances a similar distinction, as he employs 'technics' for tools and methods not based on systematic knowledge, and 'technology' for those which are. Yet, for Cardwell technics consists of tools and methods before the development of systematic knowledge about them, while I define technics as tools and methods regardless of the existence of the systematic knowledge. Donald Cardwell, *Wheels, Clocks, and Rockets: A History of Technology*, rprt edn (New York: W. W. Norton, 2001), 4–6.

[4]In discussing engineering in the context of research and development and in comparing it to science, the literature virtually always refers to the small minority of engineers working at institutes of higher education and research laboratories rather than to the majority of shop-floor and maintenance engineers. I refer to the former as research engineers. They are the engineers discussed in this book. For a critique of neglecting the work of most engineers, see David Edgerton, *The Shock of the Old: Technology and Global History since 1900* (Oxford: Oxford University Press, 2011), xvii. Edgerton's emphasis on the importance of examining the uses of technics highlights a clear difference between the majority of engineers, who characteristically work to make devices and methods function properly, and scientists.

[5]There are some differences in their views, among others Vincenti points to a few characteristics of engineering thinking that differ from the common practices of scientists. These differences, however, are not important in this context. Edwin T. Layton, 'Through the Looking Glass, or News from Lake Mirror Image', *Technology and Culture* 28 (1987): 594–607; Walter G. Vincenti, *What Engineers Know and How They Know It: Analytical Studies from Aeronautical History* (Baltimore: John Hopkins University Press, 1990).

societal needs, rather than a disciplinary one.[6] The perspectives advanced by the historians of technology and the concepts suggested by the sociologists of science can help us to distinguish science from technology, pace contrary claims that they are indissociable in one 'technoscience'. Deviations of scientists from topics and questions that stem from the inner logic of the discipline indicate an effect of external influences on their research. Similarly, deviations from topics that can help technology suggest the power of disciplinary science. As will be exemplified later, in both cases one does not need to enter scientists' inner thoughts to identify such deviations.

The technical application of piezoelectricity was part of a more general trend towards tighter connections between science, particularly physics, and technology, at the beginning of the twentieth century. The view that research in science could and should foster technology became more common among both policy makers (who had earlier often doubted the practical value of research) and practicing scientists (who became more inclined to assist technology). Already before the First World War this view led governments, commercial companies, and partnerships between them and the public to establish research institutions aimed, at least partly, at fostering technics through a better interaction with science. The war strengthened that trend when scientists and their scientific expertise were mobilized to the war technical effort. By its aftermath, research in physics proved indispensable for developing technics like radio, telephony, ultrasonic detection, and even more efficient incandesces lamps, arguably an emblematic product of the professional independent inventor. Consequently, social support for scientific research connected to technology expanded: extant institutes were extended, new ones were open, and new forms of funding were set up.[7] The connections between science and technology became tighter and continued to be so ever since.

Some of the new kinds of institutions established to harness scientific research for the improvement of technics played a central role in the technical application of piezoelectricity and its study after the First World War. Countries such as Germany and the UK were the first to found institutes to facilitate an exchange between research in physics and technology in the belief that it

[6]Terry Shinn, *Research-Technology and Cultural Change: Instrumentation, Genericity, Transversality* (Oxford: The Bardwell Press, 2008), 171–82; Marcovich and Shinn, 'Regimes of Science Production'.

[7]Shaul Katzir, '"In War or in Peace": The Technological Promise of Science Following the First World War', *Centaurus* 59, no. 3 (2017): 223–37.

would foster industrial growth. The earliest was the German Physikalisch-Technische Reichsanstalt (Imperial Institute for Physics and Technology), established in 1887. The British National Physical Laboratory (NPL) and the National Bureau of Standards of the United States followed suit in 1900 and 1901.[8] Physicists working at the three institutions, like their colleagues at the newly founded Physico-Technical Institute in Leningrad (LFTI), participated in the research and development of piezoelectricity and other frequency and time standards discussed here. Researchers at industrial research laboratories also played an important part in these research and development activities. These new laboratories dedicated to research and development in industrial corporations, isolated (to various degrees) from production lines and testing, first appeared within the German chemical industry in the late nineteenth century. In 1900 and 1911, respectively, the American electric and telecommunication corporations General Electric (GE) and American Telephone and Telegraph Company (AT&T) established larger research laboratories, which studied questions related to physicists (both laboratories participated in the research discussed here, albeit to a different degree). During the 1920s extant and new laboratories grew in size, employing an increasing number of physicists, who formed a larger share of the physics community.[9] European companies beyond the chemical industry established new laboratories or expanded their small research units. Yet, historians have concentrated on the early phase of the research laboratory rather than on the interwar period,[10]

---

[8]David Cahan, *An Institute for an Empire: The Physikalisch-Technische Reichsanstalt, 1871–1918* (Cambridge, UK: Cambridge University Press, 1989); Russell Moseley, 'The Origins and Early Years of the National Physical Laboratory: A Chapter in the Pre-history of British Science Policy', *Minerva* 16 (1978): 222–50; Rexmond Canning Cochrane, *Measures for Progress: A History of the National Bureau of Standards* (New York: Arno Press, 1976).

[9]Ernst Homburg, 'The Emergence of Research Laboratories in the Dyestuffs Industry, 1870–1900', *British Journal for the History of Science* 25 (1992): 91–111; Leonard S. Reich, *The Making of American Industrial Research: Science and Business at GE and Bell, 1876–1926* (Cambridge, UK: Cambridge University Press, 1985); Lillian Hoddeson, 'The Emergence of Basic Research in the Bell Telephone System, 1876–1915', *Technology and Culture* 22 (1981): 512–44; David A. Hounshell, 'The Evolution of Industrial Research in the United States', in *Engines of Innovation: U.S. Industrial Research at the End of an Era*, ed Richard S. Rosenbloom and William J Spencer (Boston, MA: Harvard Business School Press, 1996), 13–85.

[10]A few notable exceptions include Ronald R. Kline and Thomas C. Lassman, 'Competing Research Traditions in American Industry: Uncertain Alliances between Engineering and Science at Westinghouse Electric, 1886–1935', *Enterprise and Society* 6 (2005): 601–45 and David A. Hounshell and John K. Smith, *Science and Corporate Strategy: Du Pont R&D, 1902–1980* (Cambridge, UK: Cambridge University Press, 1988), 98–326. Steven Shapin also discusses this

and rarely looked beyond the USA.[11] Here I examine both American and European laboratories. My central concern is in the kind of research done there, rather than in institutional history, or in contemporary views of the proper role of these laboratories. I look mostly at 'rank and file' researchers rather than at a few famous but exceptional industrial scientists like GE's Irving Langmuir and AT&T's Clinton Davisson, who enjoyed more freedom than their colleagues.[12] Further, I analyse their work within a larger context of similar and connected studies performed at academic and governmental institutions. The comparison shows a marked but nondeterministic effect of institutional settings on the kind of study performed.

In studying the work of these industrial researchers, as well as that of dozens of other protagonists in this history, I examine the content of their scientific and technical research and development, and their context. In particular, I look at the reasons that led the protagonists to their particular research questions, whether they chose the topic themselves or followed their supervisors' decisions. The analysis is based on their original scientific and technological writings (journal papers, patents, and reports), published and unpublished alike, and on contemporary and later reports and recollections of their works and deeds. In a few important cases (Langevin, Rutherford, Cady, and Marrison and his colleagues at AT&T) these are supplemented by their laboratory notebooks and contemporary correspondence, found in archives in

---

period but elaborates on the sociologists and managers rather than the scientists in the laboratories, in *The Scientific Life: A Moral History of a Late Modern Vocation* (Chicago: University of Chicago Press, 2008).

[11]On industrial research laboratories in Europe, see H. Schubert, 'Industrielaboratorien für Wissenschaftstransfer. Aufbau und Entwicklung der Siemensforschung bis zum Ende des Zweiten Weltkrieges anhand von Beispielen aus der Halbleiterforschung', *Centaurus* 30 (1987): 245–92; Kees Boersma, *Inventing Structures for Industrial Research: A History of the Philips National Laboratory, 1914–1946* (Amsterdam: Aksant Academic Publishers, 2002); Marc de Vries with Kees Boersma, *80 Years of Research at the Philips Natuurkundig Laboratorium (1914–1994): The Role of the Nat.Lab. At Philips* (Amsterdam: Amsterdam University Press, 2005).

[12]I look at the middle- to high-level researchers with an academic degree (almost all with postgraduate degree); most were heads of groups within their laboratories. For studies of the more famous scientists and their use by the corporations see David Philip Miller, 'The Political Economy of Discovery Stories: The Case of Dr Irving Langmuir and General Electric', *Annals of Science* 68 (2011): 27–60; Leonard S. Reich, 'Irving Langmuir and the Pursuit of Science and Technology in the Corporate Environment', *Technology and Culture* 24 (1983): 199–221; Arturo Russo, 'Fundamental Research at Bell Laboratories: The Discovery of Electron Diffraction', *Historical Studies in the Physical Sciences* 12 (1981): 117–60.

France, the UK, and the US.[13] I employed various bibliographical databases and searched within the secondary sources to gather details relevant to the protagonists' work on piezoelectricity.

While most studies of physics in the first decades of the twentieth century concentrated on quantum mechanics and relativity, these fields occupied, at most, a small part of the ongoing research at these under-examined sites. Studying a field that underwent dramatic change but was hardly impacted by these revolutions widens our picture of how physics developed in the epoch beyond quantum mechanics, the atom, and new particles.[14] That the quantum revolution was the major and most consequential process in interwar physics does not negate the significance of other developments of the time. Quantum mechanics appears only at the margins of the story told here. It helped neither the understanding of piezoelectricity nor its technical usage. This was far from unique. The new theory, which began in the study of microphysics and particular phenomena of electromagnetic radiation, remained irrelevant in the study of many physical phenomena. This did not prevent many physicists from researching these phenomena, which were explained by 'classical physics'. Research in each field had its own characteristics. Still, it seems that research in many of these less famous fields (which formed most of the research in physics) was more similar to that of piezoelectricity than to that of quantum mechanics, in the kind of experimentation, theory, and (at least in some cases) interactions with technology.[15]

Hidden within larger devices and unknown to the wider public, crystal frequency control was one of the crucial inventions in the turn to the electronic world, first in telecommunications and entertainment and later in the century in almost any sector of the economy and at home. While big cogs and wheels continued to catch the public's imagination during the interwar period, electric technics became ever more central: smaller electric machines

[13] Archives were also useful for finding contemporary reports, especially from the WWI period. See the list of archives consulted in the List of Abbreviations.

[14] This, of course, is not the first history of physics of that time period that does not deal with quantum mechanics. A few recent books that discuss other issues include Roland Wittje, *The Age of Electroacoustics: Transforming Science and Sound* (Cambridge, MA: MIT Press, 2016); Chen-Pang Yeang, *Probing the Sky with Radio Waves: From Wireless Technology to the Development of Atmospheric Science* (Chicago: University of Chicago Press, 2013); Aitor Anduaga, *Geophysics, Realism, and Industry: How Commercial Interests Shaped Geophysical Conceptions, 1900–1960* (Oxford: Oxford University Press, 2015).

[15] For examples of interactions with technology, see the papers in the issue 'Interaction of Interwar Physics: Technology, Instruments, and Other Science', *Science in Context* (2018) and further examples in the introduction to the issue.

were replacing many of the earlier mechanical ones; new electric appliances were replacing manual work; other electric gadgets were creating new demands. The radio set used for home entertainment is probably the best example of this phenomenon. It was also the first mass-produced gadget that involved electronics, i.e. components based on electrons' particular properties as the carrier of electric current. The major electronic component, before the advent of the transistor, was the thermionic tube (or valve) based on the discharge of electrons in vacuum. Crystal frequency control was developed due to the evolving need for radio communication, used the vacuum tube, and enabled the further expansion and elaboration of electronic technics in general. First implemented in large radio transmitters in 1925, crystal frequency control continues to regulate the frequencies of electromagnetic waves in almost any transmitter, including current cell-phones. A small quartz unit still controls virtually every 'smart' device. This turn to electronic devices and methods, accompanied by the process of miniaturization, was arguably one of the most important technical developments of the twentieth century, with profound consequences for the economy, society, and culture. Leading to new, more precise, cheap, and reliable technics for maintaining and distributing time and making exact synchronization necessary for telecommunication, crystal frequency control further strengthened the role of the clock in human affairs. Significantly, it spread time awareness and discipline, marked characteristics of the modern world, from the working place and public transportation into the home and leisure activities of the general population.[16]

Since crystal frequency and sonar were early technics that relied on the vacuum tube, their history provides a vantage perspective on the development of what can be termed early electronics, which still deserves further historical study.[17] This history shows that although the term came into use to denote technics around 1930, electronics had been a vibrant, creative, and expanding technical field since the First World War. Methods of submarine detection, like sonar, exemplified its usage beyond its origins in telecommunication,[18]

---

[16]On these effects see Shaul Katzir, 'Time Standards for the Twentieth Century: Telecommunication, Physics, and the Quartz Clock', *Journal of Modern History* 89 (2017): 119–50.

[17]Frederik Nebeker's attempt at a synthetic history in *Dawn of the Electronic Age: Electrical Technologies in the Shaping of the Modern World, 1914 to 1945* (Piscataway, NJ: Wiley-IEEE Press, 2009) testifies to the many gaps in the history of early electronics.

[18]The diode and the triode were invented as detectors for electromagnetic waves. The triode was further developed for the production of oscillations for radio-telephony and for amplification. AT&T's research department improved it in order to amplify transcontinental telephone

recommending the use of the general term 'electronics' rather than 'radio engineering'.[19] Still, problems in telecommunication continued to stimulate many developments in the field in the war's aftermath. In particular, this book reveals the importance of methods connected with manipulation and control of frequencies, with and without the use of piezoelectric components. How to construct a quartz clock was a problem in electronics at least as well as in piezoelectricity.

Beyond the analysis of the interactions between science, technology, and technics, this book examines a few other themes. A central theme, connected to this analysis, is the transfer and transformation of knowledge and methods among and within different scientific and technical fields and topics. I explore the conditions and resources required for a particular researcher (or a group) to adopt a piece of knowledge, experimental procedure, or method used in one context (e.g. the study of crystals) to implement it in another (e.g. locating submarines). The process is especially interesting in the more challenging cases of crossing thematic and disciplinary borders. I also examine which parts of the knowledge or methods were transferred. Sometimes one case served as an analogy for another (e.g. locating ores for detecting submarines—Ch. 2), in others a method was used almost intact, and in still others one had to modify it (e.g. adapting laboratory piezoelectric instrument for sonar). Students of science stressed the crucial role of tacit knowledge, which cannot be (or at least was not) described fully in texts and diagrams, for assimilating techniques (especially experimental ones).[20] Here I show that often researchers with technical knowledge managed to acquire the needed tacit knowledge from written reports and publications (e.g. the use of piezoelectric quartz for measurements, Ch. 2, and electronic technics for frequency-measuring technics, Ch. 7). This, however, does not negate the many advantages of direct personal contact

signals. Sungook Hong, *Wireless: From Marconi's Black Box to the Audion* (Cambridge MA: MIT Press, 2001); Hugh G. J. Aitken, *The Continuous Wave: Technology and American Radio, 1900–1932* (Princeton, NJ: Princeton University Press, 1985); Reich, *The Making of American Industrial Research*.

[19] Radio engineering is both too narrow, as it does not include such usages, and too wide, as it relates to other topics of the arts that were not electronic in character. Electronics was first used for the scientific study of electrons, also in relation to technology. During the 1920s the noun appeared in German in reference to technics as *technische Elektronik*. Its use became popular in English for technology following the establishment of a journal under that title in 1930, Charles Susskind, 'The Origin of the Term "Electronics" ', *IEEE Spectrum* 3, no. 5 (May 1966): 72–9.

[20] Harry M Collins, *Changing Order: Replication and Induction in Scientific Practice* (London: SAGE, 1985).

between the transmitter and receptor of knowledge, including through laboratory visits. The successful transformation of knowledge turns out to be based both on the body of knowledge (the possible employment or analogy between two realms) and on the social and personal conditions that allowed researchers interested in one problem (whether technical or scientific) to know the other realm well enough to appreciate its potential usage for their own problem. Here I examine both the details of the empirical and theoretical knowledge and skills and the social context of research and development. What was the role of formal institutions (designed to facilitate such transmission) in the process? And what was the role of informal connections like personal acquaintance and contingency in the process? The analysis shows that in some cases an expertise in the phenomenon and its laboratory employment was needed for its successful technical application.

The emergence and development process of knowledge, devices, and methods, which is connected to the transfer and transformation of knowledge, is another theme of this book. I examine a few cases of invention and discovery, and many cases of expanding scientific and technological knowledge, and improving technics. The book looks at the process of invention, innovation, and improvements in various settings from the work of independent inventors to that of the modern industrial laboratory and academic departments. Sonar, discussed in Chapter 2, was the joint fruit of the inventions of a semi-professional independent inventor, Constantin Chilowsky, and of an academic physicist, Paul Langevin. Temporarily mobilized for the war effort, Langevin was a kind of occasional inventor. So was his colleague Cady, who invented crystal frequency measurement and control technics (Ch. 4). Cady's invention-endeavour followed his own discovery of piezoelectric resonance at a traditional small university laboratory (Ch. 3). I explore the resources and factors that led to these inventions and discovery, understood as processes, rather than momentary achievements.

It is easier to recognize the evolutionary character of the process of invention and innovation when its results were the stated aim of a group endeavour. I discuss here two endeavours of this kind: the war development of sonar into practical technics by French, British, and American teams (Ch. 3), and the development of frequency standard systems that included clocks based on tuning-forks (Ch. 5) and then on crystal frequency control by teams at AT&T and the NPL (Ch. 6). Their accomplishments were built on contributions from many individuals working in different organizations. The creation of the quartz clock at AT&T is particularly useful for studying the process of invention and innovation within the modern science-based industrial laboratory, exploring the cognitive and experimental steps that enabled a group of researchers to collect, combine, and adjust the rich resources at their disposal for solving an important technical problem. Still other groups invented similar technics that allowed them to construct a quartz clock simultaneously. Analysis of their work reveals different styles of research and innovation (Ch. 7). Moreover, using the comparative method this case of simultaneous, multiple, or parallel invention helps the historian to identify the epistemic, material, and social contexts of the innovation and factors crucial for its accomplishment. Parallel inventions and missed parallel inventions, i.e. similar technics that failed to achieve the successful invention's practical function, recur throughout this history from early pre-sonar systems to the quartz clock. I therefore employ comparative methods when analysing the inventions of the echo-sounding method (Ch. 2), piezoelectric filter (Ch. 8), tuning-fork frequency standard system (Ch. 5), and, as mentioned, the quartz clock (Chs. 6–7), and discuss the significance of the phenomena of multiple inventions in the conclusions to Chapters 5 to 7 and in the book's general conclusions.

Parts of the history told here have been told before, particularly regarding the inventions of sonar, crystal frequency control, and the quartz clock. These accounts appeared mainly in scientific–technical reports and textbooks, semi-popular publications, participants' recollections, and companies' histories.[21] Although some of these are informative and show a sincere effort to be fair with other contributors to the field,[22] they suffer from known drawbacks of the genre: reliance on personal memory or limited sources, and an emphasis on questions of credit and priority. They do not attempt a causal or contextual historical analysis of the development of knowledge and inventions suggested here. Moreover, the narratives regarding the invention of crystal frequency control and the quartz clock suffer from a bias of exaggerating AT&T's role in their development, since most of them were written by authors working in connection with AT&T, or were based on reports written by these authors. Above

---

[21]Two major exceptions to this literature are David Zimmerman, 'Paul Langevin and the Discovery of Active Sonar or Asdic', *Northern Mariner* 12 (2002): 39–52; and Carlene E. Stephens, 'Reinventing Accuracy: The First Quartz Clock of 1927', in *Die Quarzrevolution: 75 Jahre Quarzuhr in Deutschland 1932–2007*, ed. Johannes Graf, Furtwanger Beiträge Zur Uhrengeschichte; N.F., Bd. 2 (Furtwangen: Dt. Uhrenmuseum, 2008), 12–23. I mention my additions to their works in the discussion on sonar and the quartz clock.

[22]This effort to credit other contributors is noteworthy in the account of Warren A. Marrison, 'The Evolution of the Quartz Crystal Clock', *Bell System Technical Journal* 27 (1948): 510–88. Marrison contrived the first quartz clock.

the corporation's common stakes of increasing its fame to cultivate the pride of its research staff and to support its public image as an innovative beneficial company stood two practical interests. One was its disputable claim and long litigation over the rights for the basic patent of frequency control (discussed in Ch. 4); acknowledging the originality of Cady's work would have threatened the corporation's high financial stakes. The second was AT&T's efforts, as part of its strategy to resist US federal attempts to break up its monopoly in the 1970s and 1980s, to demonstrate that it had bestowed many technical contributions to humanity. The official histories of research and development at the corporation written during that time, thus, echo earlier claims for the origins of frequency control and other technics in its laboratories.[23]

My examination of original publications and reports beyond AT&T, however, shows that many laboratories contributed to the development of piezo-electric technics. Crystal frequency control itself evidently did not originate in the giant corporation (Ch. 4). Other technics, like the quartz clock, frequency filters, and special quartz cuts, were developed both inside and outside AT&T. Transfer of knowledge and techniques between different laboratories contributed to innovation in all the participating laboratories. At stake is not merely the allocation of due credit to particular researchers, but more significantly, the understanding of the dynamics of research. Broadening the picture beyond AT&T shows that the technics originated in shared scientific and engineering research endeavours and a common reservoir of technical methods. It questions the relative significance of different kinds of laboratories in advancing technological innovation. These laboratories include those of smaller, but still large, corporations like the Radio Corporation of America (RCA) and Philips; military ones like the US Naval Research Laboratory and the Italian Naval electro-technical and radio-telegraphic laboratory; governmental ones like the NPL and the American National Bureau of Standards; those of small companies like General Radio; and academic ones like the shortwave research laboratory of MIT, and those of Cady and Pierce. Moreover, widening the horizon shows that this was not an exclusively American story, but

[23] M. D. Fagen, ed., *A History of Engineering and Science in the Bell System—The Early Years* (Murray Hill, N.J.: Bell Laboratories, 1975); S. Millman, ed., *A History of Engineering and Science in the Bell* System: *Physical* Sciences *(1925–1980)* (New York: Bell Telephone Laboratories, 1983); F. M. Smits, ed., *A History of Engineering and Science in the Bell System: Electronics Technology (1925–1975)* (Indianapolis: AT & T Bell Laboratories, 1985). Particular claims in these and other publications are discussed mainly in Ch. 4, but also in passim in Ch. 7.

a transnational, almost global, enterprise. This was not a simple case where Europeans produced basic scientific knowledge and Americans applied it,[24] but the technical application and development of piezoelectricity included European laboratories (in Britain, France, Germany, Italy, and the Netherlands) and at least also one Japanese. Further scientific and technical studies included additional places like the USSR and China. This history well exemplifies the need for transnational history of twentieth-century science and technology.

The context relevant for the development of piezoelectric technics should also be widened to a broader technical context of devices based on physical principles and effects other than piezoelectricity. At least in two such cases, piezoelectric components replaced other components within a technical system. The idea of sending and receiving ultrasonic waves to detect submarines preceded employing piezoelectricity to produce and receive the waves. Understanding the advantages and drawbacks of the various ways to produce ultrasonics, to detect submarines, is important for comprehending why and how piezoelectricity was adopted (Ch. 2). This is also true regarding earlier means of measuring and controlling frequencies of electronic systems based on a tuning-fork. The earlier tuning-fork systems provided means, methods, and forms of thinking for developing crystal frequency standards and the quartz clock, and are therefore discussed in details in Chapter 5.

****

The evolution of the research on piezoelectricity and the development of technics based on it can be divided into five phases. As a scheme it simplifies a more complex and continuous evolution, and thus can help us recognize the process's main characteristics as a whole and its different phases. Moreover, with necessary adjustments, it can be applied as a model for the evolution of science-based technics more generally. Although each technical evolution is contingent on its particular historical and material conditions, there are some structural similarities in science-based technics' development that could

---

[24]The claim that Americans had applied fundamental knowledge of European origins was suggested by a committee headed by Isiah Bowman nominated in 1944 by Vannevar Bush, who repeated the claim in his famous 1945 report *Science—The Endless Frontier* (National Science Foundation, 1990), on pp. 6, 22, and 78 (of the Bowman Committee); Daniel J. Kevles, 'The National Science Foundation and the Debate over Postwar Research Policy, 1942–1945: A Political Interpretation of Science—The Endless Frontier', *Isis* 68 (1977): 17–19.

help in understanding their histories.[25] More specifically, four of the phases in piezoelectric technics' evolution seem characteristic of that of other technics in which scientific study precedes practical applications.

The first phase is characterized by a scientific–disciplinary study (Ch. 1). Scientists identify phenomena and questions of interest in the field and seek a theoretical account, without much thought about possible application. Specific practical applications are beyond scientists' expectations, as they literally do not imagine them; e.g. sonar and frequency control were unforeseen by the students of piezoelectricity. Discovered in 1880 and receiving a general phenomenological account in the early 1890s, research on piezoelectricity after 1900 concentrated on a few questions left open by theory, on further measurements, and on its connections to current developments in physics. Within the discipline, the phenomena found a laboratory use in experimental settings and in more stable scientific instruments. This phase can be long; in the case at hand, it took about thirty-five years, but in other cases it has been much shorter.

In the second phase, discussed in Chapter 2, the phenomenon is applied to practical technics, useful beyond laboratory and field research, for the first time. In this case, Langevin exploited piezoelectricity to convert mechanical oscillations to electrical ones and vice versa for the new methods of submarine detection. The transition to this phase seems the most precarious. It depends on, in principle, the applicability of the phenomena to a particular technical problem, in this case utilizing a piezoelectric vibrator as an ultrasonic-electric transducer. Moreover, its use for that end depends, on the one hand, on a good familiarity with the scientific phenomena and, on the other, on a clear understanding of the technical problem and its definition in a way that could lead the hitherto unapplied effect to its technical solution. Acquaintance with a phenomenon that previously had no practical use is rare among inventors and engineers, except for very famous scientific phenomena, like electromagnetic waves. The transition thus usually requires a person familiar with both realms. In the case of piezoelectricity, it required the exceptional mobilization of physicists to a technical project for the sake of the war effort and the contingent expertise of Langevin in piezoelectricity. It was also contingent on the earlier development of an ultrasound echo system for locating submarines to which piezoelectricity provided a possible solution. Chapter 2 begins with the

[25] The model sketched here is valid only for technics that followed science; many technics, including some based on elaborated technology, did not. The model is also different from cases of research-technologies, in Shinn's terms, even when they found general usage, like lasers.

development of these methods, before it examines how piezoelectricity was applied to submarine detection for war research and what allowed its successful use. In other cases, a scientist may be allured by the potential utility of a studied phenomenon in order to pursue its technical development, without having been engaged in a particular technical quest previously. This phase is precarious also because identifying a problem and a potential solution does not guarantee its success in practice.

Indication that a new technic is practical leads to a research and development phase to improve the device and its methods of use, which is the model's third phase. Under the emergency of the First World War, physicists and engineers of the major Allies powers (France, Britain, Italy, and the US) engaged in improving piezoelectric sonar. They examined questions ranging from the advantages of different ways to fix crystals to metallic electrodes, to the behaviour of vibrating piezoelectric crystals under different physical conditions, through the properties of ultrasonic waves in seawater (Ch. 3). In more abstract terms, they studied both the devices and general natural phenomena related to their design. As research and development grew, this phase required more extensive resources than the initial application of the phenomena in the second phase. It, thus, relied on larger social support, provided when the technics in question were deemed of important social interest. This was clearly the case with countering the threat of U-boats during the war. In other cases, potential material gain often secured financial support for research and development. The intensive research on piezoelectric vibrators for sonar led to some unexpected findings. In particular, relieved from the war mission of improving submarine detection, in early 1919 Cady returned to open-ended scientific research without aiming to improve the technics, which allowed him to turn a few earlier observations into the unexpected discovery of an abrupt change in the piezoelectric vibrators' electric properties near resonance frequency, as discussed at the second part of Chapter 3.

The newly discovered effect quickly became a basis for new applications by Cady, first of laboratory instruments—a frequency measurement device (1919) and then a method of crystal frequency control (1921). Soon it became clear that the new method and the effect of resonance eclipsed sonar in their usefulness and significance for other technical fields and therefore for society. Consequently, the field entered a forth phase of its history, characterized by efforts to apply the newly discovered properties of crystal resonance to useful technics (Ch. 4). Familiarity with the details of piezoelectric resonance seems important also in this transition to an applied phase (as with the transition from the first to the second phase). Yet, after piezoelectricity had been used for one technics, the idea of using it for another was clearer on the horizon. Unlike Langevin, Cady did not have a particular need for frequency control in mind. His general knowledge of wireless technics and their needs sufficed to suggest the method's usefulness.[26] The value of controlling frequencies was also clear to researchers less familiar with piezoelectricity. These included (as discussed in Ch. 4) the physicist Georges Pierce, who improved on Cady's method, and researchers at AT&T's research laboratory, who claimed rights on Cady's patent, and the Naval Research Laboratory, whom Cady assisted. Another difference between this move and the move from the first to the second phase might have originated in the wider possible application of the effect to contemporary technics. Langevin's employment of crystals as transducers answered a specific need of a particular method, whereas controlling frequencies had a wider range of possible uses in the evolving electronic technics.

Discovering an unexpected effect with a major technical application in the research and development phase (third in the scheme), as was the case with piezoelectricity, is probably a rare event. Minor discoveries of limited use (like Meissner's finding that piezoelectric resonators produce a blast of air, as described in Ch. 9) are probably much more common. Thus, it seems that in other cases of science-based technics, the research and development of a particular technics (third phase) led directly and more smoothly to the last phase (fifth here) characterized by general research and development in the field also beyond any particular application.

For physicists and engineers alike, the successful employment of piezoelectricity for sonar and frequency control established its technical utility and usefulness, and recommended its expanding study and usage in the fifth phase of its evolution—the phase of a field recognized to be technically useful. The number of technical and scientific researchers in the field increased a few fold. This phase is characterized by three parallel enterprises: (a) further research and development of the technics invented in the earlier phases and new and improved ways of employing them, in this case especially of crystal frequency control; (b) development of new applications of the phenomena to other devices and methods; and (c) an extensive study of the phenomenon within physics (in comparison with its earlier study). The move between physics and

---

[26] I discuss the contribution of his exchange with members of AT&T research laboratory in Chapter 4.

engineering became easier in this phase. Above the reliance of both science and academic engineering on propositional knowledge, they now used the same object—the piezoelectric resonator, which formed a kind of boundary object in the analysis suggested by Hong.[27]

Research and development of the technics in the fifth phase was directed at improving the frequency control units in terms of stability, range of controlled frequencies, strength of signals, etc., according to the particular needs of their uses by corporations, military arms, regulatory agencies, and radio operators. A related effort was directed at employing the technics in larger technical systems. Central among these was the use of crystal frequency control to establish frequency-measuring standards. Such standards relied directly on time measurements and thus led to the development of the quartz clock. Due to the importance of the resulting technics and the considerable resources invested in this research by a few groups, I expound and analyse the developments of these systems in Chapters 5 to 7, from their origins in the tuning-fork system to the quartz-based technics. This history is examined here as a prime example of the development of knowledge-based technics by physics- and engineering-trained researchers at industrial, national, and academic laboratories. It also makes it possible to compare a parallel process of inventions at different settings, discussed at the conclusions of this part.[28]

Following the early application of piezoelectricity, scientists, engineers, and inventors suggested quite a few novel usages of the phenomenon, forming the second enterprise in the fifth phase. The commercial suggestions were based on the properties of the resonators, like frequency filters, which found a considerable market. Yet, many measuring devices were based on knowledge extant before 1915. Apparently, the major factor that led to the new suggestions after the war was that researchers were better acquainted with the phenomenon and to its possible usefulness. These new technics required their own research and development efforts. Yet, to keep the main thread of the narrative, I discuss them only briefly in Chapter 8.

The technical use of and interest in piezoelectricity stimulated also research of piezoelectricity within physics. The annual number of papers regarded

[27]Sungook Hong, 'Historiographical Layers in the Relationship between Science and Technology', *History and Technology* 15 (1999): 289–311.

[28]Of course, research and development do not end with the construction of the first working devices; this is clearly true for precision instruments and measuring standards like frequency control units and the quartz clock. Yet, for practical reasons it remains beyond the scope of this study.

as contributions to its physical study multiplied a dozen times from the 'pre-applied' phase. Above the general expansion of physics, its technical application transformed the research of piezoelectricity within the discipline. At one level, technological research revealed unknown effects that posed challenges to the physical theory of the phenomenon and required further experimental elucidation from a disciplinary perspective. At a subtler level, the interests of users and developers of technics directed a large share of the research in the field to topics that would plausibly help technology, i.e. knowledge concerning technics. I show that the interests of users of technics, 'technological interests' channelled physicists to study questions relevant to improving technics of interests to influential users, such as state agencies and corporations. These physicists often chose problems that had a low priority according to disciplinary logic but were important for technical applications. This hidden process often evaded contemporaries and has not received due recognition in the historiography. It seems to be central to the study of a field that evolved into an applied field of technical importance like piezoelectricity. I therefore closely analyse the process in Chapter 9. I compare the effect of its technological interests with other factors fitting the disciplinary logic. These included the effect of the inner disciplinary employment of piezoelectricity in experiments, especially in ultrasonic transducers, and the employment of new experimental technics, like X-ray diffraction to study piezoelectricity.

1

PRELUDE: PIEZOELECTRICITY BEFORE ITS
TECHNICAL USE

In 1905 Woldemar Voigt, the 'dean of piezo-electricians', described crystal physics, the field that included piezoelectricity, as 'old-fashioned physics in the stronger sense; its laws are hardly poor starting points for technical application, and only the quest for scientific knowledge drives and guides it.' Moreover, he observed that the field had been 'far from the problems that have occupied the larger number of physicists'.[1] Indeed, piezoelectricity in particular, and the physics of crystals in general, attracted little attention during that period. Discovered in 1880 by Jacques and Pierre Curie, by the mid-1890s piezoelectricity had acquired an accepted body of experimental facts regarded as its core phenomena and a phenomenological (i.e. descriptive) theory that accounted for most of them, which was formulated by Voigt and elaborated for the converse effect by his former student Friedrich Pockels. Elsewhere, I examine in detail the discovery and study of the phenomenon in the disciplinary phase of research.[2] Here I will point out the main questions that occupied researchers at this phase, present the state of research on the phenomena at the later part of the phase, i.e. in the decade before the First World War, and look at the few laboratory applications of the effect. As such, this chapter is a prelude, providing background and a basis for comparison for examining the later technical application of piezoelectricity and its subsequent extended disciplinary research in the following chapters.

A mechanistic molecular hypothesis led the Curies to their discovery that pressing asymmetric crystals produces electric polarization on their surfaces. They assumed that crystals are composed of discrete electrically polarized molecules, and that therefore a change of stress would vary the distances

---

[1] Woldemar Voigt, 'Rede', in *Die physikalischen Institute der Universität Göttingen* (Leipzig and Berlin, 1905), 39.

[2] Shaul Katzir, *The Beginnings of Piezoelectricity: A Study in Mundane Physics* (Dordrecht: Springer, 2006).

Sonar to Quartz Clock. Shaul Katzir, Oxford University Press. © Oxford University Press (2023).
DOI: 10.1093/oso/9780198878735.003.0002

between them and generate an electric effect. Following their discovery, they examined the appearance of the effect in different crystal species, its directions, and magnitude, and determined the coefficients of what they found as a linear effect in quartz and tourmaline. Pursuing a theoretical inference from Gabriel Lippmann, they found the converse effect in which electric field induces mechanical stress. Other researchers soon joined the Curies, studying the relation of the phenomenon to optical effects in crystals (August Kundt and Wilhelm Röntgen) and the geometry of the effect, i.e. its dependence on various directions of strain especially in quartz. A few preliminary theoretical accounts were suggested. Yet in 1889 Röntgen found that torsion induces electric effect in quartz in a way that could not be squared with the Curies' simple polar molecules and the structure of quartz, or with any other theory hitherto suggested. A year later, that finding provided Voigt with a good reason to suggest a phenomenological theory, which avoided any assumption about the mechanism that produced the effect.

Instead, Voigt based his theory on the assumptions that the effect is linear and that it obeys the principle of symmetry. A strain, he suggested, induces electric polarization linearly proportional to its magnitude in each differential volume of the piezoelectric material. A year later, Pockels assumed analogously the linearity of the effect of electric voltage on strain. Following his earlier work on crystal physics and the approach of Franz Neumann, his teacher, and his school, Voigt asserted that 'the symmetry of the crystal's structure [is] always lower or equal but never higher than the symmetry of the physical behavior'.[3] In other words an effect can appear only when the physical action (strain or polarization) and structure produce some kind of asymmetry. Thus, Voigt deduced from the known macroscopic symmetry of the crystal classes which strains (including shear strains or torsion) induce electric polarization in which directions in each crystal and vice versa. It showed in which directions and crystals there is no effect at all, and some relations between effects in different directions (usually their equality). It did not reveal anything, however, about the absolute strength of the effect. Over the next four years, Voigt and a few colleagues elaborated the theory and formulated it into general thermodynamic formalism. This formalism helps to account for so-called secondary effects, such as the outcomes of the electric field generated by strain in the 'primary' effect and to show relations between piezoelectricity and other characteristics of the crystals such as their dielectric, elastic, and optical

---

[3] Woldemar Voigt, 'Allgemeine Theorie der piëzo—und pyroelectrischen Erscheinungen an Krystallen', *Göttingen Abhandlungen* 36 (1890): 8.

properties. The theory accounted for the observations known at the time, including those like Röntgen's that eluded earlier molecular explanations. The phenomenological approach continued to characterize most theories in elasticity and piezoelectricity until the 1990s.[4] Only a few experiments in the early 1890s were designed to examine the predictions of the theory and some of its consequences. Two experiments confirmed the relations between the effect in different directions predicted by the theory. They also rechecked, inter alia, the linearity of the effect, needed in determining piezoelectric coefficients in a few species. Voigt's theory was not only accepted but also defined the main phenomena relevant for the study of piezoelectricity.[5]

The phenomenological theory left a few open questions. Some of them followed the theory and its assumptions. For example, the theory and the earlier molecular assumptions predicted that due to thermal expansion a change in temperature would produce a voltage difference in crystals of axial asymmetry, a phenomenon known as pyroelectricity. Although piezoelectricity suggested a mechanism for the production of pyroelectricity, which had been known for more than a century, it was unclear whether there is an additional 'real' pyroelectricity produced directly by thermal motion beyond the electric polarization due to thermal expansion—a question that could be answered only experimentally. From the theory, it followed that inner strain in crystals of axial asymmetry (which are also pyroelectric) produces a continuous internal electric polarization, which is unobserved due to compensating surface electricity. Whether that was really the case was also an experimental question. Experiments on both cases, however, did not allow compelling results. A third open issue was whether the effect is indeed linear under extreme conditions of high pressure.[6] A connected question, which would be raised only following the WWI's application of the effect, was the influence of high-frequency oscillations on the effect. This was a classic question of disciplinary science: how changes in physical conditions influence a known effect.

An open question left unanswered by the phenomenological theory was the cause of piezoelectricity and its connection to the assumed atomistic–molecular structure of the crystals. A few causal theories that attempted to yield the equations of the phenomenological theory on a micro-physical

---

[4] An interest in 'first-principles-derived approaches to investigate piezoelectricity' grew during the 1990s: Laurent Bellaiche, 'Piezoelectricity of Ferroelectric Perovskites from First Principles', *Current Opinion in Solid State and Materials Science* 6 (2002): 19–25.

[5] Katzir, *B/eginnings of Piezoelectricity*.

[6] Ibid., 205–10, 222.

mechanism were suggested in the 1890s and the early 1900s. Although some were able to regain the equations, these theories suffered from a speculative basis, sometimes with assumptions or consequences that contradicted evolving views about crystal structure. They also introduced *ad hoc* assumptions tailored to yield the sought-after results. They further failed to suggest predictions beyond those offered by the phenomenological account, e.g. on the magnitude of the effect. Therefore, none of them were regarded as more than a tentative attempt to envisage a possible mechanism.[7]

In the decade before WWI, piezo- and pyroelectricity were studied almost exclusively by Voigt and his students and collaborators in Göttingen. Wilhelm Röntgen, who studied the field in the 1880s, returned to it in 1913 with a junior collaborator at his laboratory in Munich.[8] All in all, the field attracted less than twenty publications. Pyroelectricity received a central place among them. Experiments by Röntgen and Kamerlingh Onnes and Anna Beckman showed its possible appearance in quartz, a crystal of central symmetry, in apparent contradiction with Voigt's theory. Voigt tried to explain the conflicting results. Other questions were the existence of 'real' direct pyroelectricity, a subject of some controversy between Munich and Göttingen, and variations in the effects at low temperatures. A few researchers examined piezoelectricity in specific organic crystals (tartaric acid and sucrose) and under particular strains (torsion, non-uniform deformation). In these studies, the researchers examined details unaccounted for by the theory and some questions about its domain of validity as suggested by the disciplinary logic. The few researchers that did not belong to either Voigt's or Röntgen's school reached the subject by following the relevance of their own research interests to piezoelectricity. Kamerlingh Onnes and Anna Beckman included the phenomenon in their general study of changes in physical properties at low temperature, a field which Onnes dominated. As described below, Erwin Schrödinger and Max Born integrated piezoelectricity into their theories of dielectrics and crystalline

---

[7] Ibid., Chap. 3.

[8] From a list of eighteen publications published between 1905 and 1915, Voigt and his collaborator authored thirteen. Röntgen and his students contributed another three (one by a student in 1906). Six of the students and junior collaborators were foreigners and two probably Germans. Among them Jofé and Fréedricksz would continue to study piezoelectricity in the USSR, and are discussed in Chapter 9. The list is drawn from the volumes of *Science Abstracts*, the Web of Science database, and a few additions are from contemporary journals (see further information on these sources in Ch. 9, fn 1).

structure, which followed new ideas and empirical findings about the atomistic constitute of matter.

In 1912 Schrödinger advanced a new theory of solid dielectrics, which among other phenomena explained piezoelectricity. His main goal, however, was the explanation of solids. In this work, he extended to solids the assumption that dielectrics are made of polarized molecules, which Peter Debye had advanced a few months earlier for liquids to explain discrepancies in their electric behaviour. He further employed an analogy between dielectrics and magnetism, which allowed him to rely on Pierre Weiss's recent theory. Schrödinger assumed that these polarized molecules form 'elementary crystals', which although microscopic were large enough for statistical description. Due to their structure, the elementary crystals are also inherently polarized in all crystals. Their electric forces explain solidification. These elementary crystals are all piezoelectric since their polarization depends on the distance between their molecules. Macroscopic symmetric crystals, however, consist of a few elementary crystals whose effects neutralize each other. In their microform, however, they all possess permanent electric polarization, similar to that assumed by the theory of pyro- and piezoelectricity only for pyroelectric crystals. On these assumptions, Schrödinger managed to regain the results of Voigt's phenomenological theory (actually by employing symmetry considerations himself) and to estimate for the first time the order of magnitude of the piezoelectric coefficients. The theory's accomplishments, however, were marred by its speculative assumptions, and even more so by the finding of X-ray diffraction beginning in 1913, which discredited the hypothesis of non-deformable molecules while supporting the view that crystals are formed by individual atoms moveable in relation to each other, rather than by rigid (or semi-rigid) molecules.[9]

The view of crystals as an atomic lattice, i.e. a repeated orderly arrangement of atoms (or ions) rather than molecules that form the lattice points, was the departure point of Born's 1915 theory of crystals. Born had adopted the atomistic lattice view (rather than the molecular) three years earlier in a joint work with Theodore von Kármán on the specific heat of solids. There they employed the assumption of atomistic lattice to elaborate on Albert Einstein's

[9]Erwin Schrödinger, 'Studien über Kinetik der Dielektrika, den Schmelzpunkt, Pyro— und Piezoelektrizität', *Sitzungsberichte der kaiserlichen Akademie der Wissenschaften in Wien. Mathematisch-naturwissenschaftliche Klasse (IIa)* 121 (1912): 1937–72; Christian Joas and Shaul Katzir, 'Analogy, Extension, and Novelty: Young Schrödinger on Electric Phenomena in Solids', *Studies in History and Philosophy of Modern Physics* 42 (2011): 43–53.

1907 use of the quantum assumption to explain the specific heat of solids at low temperatures. In the following years, Born articulated a general theory of crystals to account for their main phenomena based on a few common hypotheses about lattice dynamics, without assuming special hypotheses about particular effects. His theory assumed the existence of central forces between moveable ions and electrons and atomic lattice whose structure was known from the particular crystal class's macroscopic structure. The quantum condition of the atoms' vibrations appeared only when it became relevant, i.e. in the account of specific heat. Like Schrödinger, Born included piezoelectricity in his theory of solids. Yet, the effect did not play a similar pivotal role in his theory, but was discussed as a characteristic crystalline phenomenon, well known to Born, a former student of Voigt. As its author declared, this was the first molecular theory that succeeded in accounting for piezoelectricity without forming any special assumptions to account for the effect, like that of permanent polar molecules. Yet, similar to previous (and later) microscopic theories, it employed macroscopic notions like crystal symmetry and assumption about linearity, which it shared with Voigt's theory. These notions allowed Born to recover Voigt's equations, but in 1915 Born did not reach conclusions beyond those given by the phenomenological theory. He continued to develop his matter theory, taking into account theoretical and experimental advancements in atomic theory. In a 1920 joint publication with his assistant Elisabeth Bormann, they managed to surpass the phenomenological theory by deducing the order of magnitude of the (one) piezoelectric coefficient of zinc sulfide, a relatively simple piezoelectric crystal. Born's main interest remained the atomistic-dynamic theory of crystals, rather than piezoelectricity.[10]

In Schrödinger's and Born's accounts, piezoelectricity was a means or a consequence of a general theory of solids. It did not motivate their research. Unlike them, Sergei Boguslawski, articulated a new theory of pyroelectricity to account for unexplained experimental results. Yet, he also reached the subject from a study of current theoretical developments in dielectrics. A native Russian, Boguslawski (also spelled Boguslavsky and Boguslavskii)

[10]Max Born, *Dynamik der Kristallgitter* (Leipzig und Berlin: B. G. Teubner, 1915); Max Born and Elisabeth Bormann, 'Zur Gittertheorie der Zinkblende', *Annalen der Physik* 62 (1920): 218–46; Nancy Greenspan, *The End of the Certain World: The Life and Science of Max Born: The Nobel Physicist Who Ignited the Quantum Revolution* (New York: Basic Books, 2005), 55–6; Jagdish Mehra and Helmut Rechenberg, *The Historical Development of Quantum Theory. Vol. 1, Part 1* (New York: Springer-Verlag, 1982), 300–2; Walter G. Cady, *Piezoelectricity: An Introduction to the Theory and Applications of Electromechanical Phenomena in Crystals* (New York: McGraw-Hill, 1946), 742.

studied in Freiburg and Göttingen, where he submitted a dissertation on crystal optics under Voigt's supervision in 1914. In the same year, his interest in crystal physics led him to examine possible implications of a theory of liquid dielectrics for pyroelectricity. Debye's theory of dielectrics that inspired Schrödinger was also the starting point for Boguslawski. Debye's hypothesis of polar molecules accounted for an observed increase in the dielectric constant at low temperatures in contradiction to its dependence only on the density as stated by the Lorenz–Lorentz law that followed common electron theory. Boguslawski expanded the theory to explain newer findings on liquid alcohols that contradicted the Lorenz–Lorentz law also at higher temperatures, unaccounted for by Debye. He assumed that the forces between the molecules depend not only on the first order of distance (elastic effect) but also on the second and the third. This assumption allowed him to move from liquids to solids and to consider the case of pyroelectric crystals, that is, asymmetric crystals for which he could assume that the coefficient of the forces of second order is not null. Boguslawski could show that his theory leads to the known linear relation between their electric polarity and the temperature. It also supports the assumption that these crystals possess permanent inner electric polarity masked by surface charges.[11]

Recent experiments on tourmaline by another student of Voigt's, Walter Ackermann, however, showed that the linear relation between electric polarity and temperature does not hold at low temperatures. Boguslawski, therefore, contrived a new hypothesis that would account for this discrepancy. He kept the assumption that forces depending on the second order of distance between particles are important for pyroelectricity. Yet, when discussing solids rather than liquids, he referred to atoms rather than molecules. The crystals were not electrically neutral due to a difference between the motions of ions and of electrons. He further added a quantum condition on the change in the magnitude of work (product of the change of place by change of momentum of the particle) as an integer multiplication of Planck's constant h. Using statistical reasoning he found the average displacement of atoms as a function of temperature (due to asymmetry, it is not zero) from which he deduced the

---

[11]Sergei Boguslawski, 'Zur Theorie Der Dielektrika. Temperaturabhängigkeit Der Dielektrizitätskontante. Pyroelektrizität', *Physikalische Zeitschrift* 15 (March 1914): 283–8; Karl Hall, 'The Schooling of Lev Landau: The European Context of Postrevolutionary Soviet Theoretical Physics', *Osiris* 23 (2008): 245–6. Joas and Katzir, 'Analogy, Extension, and Novelty', 45–6. Helge Kragh, 'The Lorenz-Lorentz Formula: Origin and Early History', *Substantia* 2, no. 2 (Sep 2018): 7–18.

polarization and thus the pyroelectric coefficient. In a first publication, he showed a general agreement with Ackermann's results. In a second communication, he provided a more thorough discussion of the limitations of his equations, both at very low and at high temperatures. He further indicated an analogy between pyroelectricity and specific heat, showing, among other things, that both can provide some means to calculate the frequency of the atoms' vibrations in the solid, that they yield similar values, and that theories of both phenomena reached their limit of validity at similar temperatures, which can be inferred from the theory itself. He, thus, succeeded to account for new findings and to connect his theory to the central development of quantum physics. This, however, was an isolated achievement that did not lead to further research on the relevance of quantum assumptions to pyro- and piezoelectricity.[12] The study of the phenomena in the fifth phase, discussed in Chapter 9. hardly included atomic considerations and no quantum rules.

Employing piezoelectricity in measuring instruments was another source of some interest in the phenomenon. With three instruments invented between 1881 and 1915, piezoelectricity was far from a major source for scientific instruments. Yet, one of these instruments would play a role in the application of the phenomenon to a mass technics. The Curie brothers devised the first laboratory instrument based on the effect—an electricity meter, already in 1881 a year after they had discovered piezoelectricity. They continued to develop it with the help of an instrument maker to measure various magnitudes including pressure., and it was in use probably only within their scientific circle.[13] Due to its role in the research on submarine detection I will return to the device and describe it in Chapter 2. In 1897, another student of piezoelectricity, Friedrich Pockels, suggested a different electrometer based on a related electro-optics phenomenon, the changing of the optical coefficient of double refraction by electric tension. Reasoning that contracting quartz by converse piezoelectricity would change its optical properties, Kundt and Röntgen had predicted and then observed an electrical effect on the secondary refractive index, leading to a change in birefringence in quartz. In the early 1890s Pockels developed a theory of the effect and carried out more exact measurements of its magnitude in quartz, concluding that it is probably due to a combination of

[12]Sergei Boguslawski, 'Pyroelektrizität auf Grund der Quantentheorie', *Physikalische Zeitschrift* 15 (June 1914): 569–72; Sergei Boguslawski, 'Zu Herrn W. Ackermanns Messungen der Temperaturabhängigkeit der Pyroelektrischen Erregung', *Physikalische Zeitschrift* 15 (September 1914): 805–10.

[13]Katzir, *Beginnings of Piezoelectricity*, 23.

secondary effect through piezoelectricity and a direct effect of the electric field. Since the change in the double refraction was small, Pockels contrived its use for measuring high electric tensions between 4,800 and 24,600 volts. The electrometer was composed of a quartz plate put between two charged electrodes and a common 'Babinet (Soleil) compensator'. The electric tension between the electrodes was measured by a light beam that propagated through the plate to the compensator, which made it possible to calculate changes in the crystal plate's optical index by observing optical path differences. Apparently, the electrometer was, at best, sparsely used.[14]

In 1915, just at the end of the period discussed in this prelude, Boris Galitzin employed piezoelectricity in a device for measuring accelerations. Galitzin, a Russian geophysicist, was interested in measuring ground acceleration mainly for seismic studies and also for guiding building construction rules. The device that he contrived was based on a pendulum whose motion made it possible to determine the acceleration. To know the pendulum's motion, however, Galitzin needed a means to measure the pressure that it exerts. He judged that piezoelectricity was the only phenomenon that made it possible to measure pressure according to the requirements of his devise, i.e. that it 'is not subject to any appreciable displacement and does not introduce a new period of oscillation'. A quartz or tourmaline plate inserted between two metallic electrodes connected to an electrometer satisfied his needs in this case. Apparently, Galitzin was the first researcher not involved in the study of piezoelectricity to suggest its application. This was probably due to two factors: (a) he defined clearly the requirements sought from a component in his invention, and piezoelectricity well answered them; (b) he was familiar with the effect. As a student in Strasbourg in the late 1880s, Galitzim should have known Kundt, its leading physics professor, and was probably exposed to piezoelectricity, which Kundt had studied a few years earlier.[15] As I show in Chapter 2, a personal exposure to piezoelectricity and well-defined requirements were also important for Langevin and Rutherford, who applied piezoelectricity for military-aimed

[14] F. Pockels, 'Ueber ein optisches Elektrometer für hohe Spannungen', in *Verhandlungen der Gesellschaft deutscher Naturforscher und Ärzte, 69 Versammlung 20–25 September 1897* (Leipzig: Vogel, 1898), 56–7; Katzir, *Beginnings of Piezoelectricity*, 44–8, 151–6, 200–5.

[15] B. Galitzin, 'An Apparatus for the Direct Determination of Accelerations', *Proceedings of the Royal Society of London A: Mathematical, Physical and Engineering Sciences* 95 (1919): 492–507, quote on 496; this is a translation of a paper published in Russian in 1915. Galitzin studied in Strasbourg from 1887, when Kundt was still there. He thanked Kundt's student Otto Wiener in his 1890 dissertation: B. Galitzine, 'Ueber das Dalton'sche Gesetz', *Annalen der Physik* 41 (1890): 588–626.

technics about the time Galitzim presented his device. Still up to 1915, piezo-electricity found only limited applications in scientific instruments and did not interest professional inventors and engineers. The few researchers who examined it were either connected to its earlier researchers and followed the dynamics of the small field or reached it due to experimental and theoretical interactions with other developments in physics. Its technical applications to devices of societal interests would change these characteristics.

2

KNOWLEDGE, EXPERTISE, AND TECHNICAL AIMS IN THE CREATION OF THE PIEZOELECTRIC SONAR

Submarine boats and mines posed a new threat to warships and marine supply routes in the World War I. During the war, the submarine warfare that
began as a menace to traditional naval vessels turned into a strategic concern for the Allied powers as it threated to counterweigh their control of
the seas. Regarded as strategically crucial yet technically demanding, submarine detection gained a top priority in military research. Among others, it
became a central topic for new researchers—academic physicists and electrical engineers—mobilized to the war research. Their contribution was crucial
for the development of underwater detection and for the first practical employment of piezoelectricity in the most important of these—ultrasonic echo
detection technics, known today as sonar. Thus, the mobilization of scientists led to the transition from the first to the second phase in the evolution
of piezoelectricity research, i.e. from the disciplinary, pre-applied phase to the
applied phase. As will be shown throughout this chapter, technical success followed the employment of knowledge and expertise acquired in the academic
research. Notwithstanding their contribution, academic researchers did not
work alone but worked together with military, and independently employed
and industry-employed inventors and engineers. It was a science-educated
independent inventor, Constantin Chilowsky, who initiated the method of ultrasonic echo detection. This chapter follows the invention of sonar from its
origins in non-piezoelectric echo-searching methods, through the understanding of the role of ultrasonics to the successful employment of piezoelectricity,
tracing the origins of the methods and intellectual and material resources
needed for its development in France, Britain, and Germany and the technical problems that drove it. Thereby it examines the roles of scientific and
technical knowledge and experience and of novel research in the invention of
sonar.[1]

[1] The chapter discusses the developments of other methods only to the extent that they per-
tain to the work on sonar and piezoelectricity. On the research on submarine detection in

Sonar to Quartz Clock. Shaul Katzir, Oxford University Press. © Oxford University Press (2023).
DOI: 10.1093/oso/9780198878735.003.0003

Sonar was one among the most novel technics developed by scientists during World War I, along with the infamous poisonous gas with the means to mitigate its effect, sound-ranging technology for locating enemy artillery, and passive means of detecting submarine objects (explained in Section 2.3). Engineers and inventors contrived other important military innovations not based on science, such as the tank. Earlier innovations, like the production of chemical substitutes and improvements in wireless communication technics, were just as important for the fate of the war as the new ones.[2] Scientific knowledge and expertise contributed to most of these innovations, yet, apparently, sonar was unique in relying on hitherto unapplied scientific phenomenon. In this sense, it was more innovative than most war-related innovations which originated in pre-war technics.[3]

The turn from unpractical to applicable phenomenon was an unexpected and complex process, and is the subject of this chapter. Piezoelectricity itself was not considered when the idea of detecting underwater obstacles by producing sound and receiving the reflected echoes was first suggested in 1912, nor two years later when it was suggested for war needs. A few inventors suggested the principle of echo detection independently. Yet, this was not a case of multiple inventions but of a *missed* multiple, since the technics suggested could not lead to the detection of U-boats. Only Chilowsky realized that ultrasonic waves are crucial for the echo method to succeed. Nevertheless, examining

general in Britain see Willem D. Hackmann, *Seek and Strike: Sonar, Anti-Submarine Warfare and the Royal Navy, 1914–54* (London: Her Majesty's Stationery office, 1984).

[2]Guy Hartcup, *The War of Invention: Scientific Developments, 1914–18* (London: Brassey's Defence Publishers, 1988) is still the most comprehensive summary of war-related science and engineering technics (including the tank). A more recent discussion of the research carried out by a few individuals is William van der Kloot, *Great Scientists Wage the Great War* (Oxford: Fonthill, 2014). On specific technics and scientific fields see David Aubin and Catherine Goldstein, eds., *The War of Guns and Mathematics: Mathematical Practices and Communities in France and Its Western Allies around World War I* (Providence, RI: American Mathematical Society, 2014); Jeffrey Allan Johnson, 'Chemical Warfare in the Great War', *Minerva* 40 (2002): 93–106; Arne Schirrmacher, 'Die Physik im Großen Krieg: Warum wissen wir so wenig über den Einfluss des Ersten Weltkriegs auf die Forschung, technische Anwendungen und Karrieren in der Physik?', *Physik Journal* 13, no. 7 (2014): 43–48; Arne Schirrmacher, 'Sounds and Repercussions of War: Mobilization, Invention and Conversion of First World War Science in Britain, France and Germany', *History and Technology* 32 (2016): 269–92; William Van der Kloot, 'Lawrence Bragg's Role in the Development of Sound-Ranging in World War I', *Notes and Records of the Royal Society* 59 (2005): 273–84; Roland Wittje, *The Age of Electroacoustics: Transforming Science and Sound* (Cambridge, MA: MIT Press, 2016), 67–114.

[3]Hartcup, *War of Invention*, 189.

parallel efforts in underwater detection will help us understand Chilowsky's contribution and its origins. This chapter's first section will look at his early description of his invention and contemporary patents, in comparison to those of other inventors. The comparison highlights the significance of Chilowsky's insight, and directs my historical analysis to its origins, found in his former experience as a student and inventor.[4]

Chilowsky, however, provided only a general scheme. This chapter's second section turns to its early development into a working device by a French group headed by Paul Langevin. To this end the group examined a few different technics for emitting and receiving ultrasonic waves. The second section traces the development of these promising methods. Although the technics were good enough to be shared with the British, in early 1917 Langevin began to consider and later adopted another detector based on piezoelectricity in quartz, resulting in an improved method for submarine ultrasonic echo detection, i.e. sonar. Interestingly, Langevin was one of two physicists who independently began using piezoelectricity for submarine detection, but with different results. In Britain, Ernest Rutherford utilized the phenomenon to examine the sensitivity of underwater sonic detectors. Following local lore, however, Rutherford's biographers have claimed that he was 'at least the co-inventor of sonar',[5] an assertion repeated in more general histories.[6] This, however, was not the case. Using archival sources including secret reports, laboratory notes, and letters, I show that Rutherford did not invent sonar. My conclusion concurs with the 1926 judgment of the 'Royal Commission on Awards to Inventors' settling a financial dispute between Langevin and the British Admiralty, to which I was unaware when first publishing my findings.[7] The rumour that Rutherford

[4]The origins of Chilowsky's ideas have not been analysed by historians. The physicist and engineer Frederick Hunt provided interesting insights into the process of the invention, relying on Chilowsky's own recollections and reports, which appeared also in his semi-popular Russian biography.Frederick V. Hunt, *Electroacoustics: The Analysis of Transduction, and Its Historical Background* (Cambridge MA: Harvard University Press, 1954); Igor I. Klyukin and E. N. Šoškov, *Konstantin Vasil'evič Šilovskij: 1880–1958* (Lenigrad: Nauka, 1984).

[5]David Wilson, *Rutherford, Simple Genius* (Cambridge, MA: MIT Press, 1983), 373–5); John Campbell, *Rutherford: Scientist Supreme* (Christchurch: AAS Publications, 1999), 371.

[6]John L Heilbron, *Ernest Rutherford and the Explosion of Atoms*, Oxford Portraits in Science (Oxford: Oxford University Press, 2003), 94; Helge Kragh, *Quantum Generations: A History of Physics in the Twentieth Century* (Princeton: Princeton University Press, 1999), 134.

[7]On the 1920s legal procedure see David Zimmerman, '"A More Creditable Way": The Discovery of Active Sonar, the Langevin–Chilowsky Patent Dispute and the Royal Commission on Awards to Inventors', *War in History* 25 (2018): 48–68. For my earlier claim see Shaul

invented sonar might have originated in or was spread by these efforts of the Admiralty to protect its financial stakes in sonar.

At stake is not merely a question of credit or priorities but the understanding of the process of invention. For this end the historian should understand what the invention was, who made it, and how. Examining the details of researchers' work exhibits the gap between suggestions and working prototypes, and what was needed to bridge it. Thus, the employment of piezoelectricity by Rutherford and Langevin and the reasons for the divergence in its use for the shared general aim of underwater detection are examined in the third and the fourth sections of this chapter. In particular, the sections look at Rutherford's and Langevin's specific technical aims, the role of their prior knowledge and experience with the phenomenon, and how they acquired it, and analyse how these manifested in their practice. The sections examine to what extent researchers could construct devices from written descriptions and the advantage of laboratory experience. The archival sources allow for a closer reconstruction of Rutherford's than of Langevin's research, especially regarding their early use of piezoelectricity. Consequently, I reconstruct the work of Rutherford and his collaborator Robert W. Boyle in more detail in order to learn about Langevin's use of piezoelectricity.[8] The similarities between the two cases help suggest what were the crucial ingredients that enabled Langevin's invention of sonar and turned piezoelectricity into an applicable field.[9]

Katzir, 'Who Knew Piezoelectricity? Rutherford and Langevin on Submarine Detection and the Invention of Sonar', *Notes and Records of the Royal Society* 66 (2012): 141–57.

[8] Material on Rutherford's and Boyle's works include contemporary letters, reports, and a research notebook. On Langevin there are reports written after his initial success, reports of his earlier non-piezoelectric work, and his own retroactive account from late 1918. For details and locations of archival sources see their listing in the Bibliography.

[9] My analysis of Langevin's research goes beyond the literature in examining its technical details and the resources that enabled him to invent piezoelectric sonar. Langevin himself gave a secret retrospective report on his work in 1918: Paul Langevin, 'Conférence Interalliée sur la recherché des sous-Marine par la méthode ultra-sonore: Historique des recherches effectuées en France' (19/10/18), L196/16; partial English translation in David Zimmerman, 'Paul Langevin and the Discovery of Active Sonar or Asdic', *Northern Mariner* 12 (2002): 39–52, 44–5. Zimmerman discusses Langevin's work as seen through the proceedings of the Royal Commission: Zimmerman, '"A More Creditable Way".' On Langevin's research during and after the war see Benoit Lelong, 'Paul Langevin et la détection sous-marine, 1914–1929. Un physicien acteur de l'innovation industrielle et militaire', *Épistémologiques* 2 (2002): 205–32.

2.1 The Origins of the Echo-Sounding Method

Inventors had suggested methods for detecting submarines long before their strategic importance was recognized. Raising such suggestions did not require an extensive research and development effort. A concentrated effort, feasible only under the crisis of war, however, was needed to transform a suggestion into a practical method. In the case of sonar it required also the application of expertise and knowledge from different sources. Sonar originated in the idea of Chilowsky[10] to locate submarine objects by the echo of a ultrasonic wave. In autumn 1914, Chilowsky, a Russian émigré living in Switzerland, approached the French authorities through Edgar Milhaud, a French professor of political economy in Geneva. In December, he sent them a short preliminary report, similar in structure to a patent application, 'about the possibility to see under water', in which he described the new method.[11]

The leading idea, as the title indicates, is that ultrasonic waves can be used as 'mechanical light' to replace light waves in searching for submarine objects. Unlike electromagnetic waves, which are absorbed in water, sound (i.e. mechanical) waves propagate to a long distance. A beam of mechanical waves, thus, could replace the light beam of a searchlight. As with the light beam an observer would receive the beams reflected from obstacles, i.e. the sound echo, and detect its existent and direction. By measuring the time it took the echo to return, one could measure the distance to the object, and by measuring the change in its frequency infer its velocity through the Doppler effect. The analogy with a searchlight suggested that the sound beam should be narrow, so as to avoid spreading the energy and maintain an intense beam over relatively long distance of a few kilometres. According to wave theory, to produce narrow pencil beams of sound, the diameter of the emitter should be much larger than the sound wavelength. This reasoning led Chilowsky to the most original ingredient of his suggestion—the use of ultrasonic waves, up to 100 kHz (audible are up to 20 kHz), the wavelength of which is considerably smaller (down to 1.5 cm) than that of audible waves.[12] These waves would be emitted from a rotating flat surface of 1.5 on 1 m, a rather bulky plate to be submitted

[10]Šilovskij in the international transliteration of Cyrillic is variously spelled among others as Schillowsky, Chilowski, and Shilovskiy.

[11]Hunt, *Electroacoustics*, 46. Hunt's book is based on discussions with Chilowsky and the original document held by the inventor.

[12]The unit hertz was introduced in the late 1920s, but researchers in the field had already used other names for the unit throughout the period discussed here. For simplicity I use hertz throughout this book.

underwater. He believed that this would make it possible to detect mines at a distance of half to one kilometre and submarines at 4–5 km. The shortwaves had the additional advantage, not mentioned by Chilowsky, of being reflected and not deflected from small obstacles like U-bouts and mines.[13]

While Chilowsky was quite clear about the general principles of the ultra-sonic echo detection method, he was vaguer about the devices for producing and receiving the mechanical waves. He proposed producing mechanical os-cillations by a magnetic field of a flat solenoid powered by high-frequency electric oscillation. As Langevin later commented, the device was analogous to a telephone receiver.[14] To reduce the water resistance of the suggested emit-ter he thought of dividing it into smaller plates that work simultaneously, but it was not clear whether they would produce the same effect as a large plate. To increase the sensitivity of the receiver he suggested focusing the 'ultrasound light' with special mirrors or lenses. It is unclear where these lenses would be located. Chilowsky acknowledged that his proposal should be developed fur-ther, and suggested that the new wireless technics, probably referring to those based on the triode valve, would be of help. Like many inventors, he underesti-mated the technical difficulties in realizing his method, but he was completely correct in pointing out the relevancy of radio technics to its development.

In February 1915, the mathematician and politician Paul Painlevé, head of the French *commission supérieure des Inventions* and the central figure in organizing war research, received Chilowsky's proposal and appreciated its potential. He invited Chilowsky to Paris and asked physicist Paul Langevin to join the inventor in an effort to produce a practical device.[15] Despite serious technical, scientific, and personal difficulties their research eventually led to a working sonar system. But why did Chilowsky, rather than someone else, come up with the basic idea of sonar?

Born in 1880, Chilowsky had just begun his career as an inventor, hav-ing patented three inventions prior to the war. Apparently, none of them were profitable, so his Russian family continued to support him. He began studying law in Moscow, but due to his political activities had to quit; af-ter some adventures including time in jail and a stay in a remote town, he

---

[13] Kljukin and Šoškov in *Šilovskij* provide a Russian translation of most of Chilowsky's report on pp. 46–50. I thank Oksana Kuruts for translating it into English and summarizing other parts of the book, and Alexei Kojevnikov for his corrections and comments.

[14] Paul Langevin, 'The Employment of Ultra-Sonic Waves for Echo Sounding', *Hydrographic Review* 2 (1924): 57–91, 72.

[15] Kljukin and Šoškov, *Šilovskij*, 50; Hunt, *Electroacoustics*, 46–7.

left for Germany in autumn 1906. There he turned to the natural sciences, in which he had shown interest since youth. He studied for a year at Darmstadt Technische Hochschule (technological college or institute) and another five years at the German University of Strasbourg, an uncommonly long time. Although suffering from tuberculosis, apparently during his time in Strasbourg he seemed to have enjoyed both his studies and the companionship of Nikolai Dmitrievich Papaleksi and Leonid I. Mandelstam, two Russian physics post-docs of his age who maintained their association with the university.[16] A patent he filed in 1913 after completing his studies suggests an interesting connection between the ideas that occupied his mind in Strasbourg, on the one hand, and the ultrasonic echo detection method, on the other hand.

Like his later suggestion to the French authorities, his earlier patent proposed a method for producing waves to scan an area for hidden objects. In 1913 he suggested the use of electromagnetic waves to detect subterranean ore deposits.[17] In 1914 he replaced the earth with water and the ore with submarines. Chilowsky's novelty in the ore patents was in the use of long electromagnetic waves. The use of short waves, which are reflected by the ore deposits, was a known method, but was often impractical due to humidity in the ground that blocked the waves. Instead, he suggested using long waves on the order of magnitude of the ore deposits, which are then neither reflected by the deposits nor blocked by humidity. With proper settings their impact on ore deposits would lead to a magnetic field that could be measured over the ground. This is quite different from the principles Chilowsky suggested for locating submarines. Yet, his two inventions share an important property: both depend on the specific wavelength in use. Other inventors suggested using reflecting electromagnetic and sound waves to locate subterranean and submarine objects. Chilowsky realized that long waves in the one case and short waves in the other could overcome the disadvantages of commonly used waves. The significance of the wavelength in the deposits patent probably encouraged him to examine the implications of using mechanical waves of different frequencies when seeking out a method for locating submarines.

Chilowsky's interest in methods of geophysics probably originated in his time in Strasbourg. The city hosted the central German seismologic station,

[16] Kljukin and Šoškov, 41–2; *Amtliches Verzeichnis des Personals und der Studenten der Kaiser-Wilhelms-Universität Strassburg* for the years 1908–13.

[17] Konstantin Schilowsky, Verfahren und Vorrichtung zum Nachweis unterirdischer Erzlager oder von Grundwasser mittels elektrischer Schwingungen, DE322040 (C), filed 16 November 1913, and issued 19 June 1920.

whose head was a professor at the university. According to his recollections, Chilowsky, Papaleski, and Mandelstam shared an interest in technics for examining the Earth's crustal layers. The encounter probably left a stronger impression on Chilowsky than on the two physicists, who pursued an academic career. In his patents the Russian inventor employed a knowledge of physics and geophysics that he probably acquired at the university. His scientific education provided him essential tools for his invention. Among others he used physics to infer the expected behaviour of electromagnetic and mechanical waves of different wavelengths in different media (ores, water, earth, etc.). Still, he worked as a professional inventor rather than a scientist: he first identified a technical problem and then looked for ways to solve it.

Other Echo-Method Proposals

The problem that first attracted Chilowsky's attention, according to his later recollection, was how to identify icebergs and similar obstacles. The sinking of the *Titanic* in April 1912 brought this menace to the attention of the public and inventors like Chilowsky. He thought of a solution, but it is unclear how far he developed it before returning to the problem in the different context of locating submarine boats and mines.[18] The famous accident inspired simultaneous invention. Three other inventors suggested more elaborate technics based on similar ideas.

Ten days after the sinking, Lewis Fry Richardson, a physicist famous today for his later work in meteorology, filed a provisional application 'for warning a ship of its approach to large objects in a fog'. He suggested emitting sound waves from a moving ship in air and receiving their echo from the object. Due to the Doppler effect the echo pitch would be higher than the original sound, an effect which Richardson employed to differentiate between the emitted and returning waves. A month later he suggests a modified version for underwater use. Since 'the absorption of sound by water is less than that by air' he recommended the application of waves of higher frequencies, from about 5,000 Hz into the ultrasonic range, which are 'desirable on account of freedom from diffraction'. Whistles or resonating cavities in the proper dimensions, filled with the right fluid, would produce the mechanical waves in the desired frequency. A mechanical receiver would focus the echo to an electromechanical

[18]Kljukin and Šoškov, *Šilovskij*, 44–5.

microphone.[19] In his use of mechanical methods Richardson differed from the other inventors of echo-systems (discussed in the following). He seemed to cling to known and tested methods and physical theories of mechanical waves, while others adopted the newer electromagnetic and newest electronic technics, which eventually proved to be more powerful.

Chilowsky and Richardson shared the idea of using the echo of ultrasonic waves for locating objects. Still there is one central difference, which can be traced back to their different aims. Richardson was interested in locating large objects (mainly icebergs) from a moving ship, whereas for Chilowsky, it was much smaller objects (mainly U-boats and mines). The spread of the sound beam and its energy was, thus, of a much higher concern to Chilowsky. Large surfaces could reflect stronger, more energetic beams. Chilowsky, thus, needed short waves to keep the energy, rather than to prevent diffraction, and therefore had to keep to high frequencies.

In his longer specification, Richardson mentioned that short underwater waves had already been suggested for signalling. The most developed of these was the system suggested in 1912 by the Canadian inventor-entrepreneur Reginald Fessenden for his 'submarine signal company'. Fessenden designed an oscillator for sending and receiving underwater soundwaves for the dual purpose of signalling and detecting underwater obstacles by the echo method. The oscillator sent rather long waves (540 Hz, i.e. 3 m). With its modest dimensions of 75 cm in diameter (compared with that in Chilowsky's proposal), it could not send a narrow beam in that wave range, as the Russian inventor would recommend. Still, the waves sent in this system were more energetic than those of earlier signalling systems since the oscillator produced continuous rather than damped waves produced by bells. Fessenden's experience with continuous radio waves, his field of speciality, probably contributed to this move. Yet, Chilowsky and Richardson also proposed continuous waves, as was common in the discussion of oscillators in physics. At the beginning of the

[19] Lewis Fry Richardson, Apparatus for Warning a Ship of its Approach to Large Objects in a Fog, GB191209423 (A), filed 20 April 1912, and issued 6 March 1913; Lewis Fry Richardson, Apparatus for Warning a Ship at Sea of its Nearness to Large Objects Wholly or Partly under Water, GB191211125 (A), filed 10 May 1912, and issued 27 March 1913. Quotations from the latter p. 1. In the provisional specification, Richardson refers to vibrations of 100,000 Hz as optimal but in the longer patent to the use of 'a frequency of 4780 or more complete vibrations per second'. From 1903 to 1913 'Richardson held a series of research and teaching posts, twice at the National Physical Laboratory, twice in industry, and twice in university physics departments.' Oliver M. Ashford, 'Richardson, Lewis Fry', in *Complete Dictionary of Scientific Biography*, vol. 24 (Detroit: Charles Scribner's Sons, 2008), 238–41, on 238.

war, Fessenden tried to interest the British Navy in the potential of his invention for detecting submarines, but to no avail. The characteristics of his early
sound–echo system suggest that, like Richardson's, it was planned for locating
icebergs rather than submarines. He might have made significant modifications to fit the new aim, but his system did not seem promising at least in the
eyes of a few experts. In a general report about the various methods for locating submarines in September 1915, Rutherford (whose work is discussed in
Section 2.3) concluded that 'the amount of sound returned [from submarines]
in this way . . . is very small and would be difficult to detect except at short distance. . . It appears to me that it would be more practical at such short ranges
to detect the submarine by its own characteristic sounds when in motion.'
When writing this, Rutherford did not know of Chilowsky's suggestion.[20]

A fourth contemporary method for detecting underwater objects based on
the echo of soundwaves was developed by the German Alexander Behm. In
later recollections, he attributed his inspiration to the sinking of the *Titanic*,
but there are signs that he might have begun considering the method much earlier. He had developed an interest in acoustics during his studies of physics and
electric technology at Karlsruhe Technische Hochschule in 1902–4 and continued working in the field as an employee of a few companies. He thought that
the rough surface of an iceberg would not return a detectable wave and instead
suggested a system for measuring the depth of a seabed, which he patented in
1913. The idea was not new, but previous methods based on measuring the
very short time interval that it takes an echo to return from the bottom were
impractical. Instead, Behm estimated the depth by measuring the decrease
in the strength of the reflected wave, which is proportional to the square of
the distance that it travelled. For this end he utilized his 1906 invention of
a 'Schallstärkemesser' (sound strength meter). Yet, experiments showed that
the intensity of the reflected wave depends on too many variables to allow a
depth estimation. In 1915 he devised a new method for measuring very short
time intervals based on a tuning-fork, which made it possible to estimate depth
and distance by the delay time of the echoes. Aimed at finding large objects
both methods employed sound in the audible range. Yet, due to the request of
the German Navy, Behm developed the method for also locating submarines
but did not move to ultrasonics and the device was quite cumbersome for

[20]Gary.L. Frost, 'Inventing Schemes and Strategies: The Making and Selling of the Fes-
senden Oscillator', *Technology and Culture* 42 (2001): 462–88. Ernest Rutherford, 'Report on
Methods of Collection of Sound from Water and the Determination of the Direction of Sound',
BIR report 30/9/1915, papers of John William Strutt, Lord Rayleigh, CUL MS.Add 8243/2, 8.

use on ships. His company delivered three prototypes of the device before the end of the war.[21]

The multiple suggestions of Behm, Fessenden, and Richardson show that the basic idea of using echo sounding as a submarine 'searchlight' was not unique to Chilowsky. Yet, none of their initial proposals were directed at locating small U-boats and consequently none emphasized the role of ultra-sonics as Chilowsky did. The initial advantage of Chilowsky's proposal was small but significant enough that the French, and later the British and Americans, decided to examine and develop it over that of the other proposals (naturally, they could not develop Behm's technics). Their efforts made the method that worked in principle into a feasible, even if not yet really practical, technics.

2.2 Early French Development of the Echo Method

Until the outbreak of the war Paul Langevin stood for disinterested scientific research. The technical challenges of the war, however, changed that. Like many scientists on both sides, he participated enthusiastically in research to improve military technics.[22] Already before Painlevé forwarded him Chilowsky's proposal, Langevin had studied ballistics for practical military ends. Beginning his academic studies in 1888 at the age of 16 at the École municipale de physique et de chimie industrielles de la ville de Paris (EPCI) with Pierre Curie, Langevin continued his studies at more prestigious Parisian institutions. In 1896, he spent a year in Joseph J. Thomson's Cambridge laboratory, training, among others, with Rutherford. Returning to Paris Langevin obtained a PhD in 1902 and taught physics at the Collège de France and the EPCI. With a series of successful experiments on ions and discharges in gases and a series of acclaimed theoretical mathematical papers, most famously on

[21] Alexander Behm, Einrichtung zur Messung von Meerestiefen und Entfernungen und Richtungen von Schiffen oder Hindernissen mit Hilfe reflektierter Schallwellen, DE282009 (C), filed 22 July 1913, issued 13 February 1915. Behm had applied for a similar patent in Austria on September 1912, which did not qualified there. Jörg Schimmler, *Alexander Behm (1880–1952): Erfinder des Echolots—Eine Biographie* (Norderstedt: Books on Demand, 2013), 28–73. Werner Schneider, 'Alexander Behm und 100 Jahre Echolotpatente', *Hydrographische Nachrichten*, no. 96 (Oktober 1996): 11–14; Günther H. Ziehm, 'Kiel—Ein frühes Zentrum des Wasserschalls', *Deutsche Hydrographische Zeitschrift*, Ergänzungsheft Reihe B, 29 (1988): 42.

[22] Unlike some of his colleagues (on both sides) Langevin did not express hate toward his colleagues on the other side of the battle field. Bernadette Bensaude-Vincent, *Langevin, 1872–1946: Science et Vigilance* (Paris: Belin, 1987), 85–6.

magnetism and electrodynamics, he established himself as a leading physicist of his generation. His wartime tasks followed military needs and his reputation as a physicist rather than a specific connection to his earlier scientific research (unlike the case of Chilowsky). Shortly after receiving Chilowsky's proposal, Langevin recognized that it had practical potential, although he was sceptical about its details. By March 1915 he recommended that research and development should begin without delay in his laboratory at the EPCI in collaboration with the French Navy. Officially he worked there alone until September 1915 when a former student, Marcel Tournier, joined him as a full-time assistant, but he probably enjoyed the laboratory assistance of his own staff from the beginning.

Langevin realized that in order to allow for a more energetic beam, the emitted ultrasonic waves should be continuous rather than damped. The realization of this idea was a challenge, since the production of energetic, high-frequency continuous waves was a new non-trivial art developed for wireless telephony. Chilowsky, who joined Langevin in Paris, had already suggested adopting methods from radio to this end. Langevin asked the Navy department for assistance in installing state-of-the-art devices in the ultrasonic detection system. Apparently, this was not a simple one-time installation, as the physicist might had thought. The co-developer (with Maurice Jeance) of the French arc radio transmission instruments, the Navy's Captain Victor Colin, installed the equipment and continued working on the production of the electromagnetic oscillations for development in the devices used by Langevin and Chilowsky.[23] Wireless technics played a crucial role in the ultrasonic devices and Colin's expertise was highly valuable to the team. Colin and Jeance's instrument was an early application and a variation on Valdemar Poulsen's arc transmitter, which, like Fessenden's high-frequency alternators, generated continuous waves. The transmitter was based on alternating discharge current between the arc's two electrodes, and the capacity and the inductance of the outer circuit determined its frequency; energy was provided by a source of direct current.[24] Colin transferred the Poulsen arc (without

---

[23] Colin continued to work with Langevin in the later stages of developing and implementing the ultrasonic devices until the end of the war. Their collaboration was dubbed 'mission Langevin–Colin' (L194 Dossier 1).

[24] In Poulsen's device, the cathode is from copper, and the anode from coal; the arc is full of hydrocarbon gas, and put under a magnetic field.

the system's other components for transmitting voice) from radiotelephony to ultrasonics.[25]

Although optimistic about Chilowsky's general idea, Langevin dismissed his particular suggestion of a magnetic generator of ultrasonic waves and began contriving an alternative. He suggested an electric method of a 'singing condenser': a large plain condenser vibrating by an alternating electric voltage on one the of plates connected to a Poulsen electric arc (developed by Colin). The other plate touched the water (through electric insulation) and transformed its mechanical vibrations. To attain strong vibrations Langevin planned to place the plates very close to one another. Although he produced vacuum between the plates, and theory allowed for a smaller distance, in practice he could not place them closer than 20 microns due to disruptive discharge. Consequently, Langevin inserted a sheet of mica, a symmetric crystal, between the plates. After further experiments he realized that he could dispense with the outer metallic plate, since the conducting seawater behaves like the outer plate of the condenser. The mica is moved by electrostriction, which is the deformation of all dielectrics due to electric voltage in proportion to the square of the voltage difference (unlike the linear and stronger piezoelectricity, which appears only in asymmetric crystals). Since this is a quadric effect, the mica's frequency was twice the voltage frequency. To allow for appreciable effect the mica sheet should have been in resonance, vibrating at its 'natural' mode with small loses of energy (little mechanical friction), which Langevin attained by making its width half the wavelength of sound in that medium. Encouraging laboratory tests in July 1915 showed that the transmitter indeed produced measurable elastic waves. The torsion pendulum used to measure the elastic pressure were unpractical for general use, so the next step was to develop a practical receiver (see Figure 2.1).[26]

From theory Langevin inferred that one could use a polarized mica condenser as a receiver, but the idea failed in practice and the group resorted to using regular carbon microphones, which were known to be unsteady in water. To detect the very weak electric output released from the microphones, however, one needed a means for amplification. The electric arc could not

[25] Paul Langevin, 'Historique des recherches', 7–8; English translation in Zimmerman, 'Langevin Discovery Sonar', 44–5. Victor Colin, 'La téléphonie sans fil', *Bulletin de la Société Internationale des Électriciens* 9 (1909): 427–50.

[26] Langevin, 'Historique des recherches'; Constantin Chilowsky and Paul Langevin, Procédés et appareils pour la production de signaux sous-marins dirigés et pour la localisation á distance d'obstacles sous-marinsFR502913, filed, 29 May 1916, and issued March 1920.

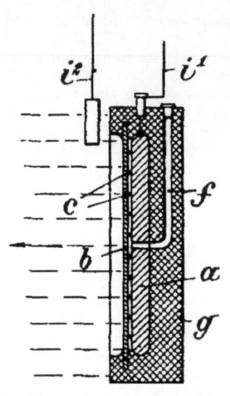

Figure 2.1 Langevin's mica transmitter: *b* is a thin mica diaphragm in contact with the water, *a* is a metallic plate placed in insulating support *g*, *f* is a tube which can produce vacuum between the metallic plate and the mica, i^1 is the pole of the electric circuit connected with the metal, and i^2 is the other pole immersed in water, *c* are points of support. *Source*: Chilowsky and Langevin's 'Procédés et appareils pour la production de signaux sous-marins'.

help here, but another method of wireless technology, which had just become available, provided a solution. The new vacuum tube amplifiers were based on the triode or audion—an electronic valve with an additional third electrode, called a 'grid', whose voltage controls the electric discharge between cathode and anode.

Circa 1912, six years after its invention by Lee de Forest for detecting radio waves, a few individuals realized that by feeding current that travels from the 'plate' (the anode) back to the grid electrode the audion could powerfully amplify weak alternating signals in the grid branch of its electronic circuit (see Figure 2.2). The resulting 'feedback' circuit led to electric oscillations that could also be used for sending continuous radio waves. Yet, research and development of the art concentrated on improving amplification, due to

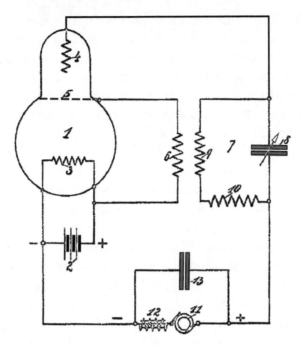

Figure 2.2 An early amplification circuit, by Alexander Meissner: 5 is the grid; 4 is the anode which feeds back the grid through the coils 9 and 6. *Source: Einrichtung zur Erzeugung Elektrischer Schwingungen.* DE291604 (C), filed 10 April 1913, and issued 23 June 1919.

the interests of the two major investors in the field—the American AT&T and German Telefunken, in wired telephony.[27] Amplification of weak electric currents was also a main interest of research on submarine detection. In autumn 1914 the French Radiotelegraphie Militaire received samples of Telefunken's triodes and began working on their improvement and employment with the help of civilian physicists and engineers mobilized to the war effort. In summer 1915 Langevin and Chilowsky could use the valves and triodes provided by the Radiotelegraphie Millitaire.[28] They also adopted the heterodyne

---

[27] Sungook Hong, *Wireless: From Marconi's Black Box to the Audion* (Cambridge MA: MIT Press, 2001), 155–89. Hugh G. J. Aitken, *The Continuous Wave: Technology and American Radio, 1900–1932* (Princeton NJ, : Princeton University Press, 1985), 224–49.

[28] Langevin, 'Historique des recherches'; Michel Amoudry, *Le général Ferrié et la naissance des transmissions et de la radiodiffusion* (Grenoble: Presses universitaires de Grenoble, 1993), 162–94. Telefunken's valves were brought by the Frenchman Paul Pichon, who was head of the company's patents department and found himself at the beginning of the war in London. His motives are unclear.

detection method from wireless telegraphy. When two electric oscillations of close but unequal frequencies meet, they interfere and produce a beat whose frequency equals the difference of the two waves' frequencies, known as heterodyne. In Langevin–Chilowsky's system a 'heterodyne valve apparatus of a frequency nearly that of the [ultra]acoustic waves is arranged to act inductively on the receiving apparatus, so as to produce beats', which were heard through telephone after amplification by valves.[29]

In winter 1915–16, Langevin and Chilowsky managed to transmit and receive clear signals on opposite banks of the Seine in Paris. In May 1916, they filed a joint patent based on Chilowsky's original idea of ultrasonic echo-detection method and on Langevin's transmitter. As common in patents it also covers technics they did not use, such as Chilowsky's original magnetic sender and Langevin's two metals–electrodes condenser. Yet, their collaboration was 'less than entirely serene. . . As a consequence, Chilowsky detached himself from experimental program, at Langevin's request, shortly after their joint patent application was drafted.' He left to investigate other military technologies.[30] Personal discordance notwithstanding, the Navy found the results of their common work worthy of further research. In April it provided Langevin with two vessels and other means to conduct experiments in seawater. Langevin divided his time between Paris and the navy base in Toulon, where Tournier conducted the research in his absence. Their research was of an engineering character, i.e. empirically examining the efficiency of different modifications of the devices, inspired by technological and scientific knowledge and the results of earlier tests. Among other things, they used a concave mirror to raise the effective sensibility of the microphone, an idea that Chilowsky had already suggested in his initial proposal. During the next winter, their device was able to receive signals at a distance of 2 km and an echo at a distance of 200 m in sea.

In May 1916 the French revealed the ultrasonic-echo system to a few British scientists and engineers who were visiting France—an additional indication of

[29] Langevin, 'Note on apparatus for the detection of submerged objects by acoustic waves of high frequency,' communicated to the BIR by de Broglie, 12/2/1917, suggests a non-piezoelectric system (here in an English translation) (BIR 3929/17) (in Langevin's archive L194/26 and ADM 212/159 and also 293/10), 7–8.

[30] Quoted in Hunt, *Electroacoustics*, 48. On Chilowsky's later work see Claudine Fontanon, 'L'obus Chiloswki et la soufflerie balistique de Paul Langevin : un épisode oublié de la mobilisation scientifique (1915–1919)', in *Deux siècles d'histoire de l'armement en France: de Gribeauval à la force de frappe*, ed. Dominique Pestre (Paris: CNRS éditions, 2005), 81–109.

their confidence in the method. Beginning early in that year such visits were the central means for exchanging technical-scientific information between the two allies, until the appointment of formal resident liaison officers in October.[31] In February 1917 Langevin sent the British an optimistic report about his system. The British were impressed enough that they decided to copy and improve the French design rather than designing alternative methods as they had done previously.[32] Yet, Langevin was also considering an alternative detecting device based on piezoelectricity. Ernest Rutherford had already employed the effect in his research on submarine detection method, but for a different aim and in a different manner. Before continuing with Langevin and his implementation of piezoelectricity I will examine the work of Rutherford's group.

2.3 Sonic and Ultrasonic Research and the Use of Piezoelectricity by Rutherford and His Group

When the French informed the British about Chilowsky–Langevin's method, the latter were already carrying out an intensive research programme on locating a submarine 'by its own characteristic sounds when in motion'. The sounds were detected by 'hydrophones'—underwater microphones, which were studied and improved by the British. Indeed, this technic which was already employed by the British navy, was the only one that would be used in action during the war. Yet, its accomplishment was very limited.[33] This is a passive method dependent on signals emitted by the sought-for object. The echo system, in contrast, is an active method where the seeker actively emits signals for locating submarine objects. Tactically, an active system allows for more flexibility in use, particularly in moving vessels, and the detection

[31] For the status of the cooperation Hackmann, *Seek & Strike*, 39, 84, BIR 8/6/16, BIR 31/12/16: 'Report of proceedings to 31st December 1916', the Royal Naval Museum Library, MSS 252/13/62, 4.

[32] Langevin, 'Historique des recherches' (Zimmerman, 'Langevin Discovery Sonar', 46–7), P. Langevin, BIR 12/2/1917; Boyle to Rutherford, 17/4/17. All the correspondences of Rutherford used here, expect those with Bragg and Paget, can be found in Cambridge University Archives (CUL) Ernest Rutherford papers. A copy of these letters is available also in the Archive for the History of Quantum Physics. Unlike Zimmerman's claim, Boyle had not suggested an active detecting method before he learnt about Chilowsky–Langevin's technique (Zimmerman, '"A More Creditable Way"', 55.

[33] Rutherford, BIR 30/9/15, 8; Wilson, *Rutherford*, 346–7. On the very modest actual yield of the hydrophones, Hackmann, *Seek & Strike*, 69–71; and Zimmerman, '"A More Creditable Way"', 55.

of immobile objects. However, the question was whether a practical active system was feasible. In September 1915, knowing only Fessenden's technics, Rutherford thought that it was not. The French suggestion of using ultrasonics changed the opinion of many. Still, Rutherford expressed doubts. '[T]he methods of production of high frequency sound by Langevin and Chilowsky', he wrote to William H. Bragg, 'seem interesting and important, but I think it will probably take a long time to bring them to a really practical issue.'[34] Instead, he continued with the development of hydrophones, which he had been studying almost from the beginning of his involvement in the war research.

It took the British Admiralty a year of fighting in the Great War and the replacement of its First Lord to begin mobilizing scientists for war research. In July 1915 Rutherford was nominated to the general panel of the new Admiralty Board of Invention and Research (BIR) and to its subcommittee, which dealt, among other things, with submarine detection. The board saw this as a most urgent problem.[35] At the age of 44 the New Zealander Rutherford held the physics' chair at Manchester, and was one of the world's leading experts on atomic physics, earning his fame in experimental research on radioactivity and the atom and his theoretical inferences from his laboratory's findings. Like Langevin, he had not worked on practical technics before the war. Rutherford was not fully happy to suspend his atomic research for submarine detection, but willingly followed the national call. That month he recruited two young lecturers to the new research on practical technics: his former student Albert B. Wood from Liverpool and Harrold Gerrard from the adjunct department of electrical engineering. Wood and Gerrard joined Rutherford and his two graduate students James H. Powell and J. H. T. Roberts in the basement of Rutherford's university laboratory. The latter two received the BIR's financial support. Wood and Gerrard left for the naval experimental station in Hawkcraig in November but continued working under Rutherford's guidance until May 1916. At that time, Boyle, a physics professor and Rutherford's former student, arrived from Canada to assist Rutherford in the war research as a BIR employee. During this period Rutherford immersed himself and his small team in submarine research. He stopped teaching probably in summer

---

[34] Rutherford to Bragg, 20/6/16; Rutherford–Bragg and Rutherford–Paget correspondence in 'Rutherford File', ADM 212/157.

[35] Roy M. MacLeod and E. Kay Andrews, 'Scientific Advice in the War at Sea, 1915–1917: The Board of Invention and Research', *Journal of Contemporary History* 6 (1971): 3–40.

1915 and had virtually no time for his pre-war research on 'pure science' until summer 1917.[36]

Wood recalled that during summer 1915, '[w]e were experimenting in a small water-filled tank with various possible sound-receivers for use under water. . . . We used a bell-type buzzer and a continuous-wave diaphragm sounder as sound sources.' Among the methods tried, 'Rutherford was hopefully, if not very optimistically, scratching small pieces of quartz crystal (with a telephone headpiece connected) to discover if the piezoelectric effect of quartz was likely to prove useful. The result of this was inevitably disappointing.'[37] Rutherford tried to exploit the transduction of electrical to mechanical energy in piezoelectricity to convert sound wave, i.e. elastic vibration, into electric wave. This might have been the earliest attempt to use piezoelectricity for practical ends, beyond a measuring instrument. For Rutherford's biographers, however, the episode presents not only that but also the first step in his road to the invention of sonar.[38] Yet, this attempt included neither of sonar's principles: it was neither an echo system nor ultrasound (Rutherford and Wood employed sonic frequencies and the detection was based on audible vibrations transmitted through the headpiece).

From September, Rutherford and his team in Manchester and Hawkcraig experimented with different kinds of 'hydrophones'. During the second half of 1915 and 1916 they followed two major lines of research, both relating to 'passive' receivers of audible sound. In the first they examined the underwater behaviour of different diaphragms and microphones

---

[36] A. B. Wood, 'From Board of Invention and Research to Royal Navy Scientific Service: Reminiscences of Underwater-Sound Research, 1915–1917', *Sound: Its Uses and Control* 1, no. 3 (May 1962): 8–17. On the support to Powell, Roberts, and Boyle see Rutherford to Paget, 3/4/16. An early mention of Boyle's work within the group: Rutherford to Bragg, 12/6/16. Rutherford's correspondence and the large number of his reports testify to the intensity of his submarine research. See also Jeff Hughes, 'William Kay, Samuel Devons and Memories of Practice in Rutherford's Manchester Laboratory', *Notes and Records of the Royal Society* 62 (2008): 97–121, 112–14. Note that the dates given by Kay are somewhat inaccurate. Kay reported on Rutherford's attitude toward war research. In the same vein Rutherford wrote to Bragg (23/5/16), 'It is a pity that it is so difficult for us [British] now to devote our attention to the pure science problems.' See Shaul Katzir, 'Manchester at War: Bohr and Rutherford on Problems of Science, War and International Communication', in *One Hundred Years of the Bohr Atom: Proceedings from a Conference*, Scientia Danica. Series M · Mathematica et Physica 1 (Copenhagen: Danish Academy of science, 2015), 495–510.

[37] Wood, 'From Board of Invention and Research to Royal Navy Scientific Service', 10.

[38] Wilson, *Rutherford*, 373; Campbell, *Rutherford*, 371.

informed by the mathematical theory of Horace Lamb. The second line of inquiry included testing, improving, designing, and constructing particular hydrophones. Designing receivers sensitive to the direction of sound occupied much of Rutherford's attention, leading, among other results, to a joint patent with another established physicist, W. H. Bragg, who at that time headed the BIR's experimental station.[39]

The Design of Rutherford's Piezoelectric Device

During the summer and fall of 1916 Rutherford's team tried to determine the efficiency of their microphones and diaphragms, i.e. the ratio between the amplitude of the underwater sonic waves and the electric signals that they emit. Thus, they needed to measure amplitudes of vibrating diaphragms, which they assumed to be equal to those of the sound waves.[40] The assumed amplitudes of the sound waves (10^{-6} to 10^{-8} cm), however, were two to four orders of magnitude smaller than the sensitivity of common optical technics for their measurement, suggesting the need for a new method. Rutherford saw a possible solution in piezoelectricity, where a small mechanical amplitude can generate observable electric voltage. He presented his ideas in a 'preliminary notice of a new method of measuring the amplitude of vibration of diaphragm and of exciting supersonic waves in water' on 28 September 1916.[41]

At the core of Rutherford's method lay a specific device—the 'quartz piezo-electrique' (his term) whose properties shaped its later use. Originating in Jacques and Pierre Curie's 1882 instrument for measuring either electric charge or pressure, the device consisted of a long and narrow quartz plate or bar (say $100 \times 20 \times 0.5$ mm), whose two larger surfaces are metallized. The Curies utilized the particularities of 'transverse' piezoelectricity in quartz, i.e. strain due to electric voltage along a perpendicular direction, where the

---

[39] Hackmann, *Seek & Strike*, 51–4; Bragg–Rutherford correspondence; Wood, 'From Board of Invention and Research to Royal Navy Scientific Service', 12–13; Campbell, Rutherford, 369.

[40] Rutherford to Bragg 12/6/16. See also the letters of 27/9 and 5/11, and Rutherford BIR 18/12/16: 'Sensibility of diaphragms for reception of sound from water' (BIR 15239/16) ADM 218/14.

[41] A letter of Rutherford to BIR 28/9/16 (BIR 11738/16). I thank Francis Duck for a copy of the notice, kept at UK National Archives T173/161. I had formerly used a French translation from 23/1/19 of Rutherford's original report from 28/9/16 in a letter of A. Grasset to P. Langevin 23/1/19 (ESPCI L138/153).

resulting elongation of the crystal per volt is proportional to the ratio between the crystal's length and thickness,

$$\Delta l = -d_{11}\frac{l}{e}V,$$

where l is the length, e is the breadth of the bar (i.e. 100:0.5), and d_{11} the piezo-electric main coefficient of quartz (Figure 2.3).[42] Following the Curies' design, Rutherford firmly fastened one end of the crystal, leaving the other end free to move or vibrate lengthwise due to alternating current.

Rutherford suggested three uses of the quartz piezo-electrique, of which only the first would be applied. In the first, the plate's free end is attached to a microphone, or a diaphragm, and thus produces elastic vibrations on the surface of the microphone. By connecting the microphone to a separate circuit, the experimenter can determine the minimal voltage on the quartz sufficient to produce a vibration detectable by the microphone. Using piezoelectric theory and empirical data one can calculate from the voltage the expansion of the quartz, which was then assumed to be equal to the amplitude of the microphone's vibration. In the second method, the piezo-electrique is used to balance the sound produced by a source unconnected to the receiving system. As with the previous method, the known value of the piezoelectric coefficient allows one to calculate the amplitude of the crystal's vibrations and thus the amplitude of the diaphragm and from it that of the underwater sound waves.

The third method departed from Rutherford's main research. It employed the device to produce waves rather than to measure them, and it dealt with ultrasonics. Rutherford suggested generating underwater ultrasonic waves by connecting the electrodes of the piezo-electrique plate to a high-frequency source, like the common Poulsen arc. A small plate rigidly connected to the free end of the quartz, and exposed on one side to the water, would communicate the ultrasonic vibration to the medium. Since the power communicated by one plate is low, Rutherford suggested multiplying the effect by connecting a few bars in parallel to the same power source. Although he believed that this arrangement would be generally efficient, he admitted that the 'actual energy communicated to the water will be comparatively small'. This is

---

[42]P. H. Ledecoer, 'Nouveaux électromètres à quadrants apériodiques', Œuvres de Pierre Curie (Paris: Gauthier-Villars, 1908), 564–86 and J. Curie, 'Quartz piézo-électrique', ibid., 554–63, and Shaul Katzir, The Beginnings of Piezoelectricity: A Study in Mundane Physics (Dordrecht: Springer, 2006), 23.

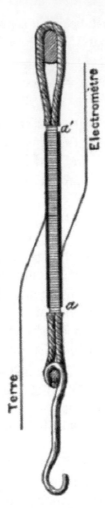

Figure 2.3 The Curies' 'quartz piézo-électrique' as used in their original instrument. *Source*: J. Curie *'Quartz piézo-électrique'*, 557.

the first recorded suggestion to employ piezoelectricity for the production of supersonic waves, but it is far from sonar. Compared to the later method, it lacked any reference to echo detection, and consequently any suggestion for a receiver. While Rutherford was probably inspired by Chilowsky–Langevin's method, his device could be used also for other ends like underwater signalling, a possibility which he had mentioned earlier.[43] This was Rutherford's second attempt to employ the effect beyond a measurement device. Yet,

---

[43] Rutherford BIR 30/9/15, 2.

as with the earlier attempt to utilize the effect for sound detection, he did not make any actual advance towards a practical device. A comparison of Rutherford's emitter with Langevin's later device, which differed in the crystal cut and oscillation modes, merely highlights the inefficiency of the former emitter, inefficiency which Rutherford had already acknowledged.

The Origins of Rutherford's Proposal

Searching for a means to measure the tiny vibrations of his microphones, Rutherford found a solution in piezoelectricity. Yet, to apply piezoelectricity, familiarity with the effect and its experimental manifestations was required. Moreover, as mentioned, Rutherford employed a particular device that utilized piezoelectricity in a specific way, rather than designing an original instrument. His application of the phenomenon, therefore, depended on his familiarity with the Curies' 'piezo-electrique balance'. However, this device was not well known in the scientific world; in technological circles it was almost unheard of.[44] Although detailed information about the device appeared in publications that were accessible to most scientists, most did not look for that knowledge.

Rutherford himself had encountered the piezoelectric measuring instrument through his research on radioactivity. In 1898 Marie Curie introduced a *quartz piézo-électrique* balance to determine the weak charges radiated from the small samples at her disposal, probably following the advice of her husband, Pierre.[45] The method had rarely been used earlier,[46] and remained a specialty of Curie's laboratory also into radioactivity research. It stayed a local practical knowledge. Rutherford probably did not use the method himself, as none of his own and his collaborators' research papers on radioactivity mention it. Yet, he followed closely the Curies' research, the main competition to his own. Moreover, he did describe 'measurement by means of the quartz piezo-electrique' in his 1904 and 1913 textbooks, where he followed Marie

---

[44]Philippe Molinié and Soraya Boudia, 'Mastering Picocoulombs in the 1890s: The Curies' Quartz-Electrometer Instrumentation, and How It Shaped Early Radioactivity History', *Journal of Electrostatics* 67 (2009): 524–30.

[45]Soraya Boudia, *Marie Curie et son laboratoire, Science et industrie de la radioactivité en France* (Paris: Editions des archives contemporaines, 2001), 58–61; Katzir, *Beginnings of Piezoelectricity*, 214–15.

[46]To my knowledge, only Jacques Curie had used it in research: Jacques Curie, 'Recherches sur le pouvoir inducteur spécifique et la conductibilité des corps cristallisés', *Annales de chimie et de physique* 17 (1889): 385–434.

Curie's publications.[47] To explain the use of the quartz piezo-electrique to his reader, Rutherford had to have a thorough understanding of the method, albeit not necessarily full mastery of the procedure. Thereby he became more familiar with the piezoelectric device and with the phenomena than most physicists.[48]

The knowledge of the quartz piezo-electrique, thus, followed a personal and contingent path from Jacques and Pierre Curie through Marie Curie to Rutherford. In her early experimental research, Marie followed instruments and methods previously employed by Pierre. This suggests that he introduced his wife, a doctoral student at the time, to the use of the piezoelectric device, which he had invented with his brother a decade earlier.[49] While the personal contact helped in directing the young researcher to this uncommon method, by reconstructing Marie's experiment on radioactivity, Boudia and Molinié have shown that it was unnecessary for mastering the quartz and the electrometer.[50] The tacit knowledge required to work with the apparatus either was shared by contemporary experimentalists or could be easily acquired by working with the *quartz piézo-électrique*. Nevertheless, contemporary researchers who lacked a personal connection to the inventors of the balance did not employ it. This suggests that, at least in some cases, obstacles to adapting an experimental method do not originate in difficulties in gaining non-verbal knowledge necessary for commanding the laboratory settings. Awareness of the possible benefits of a new method seems more important.

For his vibrating quartz Rutherford did not need to master the *piézo-électrique*, as Marie Curie did. Although some cases of complex experimental apparatuses suggest the significance of personal contact,[51] in other cases, like the one described here, one could build working devices from the circulation of

---

[47]Ernest Rutherford, *Radio-Activity* (Cambridge, UK: Cambridge University Press, 1904), 89; Ernest Rutherford, *Radioactive Substances and Their Radiations* (Cambridge, UK: Cambridge University Press, 1913), 104.

[48]Only few books provided an explanation of the instrument. In addition to Rutherford's and Curie's books on radioactivity, the quartz piezo-electrique was mentioned in Stefan Meyer and Egon R. von Schweidler, *Radioaktivität* (Leipzig: Teubner, 1916).

[49]Boudia, *Marie Curie et son laboratoire*, 47–64.

[50]Molinié and Boudia, 'Mastering Picocoulombs in the 1890s'.

[51]Harry M Collins, *Changing Order: Replication and Induction in Scientific Practice* (London: SAGE, 1985), 51–78; J. L. Heilbronan and Robert W. Seidel, *Lawrence and His Laboratory: A History of the Lawrence Berkeley Laboratory* (Berkeley: University of California Press, 1989), 317–52.

texts and figures.[52] Rutherford did not need to learn a technic from those who mastered it previously. Moreover, Rutherford did not replicate Curies' use, as he modified the device for his own aims. His originality lay in applying an alternating voltage of thousands of cycles per second to the quartz piezo-electrique. Thereby he departed from earlier static or semi-static measurements. He also departed from previous scientific study of piezoelectricity, as no experiment had been carried out on oscillating crystals. Rutherford, therefore, argued for the validity of the law, which was experimentally supported only for static cases, in the dynamic case, which was used in his apparatus. Thereby, he extended the law to the latter domain through a theoretical argument. Technical application demanded extension of the theory to the unexamined domain.

Although Rutherford's device went further than the Curies', its origins in this predecessor limited its use as a supersonic generator. While Rutherford turned a static device into a dynamic one, it remained a measuring instrument. This device was highly sensitive to changes in voltage, and thus useful for his primary aim of determining amplitudes of underwater diaphragms. High sensitivity, however, was not useful for producing underwater waves. While large amplitude is important for a measuring instrument, in contrast high energy (or more precisely, power) is needed to generate waves. In his suggestion to use the instrument as a generator, however, Rutherford kept the Curies' crystal cut, which gave high sensitivity but low power, rather than rethinking the design based on the requirements for an ultrasonic emitter, as Langevin would do. Although he acknowledged that the device could communicate only a small amount of energy, he did not depart from the Curies' basic design, and thus proposed only an inefficient generator.[53]

---

[52] Da Silva Neto and Kojevnikov concluded from Soviet research that led to the maser that 'technology transfer could take place without much personal contact and tacit knowledge, through the attentive use of formal publications and communications, however censored and restricted'. Climério Paulo da Silva Neto and Alexei Kojevnikov, 'Convergence in Cold War Physics: Coinventing the Maser in the Postwar Soviet Union', *Berichte zur Wissenschaftsgeschichte* 42 (2019): 396. Another example, 'Yasuo Koide and Hiroshi Amano, two graduate students in [Akasaki's] laboratory [at Nagoya University], built a MOCVD reactor from scratch on the basis of articles they had found in the technical literature. They were assisted in this endeavor by students from Toyohashi University of Technology, a technical university nearby.' Christophe Lécuyer and Takahiro Ueyama, 'The Logics of Materials Innovation: The Case of Gallium Nitride and Blue Light Emitting Diodes', *Historical Studies in the Natural Sciences* 43 (2013): 263.

[53] Despite the claims of his biographers, Rutherford's design indicates that he aimed to use the quartz for measurement on hydrophones, and only later did he consider producing ultrasonic waves. This is also suggested by the fact that the BIR annual report of 1916 refers to

The Use of Rutherford's Device

Contemporary documents, including a laboratory notebook, correspondence, and reports,[54] show that Rutherford and his collaborators in Manchester and London constructed two devices according to his proposal and used them to study diaphragms and microphones. However, the documents show no hint that they applied the quartz piezo-electrique to generate supersonic waves.

For about a week in October–November 1916, Rutherford, his laboratory assistant William Kay, Powell, and Roberts experimented with the quartz piezo-electrique, at the laboratory at Manchester University. They examined the minimal amplitude detectable by an underwater diaphragm and a microphone at frequencies of about 1,000 Hz (soprano pitch), following the first method suggested by Rutherford. They detected vibrations either directly, when the other side of the diaphragm was in air, or through the microphone, probably by means of a telephone earpiece.[55] Their general laboratory skills allowed them to produce the experiment without direct contact with earlier users of the piezo-electrique. Rutherford was satisfied with the results, writing Bragg that 'the quartz piezo-electrique works like a charm. At a frequency of 1000 I can detect a condensation [relative change of the wave density] of 10^{-11} while 10^{-10} gives a sound that anyone could hear at once in a moderately quiet room.'[56] The experiment supported Rutherford's belief that sensitive sonic receivers would detect submarine engines. Bragg was 'delighted to hear about the piezo-electrique, it sounds most useful', and recommended its use to other

Rutherford's notice as that on a 'method of measuring vibrations of diaphragms by the quartz "piezo-electrique"' (19).

[54] Especially Rutherford NB 19 (CUL), Rutherford–Bragg and Rutherford–Boyle correspondences.

[55] NB 19 does not mention how the quartz was connected to the diaphragm. The above description is based on Rutherford's calculations for the effect of crystal on the diaphragms, on the description in Rutherford's note, and on Boyle's report to Rutherford on his own later experiment where 'the microphone was mounted on the piezo—in a manner something like what you described' (Boyle to Rutherford 12/12/1916, and see the rest of this section). The use of a diaphragm whose other side is in air is described in J. H. Powell and J. H. T. Roberts, 'On the Frequency of Vibration of Circular Diaphragms', *Proceedings of the Physical Society of London* 35 (1923): 170–82.

[56] Rutherford to Bragg 5/11/16, similarly 2/11/16. Rutherford, thus, did not explore in these experiments the possibility of ultrasonic echo detection; thus, it is misleading to quote these sentences in the context of sonar research, as does Campbell, *Rutherford*.

researchers of the BIR for examining how electrical and sound signals are related in microphones and diaphragms.[57]

The Mancunian team did not experiment with ultrasonic frequencies. Yet following the French information about the ultrasonic echo method, in August 1916 the British began their own research, exploring alternative means of producing and receiving ultrasound. The BIR sent Boyle, Rutherford's collaborator, to conduct the research in London, in the private laboratory of the electrical engineer, inventor, and industrialist Sidney G. Brown.[58] He also carried out experiments in seawater at the Hawkcraig Naval Station. So Rutherford sent a second crystal plate (which Maurice de Broglie, the French scientific attaché to the British Admiralty, had brought from Paris following Rutherford's request) to Boyle for ultrasonic experiments.[59] In early December 1916, Boyle determined the minimal alternating voltage on the crystal that led to amplitudes detectable by the microphones. With his assistant and Brown's employee, B. S. Smith, he modified the settings for the needs of ultrasonics[60] (see Figure 2.4). Boyle and Smith could not replicate Rutherford's success. The high frequencies led to interference from the electromagnetic waves produced by the instrument. Moreover, they found 'that the piezo effect per volt $\propto$ to say $\frac{\text{currnet in galv. due to piezo}}{\text{voltage on Quartz}}$ falls as the frequency goes up'. Due to these problems, they failed to use the instruments at frequencies above 31,000 Hz. This cut-off was significantly lower than in their studies of microphones by other means.[61] There is no hint that Boyle and Smith tried to further improve the experimental setting to reach higher frequencies. They probably did not dedicate more than two weeks to the measurements with the piezoelectric device, for these pertained only to a measurement of one aspect of the echo method system they were studying. In their goal-directed research on means to emit ultrasounds they had no time to dwell on such a side issue. By January 1917

[57] Bragg to Rutherford 8/11/16, Wood to Rutherford 10/11/16.

[58] BIR 31/12/16, 21; Rutherford to Bragg 12/6/16; Boyle BIR 9/9/16: 'Attempts to transmit and receive supersonic vibrations' (BIR 10833/16), ADM 293/5; Boyle BIR 23/11/16: 'Production and reception of high-frequency sound waves by the method of the Brown grid magnetophone' (BIR 14243/16) ADM 293/5; Boyle 19/10/18: 'Conférence interalliée du 19 octobre 1918: Compte rendu officiel—expose du Docteur Boyle', L194/09; Hackmann, *Seek and Strike*, 84.

[59] De Broglie's to Langevin 23/12/18; de Broglie to Rutherford, fall 1916.

[60] Rutherford NB 19, 16; Boyle to Rutherford 12/12/16, Rutherford to Boyle 2/12/16.

[61] Boyle to Rutherford 12/12/16. Boyle and Smith BIR 28/11/16: 'Reception of High Frequency Sounds by Microphones' (BIR 14244/16) ADM 293/5.

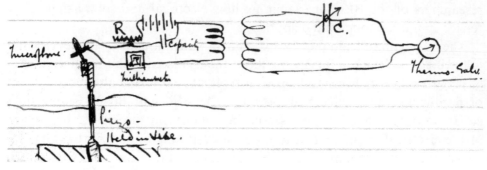

Figure 2.4 Boyle's drawing of his measurement with the piezo-electrique (on the lower left side), which is connected to a microphone. The microphone, which forms part of an RCL circuit on the left, generates alternating current in the circuit as a result of its mechanical movement. Through the coils this current induces an electric current in the CL circuit to the right; a thermoelectric galvanometer is used to measure the current. *Source*: Boyle to Rutherford, 12 December 1916.

the piezo-electrique was waiting to be sent back to Rutherford, who was in no hurry to get it.[62]

Boyle and Smith showed much interest in piezoelectricity. 'There can be quite a bit of work done on the "piezo",' Boyle wrote to Rutherford, 'but', he added, 'after the war'.[63] Clearly, Boyle did not attempt to employ the phenomenon for the very technical problem he was trying to solve: the transmission and reception of underwater ultrasonic waves, although he did examine quite a few solutions to the problem. Initially, Boyle and Smith attempted to produce and receive underwater ultrasound by producing alternating magnetic field in some resemblance to Chilowsky's initial idea. However, in November 1916 they concluded that the production of high-frequency waves 'by electro-magnetic apparatus of this design, is not possible'. Consequently, they turned to 'electro-dynamic and electrostatic methods',[64] and even examined

---

[62]No further experiments with the device are mentioned in Boyle–Rutherford correspondence. E.g. the next letter by Boyle, of 2/1/17, again discusses supersonic transmitters (not piezoelectric). On 12 January Boyle asked Rutherford, 'what you wish me to do with your quartz piezo-électrique', asking again on 13/2; on 2/3 he verified that Rutherford had received the device.

[63]Boyle to Rutherford 12/1/17, 6.

[64]Boyle BIR 9/9/16, 23/11/16 (quote p. 7). BIR report 31/12/16, 21.

mechanical methods, all, however, without much success.[65] Despite Rutherford's suggestion to employ piezoelectricity for the production of ultrasonic waves, Boyle did not explore the use of the 'piezo' for this end. Boyle would do so only following Langevin's success, and then would learn the technique in Langevin's laboratory.

Boyle and Smith's failure to explore a piezoelectric means can be explained by the low-power output of the emitter purposed by Rutherford, but also by their superficial knowledge of the effect. In addition, their problematic experience with the quartz plate at high frequencies militated against using it to generate waves at those frequencies. Even on paper, Rutherford's piezoelectric generator was inferior to other means. As a receiver it seemed much less sensitive than their microphones. They never really viewed the piezo system as a transducer. Apparently, to make it an efficient transducer required good knowledge of the phenomena and preferably experience with handling these crystals. Boyle and Smith lacked both. Boyle, the more knowledgeable of the two, had learnt the basics of piezoelectricity only following his work on the piezo and continued to show rudimentary knowledge of the phenomenon.[66] In principle, Boyle and Smith could have gained better command of the effect from printed sources, but that would have required much time for locating and learning the relevant knowledge and thus a prior belief that the phenomena might be valuable for their research. Such a belief, however, required a thorough acquaintance with the phenomenon received.

At a secret inter-allied meeting in October 1918, after experimenting for more than a year with Langevin's piezoelectric device, Boyle candidly described the British research: 'The utilization of the piezoelectric properties of quartz was introduced to England not at all for ultrasonics, but for measuring the amplitudes of certain mechanical movements . . . Rutherford followed his method of measuring amplitude and suggested the possibility of obtaining ultrasonic vibrations . . . But to be totally frank, it should be said that we undertook nothing similar [to Langevin's work] in England since we did not know the amplifiers of high power.'[67]

[65] Boyle to Rutherford 5/10/16, 22/10/16, 2/1/17.

[66] Boyle 19/10/18, 2; see also Boyle to Rutherford 12/1/17, where he showed ignorance of basic experimental findings from the early 1880s.

[67] Boyle 19/10/18, 2–3. Indeed, also in his first report on Langevin's work Boyle did not see any similarity to Rutherford's device, which he did not mention at all (Boyle, BIR 1/8/17: 'Report of Mission to France for the Admiralty Board of Invention and Research.—May 20th to July 19th, 1917' (BIR 30061/17), ADM 212/159, a copy in 293/10).

One should not, however, uncritically accept Boyle's explanation for undertaking 'nothing similar'.[68] Granting the claim that the British did not have powerful amplifiers, the above analysis shows that the British did not reach a point where better amplifiers would have made a difference. Valve amplifiers, to which Boyle referred, were unnecessary for transmitters, where other means (like the Poulsen arc) could and did produce powerful electric oscillations. While valve amplifiers played a crucial role in Langevin's receivers the British did not even suggest the use of piezoelectricity for that end. Moreover, the British, including Boyle himself, had not been so ignorant about electronic valves as he suggested. Shortly *before* their above-mentioned experiments, Boyle and Smith experimented with a circuit in which a 'valve relay can be used to rectify the signals and to produce beats of audible frequency by interference with supersonic vibrations. The same valve can also be employed to magnify the received signals, and in this way to increase the sensitiveness.'[69] Boyle and Smith probably employed either valves designed by the French military telegraphy, which had been regularly produced in Britain from 1916, or a British modification of them. The performances of the valves produced in the two countries and of multi-valve amplifiers based on these were similar, and clearly did not present a qualitative difference that would have prevented an attempt to employ the British valve for a piezoelectric receiver and emitter.[70]

The French 'hard' valve employed high vacuum, while former 'soft' valves used in Britain depended on small quantities of gas in the tube. 'Hard' valves were more stable and easier to handle than 'soft' valves, a fact highly significant in the battlefield. Soft valves, however, were more sensitive and could be used by trained technicians on board a vessel or offshore, and, thus, for detecting submarine signals. Indeed, in March 1916, the BIR panel, of which Rutherford was a member, received a report on 'valves for submarine sound amplification'. This was one in a series of reports on the use of thermionic valves discussed in the BIR from December 1915.[71] Yet, Rutherford had most probably already known about soft-valve amplifiers, among others, from J. J.

[68] Yet, even Hackmann seems to follow Boyle's explanation: *Seek & Strike*, 80–5.

[69] Boyle and Smith BIR 28/11/16, 2–3.

[70] In a letter to Rutherford from 2/3/17, Boyle mentioned 'amplifying by valve after valve', a practice common with the French valves, due to their lower amplifying power. Wood's comparison of valves after the war (BIR 12/12/18: 'Comparison of valves—French, British and German', ADM 218/282) shows that their properties were very similar. Most of these characteristics were shared already by 1916 models (but not by earlier models in Britain). Gerald F. G. Tyne, *Saga of the Vacuum Tube* (Berkeley Heights, NJ: Prompt Publications, 1977), 200–32.

[71] BIR 31/12/16.

Thomson, a member of the BIR central committee and Rutherford's former teacher, who supervised the development of soft valves at his Cavendish Laboratory.[72] Thus, it is highly unlikely that Rutherford and his associates were ignorant of strong amplifying valves in September 1916; they certainly used valves two months later.

2.4 Langevin's Piezoelectric Sonar

The sensitivity rather than the power of piezoelectricity recommended its use for Rutherford. Langevin also began using the effect for its sensitivity, for a receiver of feeble ultrasonic echo waves. In a sense the success of Langevin's mica emitter opened the way for the use of the piezoelectric method. A satisfying emitter revealed the limitations of the underwater carbon microphone as a detector, a component which until then did not receive much attention. In late 1916 Langevin's group carried out *in situ* experiments which showed that 'the microphone gave very irregular results and required delicate tuning in order to keep the sensibility of the carbon contacts approximately constant despite the variations in external pressure due to the motion of the sea.'[73] The realization of the problem followed the advancement of the French research at sea, with the collaboration of the Navy.

Looking for an alternative, in February 1917 Langevin employed a slim quartz sheet connected to an electric circuit as a receiver. Unlike Rutherford, he did not rely on the Curies' instrument. Instead, he employed a crystal of different dimensions. Most importantly, instead of a thin bar he utilized a square plate cut in a plane perpendicular to the cut of the crystal in the 'quartz-balance' (a large surface in the *yz* instead *xz* plane in common symbols) (see Figure 2.5). With this cut, he employed the longitudinal effect in the *x* direction, i.e. the generation of electric voltage along the direction of the changes of pressure (and vice versa), rather than a transverse effect. The large square surface was exposed to the water pressure due to ultrasonic wave. This considerably increased the sensitivity of the plate to the small changes of pressure as the total charge produced by the crystal is proportional to the size of the surface, and the electric detection depends on the charge,

$$Q = -d_{11} l l' P_x,$$

---

[72] G. W. White designed soft valves in Cambridge since autumn 1914; Tyne, *Saga of the Vacuum Tube* and Keith R. Thrower, *History of the British Radio Valve to 1940* (Hants: MMA International, 1992), 32–5.

[73] Langevin, 'Historique des recherches', 9.

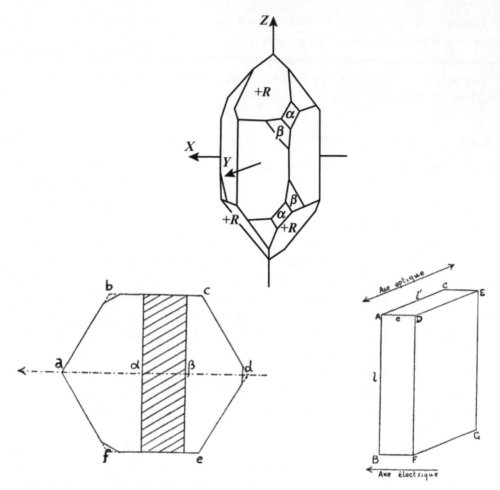

Figure 2.5 Quartz crystal with its directions and Langevin's cut. The left lower
figure is a cut in the *xy* plane. The line *ad* is in the *x* direction. The crossed area
between α and β is the ADFB section of the plate cut by Langevin seen on the
right side. The electric axis is the *x* axis, and the optic is the *z* axis. *Source*:
Upper figure is adapted from Woldemar Voigt, *Lehrbuch Der Kristallphysik* (Leipzig:
B. G. Teubner, 1910); the two lower figures are from Tsi-Zé Ny, 'Étude expérimentale
des déformations et des changements de propriétés optiques du quartz sous l'influence
du champ électrique', *Journal de Physique et le Radium*, 9 (1928): 13–37.

where Q is the total charge, l is the length in the *y* direction, l' is the length
in the *z* direction (Figure 2.5 low right side), and P_x is the pressure.[74] The
surface that was exposed to the water behaved as an electrode, and the other

---

[74] In this cut a pressure along the *x* direction can induce also an electric charge through the
transverse effect, i.e. on the *xz* surface in the same amount as the longitudinal on the *yz* surface.

was connected to an insulated metallic plate, connected to an electric circuit for detection. Unlike a claim in the secondary literature, it allowed Langevin to obtain the required quartz sheets from common crystals.[75] In addition, he benefited from French advancements in vacuum tube amplification used in radio, which facilitated detecting the feeble electric signals generated by the quartz plate. He had already used earlier models of amplifiers with the microphones. As with microphones he detected the returning waves aurally by the heterodyne method. These conditions allowed him to 'obtain promising results in reception by employing plates of quartz as receiver' during the spring.[76]

Encouraged by the success of the receiver, in April Langevin modified the device for use also as a transmitter, assuming that it might be better than the mica emitter. He employed elastic theory to devise a thicker plate that would be a better piezoelectric transmitter because it resonated at a frequency desired for submarine detection. To this end he used a 'crystal of exceptional size and purity', obtaining a few sheets 'of about a square decimetre of surface, and of fifteen millimetres thickness'. The plate surface had to be large enough to allow a narrow beam. He acquired the crystal from a highly probable supplier, the optical instrument maker Ivan Werlein of Paris, an expert in crystal cuts who provided crystals to many French scientists, including the quartz piézo-électrique used by the members of the Curie family.[77] Yet, Langevin was not dependent on a particular quartz specimen. In initial experiments, he examined the mechanical power produced at different frequencies and different cuts of 'Werlein quartz' and of a quartz from Amédée Jobin, another maker of optical instruments. French suppliers and warehouses had enough quartz crystals for research, development, and usage, and could also provide crystals to their allies, especially the British. Cutting was first done

This arrangement, however, is technically more complicated and does not seem to have a clear advantage. Later designs used also such an arrangement.

[75] For this claim see, e.g. Hackmann, *Seek and Strike*, 80; Langevin, however, explained that a special crystal was needed for his *later* work on the transmitter. In the latter case it is important that the crystal would vibrate near resonance, but this is not a prequisite for a receiver. Langevin, 'Historique des recherches',10–11, and see below.

[76] Boyle, BIR 1/8/1917.

[77] Ivan Werlein, 'Constructeur d'instruments d'Optique', was a life member of the Société française de physique (*Bulletin de la Société française de physique*, 1896, no page number). His help is mentioned among others in Jacques Curie and Pierre Curie, 'Sur un électromètre à bilame de quartz', *Comptes rendus* 106 (1888): 1287–9; Marie Curie, *Traité de radioactivité*, vol. 1 (Paris: Gauthier-Villars, 1910), 101.

by instrument makers and mineralogists and then moved to companies with expertise in marble cutting.[78]

Langevin's early experiments with the new piezoelectricity emitter showed an impressive yield in ultrasonic power; 'fish put within the beam in the vicinity of the source were killed immediately'. In the summer the group experimented in the sea with both the piezoelectric and the mica transmitters using piezo-electric receivers.[79] The electric tension employed, however, was much too high for use in a ship, and crystals of that size were too rare to provide such an instrument to every vessel. Langevin, thus, encountered an engineering problem of design: how to use thinner quartz crystals, which allow for lower electric tension and are easier to attain, and still have an emitter whose reso-nance frequency is in the useful ultrasonic range and not higher (such that it does not lose much energy on friction). One solution was examined in late 1917: 'Instead of using the quartz in a solid block, it can be divided into thin slices, the slices being placed together in a pile with positive and neg-ative faces alternately together and thin metal foil between them . . . The full available voltage can be imposed on each sheet and a high straining pressure be obtained.' Preliminary experiments gave promising results. Yet, between November and February, Langevin contrived an alternative solu-tion: a steel–quartz mosaic 'sandwich' that obtained similar piezoelectric and elastic properties from easily available crystals.[80] In this compound a thin-ner sheet of quartz (first of 4 mm) composed of pieces in a mosaic was put between two thicker plates of steel (of 30 mm) (see Figure 2.6). All crystal slabs were oriented in the same way, with an electric axis perpendicular to the steel plate, to keep the same piezoelectric behaviour. Unlike earlier de-signs, resonance frequency was determined by the property of the compound steel–quartz sandwich, which behaved elastically as one solid, and not only by

[78] On experiments on the crystals see L148/18 Notebook 'Servis US Cahier d'ordres du Di-recteur', undated by content probably from 1917. In 1923 the French turned to a company with experience in marble. A. B. Wood, 'Admiralty Experimental Station, Parkeston Quay (Har-wich) 1917 to 1919', *Journal of the Royal Naval Scientific Service* 20, no. 20 (A. B. Wood, O.B.E., D.Sc. Memorial number) (July 1965): 39; Hackmann, Seek and Strike, 80, 85–6; Benoit Lelong, 'Paul Langevin et la détection sous-marine, 1914–1929. Un physicien acteur de l'innovation industrielle et militaire', *Épistémologiques* 2 (2002): 217.

[79] Langevin, 'Historique des recherches', 11; Boyle BIR 1/8/17.

[80] Langevin, 'Historique des recherches', 11–12; Langevin, 'The Employment of Ultra-Sonic Waves for Echo Sounding'. On the pile, Robert Boyle, 'The Piezo-Electric Phenomenon, with Reference to Acoustic Vibrations' (BIR 38164/17), undated, p. 12; the list of the report states that it is from December 1917, in ADM 293/10 and ADM 212/159.

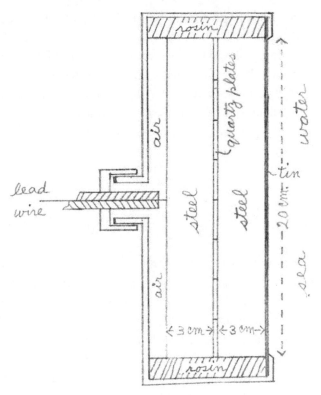

Figure 2.6 Langevin's steel–quartz mosaic as of May 1918; the width of the the quartz plates between the thick steel plates is 2.5 mm. *Source*: Drawing probably by Compton from his report from 15 May 1918, 3.

the crystal. To have a resonance in the desired frequency the sandwich width had to be half the wavelength of the emitted ultrasound in steel and quartz (luckily, they are very close). In February 1918 Langevin's team carried out encouraging experiments with the quartz sandwiches used as an emitter and as a receiver.[81] As discussed in Chapter 3, following their early success with the piezoelectric transducers, the French informed their allies about the details of the research, which was consequently pursued by British, American, and Italian groups. Still Langevin's group continue to lead, even if its transducers were not able to go into service before the war ended.

Whereas Rutherford learnt about the use of piezoelectricity from publications, Langevin enjoyed a direct personal contact with the Curies, which

[81] The mosaic is dated to November 1917–February 1918 in a historic outline prepared for the Allies conference in October 1918 (L194, 7, p. 21).

included observation and probably manipulation of piezoelectric crystals. Pierre's long and lasting influence on Langevin dated to the latter's earliest training in academic physics at Curie's laboratory at the EPCI. In 1888, when Langevin entered the school, and at least until the mid-1890s Curie still carried out research related to piezoelectricity, his first major scientific achievement. He also furnished its laboratory with related apparatuses, including a piezo-electric balance electrometer.[82] In 1905, the device, along with the physical laboratory came under the responsibility of Langevin, who succeeded Curie as a physics professor at the school.[83] Already during the 1890s Langevin had developed a close friendship with the Curies, which a few years after Pierre's death developed into a love affair with Marie that ended in a public scandal in 1911–12. Alongside their long friendship he kept up with their research on radioactivity.[84]

When the issue of priority regarding sonar and the use of piezoelectric-ity arose, Langevin recalled that back in 1915 he had already considered a piezoelectric ultrasonic sender, although, as mentioned, he did not pursue the idea until after he had discovered problems with his *receiver* in late 1916. In December 1918, he and de Broglie recalled that in January 1917 the latter had told the former about Rutherford's attempt to produce ultrasonic waves by a piezoelectric quartz and that Langevin had answered that he had already considered a similar method. Langevin added that he had already thought about the employment of 'his' cut and had carried out related calculations in late 1916. Yet, even if Langevin remembered correctly, knowing that Ruther-ford suggested utilizing piezoelectricity for a sender might have encouraged him to pursue his own more efficient piezoelectric receiver.[85] Notwithstand-ing, the comparison with Rutherford and his team suggests that Langevin enjoyed a more thorough knowledge of the phenomenon in its experimental

[82] Paul Langevin, 'Pierre Curie', *La Revue du mois* 2 (1906): 5–36; Katzir, *Beginnings of Piezoelectricity*, 95, 132; Curie's notebook in Bibliothèque nationale de France—NAF 18369.

[83] According to Frédéric Joliot-Curie, Langevin asked his assistant to cut from the crystal in that same device to make the first piezoelectric receiver. Joliot-Curie knew personally a few of the people involved but does not reveal source of the story; see Frédéric Joliot-Curie, 'Le professeur Langevin et l'effort scientifique de guerre', *La Pensée* 1 (Oct. 1944): 32–7, on 35.

[84] Bensaude-Vincent, *Langevin*.

[85] De Broglie to Langevin 23/12/18 (L194/49), Langevin to de Broglie 27/ 12/18 (L194/50). The two agreed that Langevin had not known Rutherford's report when de Broglie informed him about it in January 1917. References to the 1915 considerations appear only after this cor-respondence, e.g. 'Echo Sounding', 75. Only later Langevin mentioned that his work on the receiver had begun in late 1916, for example in 'a short history of piezoelectricity', written probably in the second half of the 1940s (L194/25).

manifestations, which enabled his success. This familiarity with various manifestations of piezoelectricity, especially with different cuts and directions of the effect, allowed him to think flexibly about its uses and to apply it for his needs.

2.5 Conclusions: Langevin and Rutherford and the Technical Application of Physics

The practical application of piezoelectricity beyond the laboratory depended on mobilizing physicists to the research and development enterprise of the First World War. Only after scientists like Langevin and Rutherford had engaged in solving a technical problem, namely locating submarines, did they suggest the practical application of the phenomenon, although they had known it for quite a few years, and sonar would have been useful for civilian and military vessels also before the war. Piezoelectricity, like many if not most phenomena known from the laboratory, did not recommend itself for technical employment, because, as shown in this chapter, making it a basis for practical devices required much time and effort. Before its concrete use was determined, scientists did not invest the resources needed for its technical development. Its transformation into an applied phenomenon required a concentrated technological study, and was not a by-product of scientific investigation. Thus, these physicists needed an external pressure to invest effort into improving technics and a specific technical problem to investigate employing piezoelectricity. Inventors, on the other hand, did not know piezoelectricity well enough to suggest using it for the problems that they formulated, such as the production of ultrasonic waves in Chilowsky's method. While conducting research within a limited time 'utilitarian regime', the physicists employed sources from their scientific knowledge and practice, like that of piezoelectricity. In a sense, they suppressed their scientific persona for a technological one (albeit reluctantly in the case of Rutherford, footnote 36). They did not differ in the strategy of relying on their expertise from inventors like Chilowsky, who employed his own experience with detectors, and technological knowledge, but they could more easily transfer scientific knowledge into technology.[86]

Rutherford considered passive reception of U-boats' noise as the most promising method for their detection, and therefore concentrated his efforts on developing such methods. Looking for ways to answer a particular

---

[86]Scientists transferred methods and knowledge from physics to technology also with other WWI technics, e.g. Tucker's microphone based on a laboratory device for a gun-locater system. Hartcup, *War of Invention*, 72–3.

question that emerged in this research, he exploited and modified the Curies' piezoelectric quartz to measure small mechanical vibrations. His piezoelectric instrument and his acquaintance with efforts to produce ultrasonic waves inspired his suggestion to employ the instrument as an ultrasonic emitter. Yet, regarding his preference for the passive system, he considered the problem of ultrasonic emission only briefly, and did not work on the realization of his proposal. Contrary to the claim of Rutherford's biographers, his suggestion was far from sonar. Langevin, in contrast, became the champion of the ultrasonic echo method and immersed himself in its development. His invention of the piezoelectric sonar resulted from a concentrated effort to solve specific problems: ultrasonic reception and then emission. His success highlights the importance of framing precise technical aims. Formulating the technical problem in specific terms fitting the problem in question was also central to Chilowsky's realization that one must employ ultrasonics in order to detect submarine objects. Those who framed it differently missed the importance of the wavelength. Langevin's work also shows the crucial role of actual research toward the technical aim, research which often reveals unforeseen problems like the effect of the sea on the carbon microphone. Inasmuch as Langevin's commitment to the echo method resulted from his personal identification with a technic in which he had invested his time and prestige, the case also points to the power of such personal interests in guiding technical research and development.

By following the Curies' measuring instrument, Rutherford succeeded in swiftly designing and constructing a dynamic measuring instrument. Considering his limited practice with piezoelectricity, restricting his efforts to the extant instrument was a wise move. Yet in so doing he limited himself to the particularities of that instrument, precluding the use of piezoelectricity for an efficient sonar system.[87] Rutherford's lack of experience with the various manifestations of piezoelectricity discouraged him from exploring beyond the restrictions of the particular device at hand, even when he did consider its utilization to a novel end. Langevin, better acquainted with the effect on

---

[87]This restrictive role of the individual instrument resembles the power of *Drosophila* as an experimental system in directing research to questions of chromosomal mechanism to which it was designed, and at the same time discouraging study of other questions like those of development and evolution. Kohler describes how during the 1910s and 1920s, the advantages of the system in the study of genetics directed experimentalists to work on that subject, and diverted them from other kinds of questions, not due to lack of interest but to the difficulties in applying their experimental system to these issues. Robert E Kohler, *Lords of the Fly: Drosophila Genetics and the Experimental Life* (Chicago: University of Chicago Press, 1994),173–207.

its experimental and theoretical aspects, more easily transcended the Curies' design and suggested a novel method that enabled a breakthrough.[88]

Notwithstanding the differences between them, that Rutherford and Langevin, two physicists with previous encounters with piezoelectricity, were the only ones who proposed its use for practical ends suggests that prior familiarity with the phenomenon enabled its novel technological application. Even as researchers could have acquired the knowledge needed for applying the effect without a direct connection with previous workers in the field, a prior encounter with the effect still seems as a prerequisite for considering its use for practical ends. Rutherford's and Langevin's acquaintance with the phenomenon put it on their horizons and thereby made it a candidate for application. Boyle's obliviousness to the potential utility of piezoelectricity well exemplifies the crucial role of prior knowledge of phenomena, not so much in acquiring knowledge to command the effects, but as in recognizing their possible use and benefits. In principle Rutherford could have examined further manifestations of piezoelectricity more useful for ultrasonic transducers. Yet the case suggests that in practice scientists (including highly creative ones) prefer staying closer to methods and knowledge with which they are more familiar, rather than looking for solutions in areas in which they are less, such as other piezoelectric cuts for Rutherford and Boyle. Langevin's earlier extensive knowledge of and practice with piezoelectricity provided him resources for considering the manipulation of the crystals needed for sonar. Similarly, Chilowsky's experience and knowledge of the method for probing by waves to discover hidden objects (unique among those who attempted echo-sounding methods) provided him resources for considering which kind of waves would be useful for submarine detection and to realize the superiority of ultrasonics. The critical contribution of prior familiarity with the phenomena points at the role of contingent and often personal routes by which knowledge moves, as exemplified in Langevin's connection to the Curies and the coincidence of their and of Rutherford's interest in radioactivity.

---

[88] Although theory was important in Langevin's work, as Zimmerman claims, it was only part of the scientific expertise that he brought with him and did not distinguish him from Rutherford. Moreover, as an inventor and developer he relied also on the results of his empirical work, e.g. his use of mica without a second metallic plate and the sandwich design, although informed from theory. See also the discussion in Chapter 3. Zimmerman, "'A More Creditable Way'", 68.

3

FROM SONAR RESEARCH TO THE DISCOVERY OF SHARP ELECTRIC RESONANCE

Langevin's piezoelectric ultrasonic receiver and his later quartz-steel sandwich for emitting and receiving waves were groundbreaking devices, which allowed for the construction of actual sonar systems for locating submarines. Still, there was a gap between these working prototypes and practical technics that could be implemented in navy vessels. Closing the gap required an intensive research and development endeavour, made possible due to the high priority given to the submarine detection problem and the promise of Langevin's technics. Allies' researchers and military personal in France, Britain, Italy, and the United States participated in this effort.[1] Their intensive research and development effort formed the third phase in the evolution of piezoelectric technics, sketched in the introduction to this book. The first part of this chapter examines the allies' extensive research on sonar, drawing mainly on contemporary reports. It presents the main research groups, their attempted improvements of the devices and their testing, and their research on piezoelectricity and related

---

[1] To give an estimate of the extent of the project, the initially approved annual budget of the American Naval Consulting Board was US$1.5 million, which is about US$500 million in economic costs in 2018. Locating submarines was the board's main task. Although some funds went to other aims, even in the United States, other agencies invested much money and materials and recruited personal, while the other powers carried out their own large projects in the field, even if the American one was larger. Daniel J. Kevles, *The Physicists: The History of a Scientific Community in Modern America*, 2nd edn. (Cambridge, MA: Harvard University Press, 1995), 107. Calculating the relative value of currency is based on the website Measuring Worth (http://www.measuringworth.com).

In 1926, Langevin estimated that the French Marine spent between £40,000 and £50,000 to support his group's research alone. This was about one hundred annual salaries of an American professor. The sum amounts to £20 million in economic costs in 2018. David Zimmerman, '"A More Creditable Way": The Discovery of Active Sonar, the Langevin–Chilowsky Patent Dispute and the Royal Commission on Awards to Inventors', *War in History* 25 (2018): 61.

Sonar to Quartz Clock. Shaul Katzir, Oxford University Press. © Oxford University Press (2023).
DOI: 10.1093/oso/9780198878735.003.0004

effects in crystals and in 'sandwich' transducers. Thereby, it explores the kind of work done in the research and development phase.[2]

Scientists were central to the work on submarine detection in general and on ultrasonic echo technics in particular. They continued to lead the development of the piezoelectric echo method even after Langevin had suggested its working principles. During World War I scientist expressed confidence in the power of scientific research to solve technical problems posed by the war and in its superiority over traditional methods of technical innovation. It was also the view of academic scientists and engineers involved in sonar research, as shown by their rejection of tinkering methods of trial and error, still often in use by professional inventors.[3] While they did examine the consequences of many variations in their devices and systems, they carried out methodological studies of their impact. Most of their research was technological in character, examining the effect of the devices' varying different parameters on their performance. Their methodology did not differ much from that of their experimental work on physics, but its subject matter and aims had changed. To paraphrase Edwin Layton, they crossed through the looking glass that separated scientists and engineers.[4] Still their research involved also more general, hence scientific, study into the properties of related materials and phenomena including properties of metals and crystals, unknown or little known piezoelectric and elastic coefficients and interactions, and the conditions of seawater and their influence on the propagation of ultra-sound.

The research on sonar drew unprecedented human and material resources to the hitherto calm field of piezoelectricity. The topics of study followed

---

[2]Research on locating submarines was not confined to active echo systems. Even larger resources were invested in improving older passive means of locating submarines by the noise that they produced. Unlike sonar, these were put in active service during World War I, but with limited success and limited future. On the organization of submarine detection research and the research of passive methods see Willem D. Hackmann, *Seek and Strike: Sonar, Antisubmarine Warfare and the Royal Navy, 1914–54* (London: Her Majesty's Stationery office, 1984), Chapters 2–3, 11–71, and Roland Wittje, *The Age of Electroacoustics: Transforming Science and Sound* (Cambridge, MA: MIT Press, 2016), 89–101.

[3]On the view that science is more potent than other methods see Kevles, *The Physicists*, 109–16 and Ian Varcoe, 'Scientists, Government and Organised Research in Great Britain 1914–16: The Early History of the DSIR', *Minerva* 8 (1970): 192–216. Similar attitude is expressed in Ernest Rutherford, 'Henry Gwyn Jeffreys Moseley', *Nature* 96 (1915): 33–4.

[4]According to Layton, engineering and science mirror each other, agreeing in their methods but differing in their aims. Layton, 'Through the Looking Glass, or News from Lake Mirror Image'.

questions related to the employment of crystals in ultrasonic transducers. They included known issues in the study of the effect, in particular the behaviour of quartz and Rochelle salt crystals under pressure and electric fields and the determination of piezoelectric coefficients. They extended also to previously unexamined issues regarding the behaviour of crystals under high-frequency vibrations. In both kinds of studies, researchers gained surprising and unexplained results that led to the discovery of two major effects: (a) an abrupt change in the electric behaviour of piezoelectric crystals near resonance, and (b) ferroelectricity—the induction of permanent electric polarization by the application of a transitory field. Yet, although originating during war research, the two phenomena were discovered only after the end of WWI. This chapter examines the main characteristics of sonar research during the last part of the war, showing how and why it led to the peculiar observations later related to these effects and why they were not further explored at the time. In the chapter's second part, I examine how and why immediately after the war Walter G. Cady followed the few peculiar observations to discover the sharp steady electric resonance of piezoelectric crystals, which would become the basis for the highly useful crystal frequency control. Scientists explored phenomena related to ferroelectricity somewhat later than piezoelectric resonance; I therefore discuss it in Chapter 9 on the scientific research at the fifth phase.

In 1918–19 Cady was the first to recognize and to further examine the significance of the abrupt change in crystals' electric properties at their resonance frequency. Based on Cady's notebooks, the second part of this chapter discusses how his work on piezoelectric ultrasonic devices, coupled with his general interest in natural phenomena, drove him from his original research programme on ultrasonics to that on crystal resonance, which in turn led to the discovery of the concomitant electric effect's sharpness and stability and then to suggest ways to utilize it. Due to the importance of this discovery I describe Cady's background and examine closely his work on crystals.

Cady's transition was also a move from technological research aimed at improving specific technics, to disciplinary or pure research lacking a clear practical aim. As discussed in the Introduction, technological or 'engineering research' shares the methods of the natural sciences, but differs in its aim. It seeks knowledge for the sake of the design and improvement of practical technics. Therefore, its questions and results differ from those of disciplinary research. The research of Cady and his colleagues included both kinds of research in the study of one and the same subject—oscillating piezoelectric crystals. It provides, therefore, an advantageous perspective on the controversy regarding the relationships between disciplinary and utilitarian research, their significance, and even their very existence.[5] In Cady's research, knowledge was sometimes its own aim, sometimes it was a means for improving design, and sometimes it was both, as questions of understanding and intelligibility were linked to possible practical usages. A close examination is therefore required to plausibly reconstruct the goals of Cady's research, whether of disciplinary knowledge or of design, at different stages in order to appreciate the contribution of and limitation imposed by each kind of research. This investigation shows that although parts of the research cannot be neatly divided between the poles of 'engineering' and 'pure' research, at other instances the two kinds of research led to investigations with significantly different immediate aims, examined phenomena, and, consequently, results.

3.1 Improving the Piezoelectric Sonar

The French were quick and generous in sharing details about Langevin's advancements with ultrasonic echo technics (the receiver, the one quartz-crystal transmitter, and the sandwich oscillator) with their allies. Researchers in Britain, Italy, and the United States (after it entered WWI) turned to copy, adapt to their own particular needs, and improve on Langevin's suggestions. The free flow of information and materials, facilitated by the war emergency, helped foreign researchers enjoy the French experience. Robert Boyle was Langevin's closest foreign collaborator. He spent six weeks with Langevin in Toulon in May–June 1917, learning about the success of the receiver and experimenting with the new piezoelectric emitter and returned there in May 1918. He paid at least two additional visits to France, whereas Langevin visited the British experimental station twice.[6] When direct meetings were impractical, the researchers learned from detailed written reports, which included also

---

[5]See the Introduction to this book and the literature quoted there. For the controversy see, for example, Sungook Hong, 'Historiographical Layers in the Relationship between Science and Technology', *History and Technology* 15 (1999): 289–311.

[6]Boyle BIR 1/8/17: 'Report of Mission to France for the Admiralty Board of Invention and Research.—March 20th to July 19th, 1917' (BIR 30,061/17), ADM 293/10, a copy also in 293/10; Boyle visited France also in May, June, and October 1918. Hackmann, *Seek and Strike*, 89; mention of Langevin's visits in his correspondence at ESPCI and A. B. Wood, 'Admiralty Experimental Station, Parkeston Quay (Harwich) 1917 to 1919', *Journal of the Royal Naval Scientific Service* 20, no. 20 (A. B. Wood, O.B.E., D.Sc. Memorial number) (July 1965): 40.

blueprints of the devices, from correspondence and in some cases from the devices themselves that were sent to them. Most of the reports were written and also orally delivered by scientific attachés. Unlike Langevin and Boyle, they reported also on work in progress and on various comments and ideas of the researchers. The experimental physicist and navy officer Maurice de Broglie was central in transferring knowledge between France and Britain, among others, through his visits in Toulon and Harwich. The head of the Italian Ufficio Invenzioni e Ricerche, the parallel to Painlevé's Ministère des inventions, Vito Volterra personally participated in Langevin's experiments in Toulon. The physicist Karl Compton, assistant for the American scientific attaché in Paris, and the engineer Samuel L. G. Knox, the scientific attaché in Rome, also visited Langevin and joined his experiments, reporting on them and commenting about their advantages and drawbacks to their colleagues in the United States.[7] On 19–21 October 1918, delegations from the four allied countries met to compare the progress on sonar research at the different sites and to plan further research. Yet, when they met, the end of the war was, at last, in sight, and the future of the research during peacetime was unclear.[8]

A few dozen American scientists and engineers learnt about the details of the ultrasonic echo methods in a special three-day conference in Washington, DC, in June 1917. British and French scientists, including Rutherford, and physics professors Henri Abraham, who followed closely Langevin's work,[9] and Charles Fabry described the devices and research on submarine detection methods (the two physicists appear again, respectively, in

[7] Reports of Compton from the Paris office in the Records of the National Academy of Science, USNA (entry 2, box 2), Elias Klein, *Notes on Underwater Sound Research and Applications before 1939* (Washington, DC: Office of Naval Research. Department of the Navy, 1967), 7–8. There are many mentions of de Broglie and his role in British reports and in correspondences of Rutherford and Langevin; see also Boyle's report at the interallied conference, Paris, 19–21 Oct. 1918: 'Conférence Interalliée du 19 Octobre 1918: Compte rendu officiel—expose du Docteur Boyle', ESPCI L194/09; Volterra's statement, there; on Volterra's role in the Italian research, Giuliano Pancaldi, 'Vito Volterra: Cosmopolitan Ideals and Nationality in the Italian Scientific Community between the Belle Époque and the First World War', *Minerva* 31 (1993): 34–5.

[8] 'Conférence Interalliée sur la recherches des sous marins par la methode ultra-sonore', Séances du lundi 21 Octobre 1918, L194/7.

[9] At least in August 1918 Abraham was familiar enough with the details of Langevin's system that his opinion about the appropriate width of the radiated beam was mentioned in the same breath as that of Langevin himself. Karl Compton, report 122, 'The Langevin apparatus for supersonic signalling and detection of submarines by echo', 28.8.1918, p. 6 in USNA, Paris Office Box 5.

Chapters 5 and 9). The French focused on Langevin's piezoelectric transmitter and emitter. During the following months, American scientists imitated the piezoelectric devices and began exploring additional kinds of crystals and transducers. They continued to follow the progress in France and Britain through written documents and from May 1918 also enjoyed informative reports of their scientific attachés.[10] American researchers believed that these means of communication left them at a disadvantage compared to their European colleagues. At the October 1918 meeting, Columbia electrical engineering professor John H. Morecroft, speaking on behalf of his colleagues, expressed 'our regret that we could not have the benefit of Langevin's advice and counsel as Professor Boyle has been able to have. I feel that our advancement in the problem would have been immeasurably hastened, had we been able to associate with him more closely.'[11] They enjoyed, however, considerably richer resources than their colleagues across the Atlantic. US researchers carried out studies relating to submarine detection in ten different sites, by about eight research groups (partly connected to each other), seven of which examined piezoelectricity. The work of the latter groups was coordinated by the National Research Council, whose director of research, Robert Millikan, a distinguished experimental physicist, chose to direct its antisubmarine committee, an indication of its deemed importance. Millikan achieved a close cooperation between academy and industry and among different corporations.[12]

[10]Walter G. Cady, 'Piezoelectricity and Ultrasonics', *Sound: Its Uses and Control* 2 (1963): 47–8. Hackmann, *Seek and Strike*, 41–2.

[11]J. H. Morecroft, 'Historical Survey of Development in the United States', presented at an interallied conference regarding ultrasonics in Paris, 21 Oct. 1918, published in Klein, *Notes on Underwater Sound Research*, 13.

[12]The groups included such academic scientists as professors of electrical engineering Michael Pupin (Columbia) and Harris J. Ryan (Stanford); the astronomer John A. Anderson (Mount Wilson Observatory); and the physicist Albert P. Wills (Columbia); as well as industrial scientists from the laboratories of General Electric, Western Electric, and other telecommunication companies; and a group from the American Bureau of Standards. See Hackmann, *Seek and Strike*, 41, 90–2; Willem D. Hackmann, 'Sonar Research and Naval Warfare 1914–1954: A Case Study of a Twentieth-Century Establishment Science', *Historical Studies in the Physical Sciences* 16 (1986): 83–110, 95–7; Cady, 'Piezoelectricity and Ultrasonics', 46–9. On the creation of the NRC and its advocacy of 'scientific' research and study of submarine detection see Kevles, *The Physicists*, 109–26.

Langevin's and Boyle's Work

Following the design of the steel–quartz mosaic 'sandwich' transducer in late 1917, and the successful experiments on its use in February 1918, Langevin's group concentrated on its improvement and its methods of use. The improvement included changes in the sandwich design, and in the electric circuits for transmitting and receiving waves. Probably the most important change was to use the same sandwich for both emitting and receiving the ultrasonic waves, while in the original design one crystal was used as an emitter and another as a receiver. Some time between February and May 1918 the group connected a crystal sandwich through a commutator to emitting and receiving electronic-valve circuits. Yet, connecting and disconnecting the receiving circuit caused distributing noises through the telephone used for detection in the receiving circuit. This was due to a technical problem with the electric system, under Colin's responsibility, and the solution was found in the same field: permanently connecting the receiving circuit to the sandwich; due to its low voltage, it did not affect the emitting circuit. Other changes were related to the electric circuits and their valves and specifications of the sandwich. For example, between May and August 1918, Langevin turned to quartz plates of 4 mm instead of 2.5 mm width (while the steel plates remained 30 mm each).[13]

Most of the questions examined by the group pertained to design specifications like the optimal mixture to cement the quartz mosaic to the steel (2/3 rosin, 1/3 beeswax, preferred by Langevin, but not by all); the optimal width of the quartz plate; the optimal width of the ultrasonic beam. The specifications were chosen due to their usefulness in practice rather than based on theoretical considerations. The latter served as a general guidance in suggesting modification in the devices. For example, since the damping due to the viscosity of water was known to be proportional to the square of the wave frequency, in May 1918 Langevin considered lowering the frequency in use to 20,000 Hz. Yet, he did not change the frequency of the transducer. The sources are silent about his reasons; it might be that the 38,000-Hz sandwich satisfied the needs, or that a 20,000-Hz transducer showed in practice other problems, like a too-wide beam.[14] This kind of questions and combination of

[13] Karl Compton, reports no. 37, 'Submarine Signaling by Supersonics, and the Detection of Submarines by Echo', 15.5.18; no. 40, 'A Report on Langevin's Work', 18.5.18; no. 122, 'The Langevin Apparatus', 28.8.18, USNA, Reports received from the committee's Paris office, 1917–1921 (boxes 2, 5).

[14] In 1924 Langevin referred to 40 kHz as the optimal frequency for additional theoretical reasons regarding the proportion between the sender surface and the wavelength, but in 1918

guidance from known scientific laws and empirical examination of their performance under the required settings characterizes also the work of British and
American researchers. The researchers did not aim at better knowledge of the
phenomena but at more efficient working devices and systems. For the latter,
they carried out also general studies of the devices and materials they used,
such as Langevin's examination of the properties of different crystal plates
at different frequencies, as mentioned in Chapter 2. Sometimes, they examined beyond the particular devices more general and previously little studied
phenomena, such as the temperature gradient of seawater with depth and the
consequent refraction of ultrasonic waves.[15]

Improvements in the transducers, in their electronic circuits, and in the
mechanism of controlling the whole apparatus (for example, allowing it to
survey 360°) and the development of appropriate tactics for their use on moving sea vessels allowed the system to be used. In July–August 1918 Langevin's
team performed trials on the sea in which the system managed to chase a submarine, first detecting its echo at a distance of 0.8 to 1.3 km. The French Navy
planned to install the device in seven ships by early 1919, but the end of the war
stopped the programme. Karl Compton recommended 'immediate adoption
of the apparatus' for use to supplement the submarine listening devices in use
in the United States. This, he wrote, 'appears to be the unqualified opinion of
all who have studied the question'.[16]

Although impressed with the use of piezoelectricity, since the British did
not have a quartz crystal with the desired properties to imitate Langevin's
early receiver, Boyle and Smith began researching piezoelectric transducers
only in October 1917, after receiving a suitable crystal from Langevin. In the
meantime, Boyle studied the basics of piezoelectricity in quartz. After making some advances with a single-crystal emitter in early 1918, in May they
followed Langevin again and adopted his successful sandwich in a somewhat
different design. Like the French team they used the same oscillator as both
emitter and receiver. Their research focused on the device's practical aspects
and how to install it on a moving vessel, such as examining suitable materials
for the dome that would protect the oscillator from the sea but would enable

---

he considered a larger surface as the problem's solution. He did not mention the use of the
20-kHz transducers in any of the documents that I have seen. Compton, report 37, 'Submarine
Signaling by Supersonics'; report 40, 'A Report on Langevin's Work'; report 122, 'The Langevin
Apparatus'. Langevin, 'The Employment of Ultra-Sonic Waves for Echo Sounding', 68–70.

    [15]Compton, report 37, 'Submarine Signalling by Supersonics'.

    [16]Hackmann, *Seek and Strike*, 83. Compton, report 122, 'The Langevin Apparatus', 6.

undisturbed motion of ultrasonic waves. Following successful experiments in seawater, a few days after the end of the war, the team, together with navy personnel, carried out sea trials on the system known in Britain (and sometimes in the United States) as 'ASDICS'. The British navy installed ASDICS on a few ships beginning in 1919.[17]

Research and Development in the USA

The USA carried out more extensive research on submarine detection in general and on ultrasonic echo methods in particular than the European powers. The larger enterprise allowed for research in more directions. Like with Langevin's research, most of the US research was related to specific design questions. Spending more resources and being less invested in the particular quartz transducers they also explored alternative designs. The extensive sources suggest that the American researchers, especially in their early phases of the research, dedicated more time to basic questions about the physical effects related to the technics than Langevin and Boyle had done.[18] The basic properties of the crystals under high-frequency oscillations were studied among others by a group around the Columbia physicist Albert P. Wills, by Cady at Wesleyan University, and by the astronomer and physicist John A. Anderson's group at Throop College of Technology (soon afterwards Caltech).[19] The California group also carried out a systematic study of choosing an optimal quartz width for the sandwiches. They examined the intensity of ultrasonic waves at different frequencies of quartz-metal sandwiches, where the thickness of the quartz plate to that of the total sandwich varied from

[17]'Conférence Interalliée du 10 Octobre 1918, Compte rendu officiel, Expose du Docteur Boyle' (L194/09), Robert Boyle, 'The Piezo-Electric Phenomenon, with Reference to Acoustic Vibrations', BIR 38,164/17 (undated, the list of report states that it is from Dec. 1917), in ADM 293/10 and ADM 212/159; Hackmann, *Seek and Strike*, 85–9. Hackmann overlooked that Boyle was informed about Langevin's sandwich only in May 1918 (according to Boyle's report in October of that year). Zimmerman, "'A More Creditable Way'", 65.

[18]The impression might be an artefact of the fact that US researchers sent more detailed reports to the NRC, which coordinated the research, since they were spread over a few sites. These reports included findings of their initial experiments as well as about the results of their trials. As mentioned in Chapter 2, Langevin's notebook reveals a basic study of the behaviour of quartz crystals. Still the amount of results reported by the Americans suggests that they carried out more research into the phenomena of the piezoelectric transducers.

[19]Although primarily an astronomer Anderson had studied tourmaline crystal before the war, and thus had relevant experience with crystal physics. I. S. Bowen 'John A. Anderson, Astronomer and Physicist', *Science* 131, no. 3401 (4 March 1960): 649–50.

5 to 100%. The California and Columbia groups also questioned Langevin's choice of longitudinal vibrations and examined the use of transverse oscillators, as they required lower voltage (the cut they used was still different from 'Curies' cut' in the 'quartz piezo-électrique', which Rutherford used). In this case the exploration of alternatives did not lead to a change in the devices that they used, which were still based on Langevin's longitudinal oscillator.[20] As a preparatory stage to this research, and to the testing of crystal specimens, the researchers studied piezoelectric theory and experimented with its phenomena.

Like their European colleagues, the American professors involved in the research took care of many practical questions related to the technics. These included not only the design of transducers, their installation in the ships, and how they were used but also about issues such as the supply and cut of proper crystals and the supporting electronic instruments. While the European professors did not perform research in the latter realms, US professors also examined the supporting electronic circuits. This was perhaps due to the higher number of academic researchers in the American enterprise, on the one hand, and to the better cooperation between scientists and military in France and between independent engineers and technical people in Britain, on the other. While navy engineer Victor Colin was responsible for the valve circuits in Langevin's system, Columbia electrical engineering professors Morecroft and Michael I. Pupin developed generating circuits, valves, and high-frequency amplifiers.[21]

After learning about Langevin's one-crystal oscillator in April 1917, American researchers examined also non-piezoelectric transducers, returning, among others, to ideas tried out by Chilowsky and Langevin. Yet they turned almost fully to piezoelectric methods later that year after learning about Langevin's success with multi-plate crystal senders.[22] The major alternative

---

[20] A. P. Wills, 'Asdic laboratory report', November 1918 to 1 June 1919, held at Elias Klein papers in AIP; J. A. Anderson, San Pedro Group, 'Behavior of Rochelle Salt Crystals', 13/3/1918 USNA Washington office (box 72, #2); W. G. Cady 'Report on Experiments with Rochelle Salt', 8/5/18, from Wesleyan University, USNA, Washington office (box 50, #15, a copy).

[21] Wills, 'Report'; 'New London Reports', March 1918–February 1919, held at the (US) National Academy of Science.

[22] Morecroft recalled that a group in Columbia began with non-piezoelectric means in April 1917 and turned to quartz oscillators in September that year. Since this is before Langevin constructed the sandwich, I assume that they received information about Langevin's emitter consisting of a few quartz plates (see Ch. 2). Anderson in California reported the use of metals to produce ultrasound in 3/12/17, USNA, Washington office (box 72). For a later reference to

that the Americans explored was the use of Rochelle salt crystal instead of quartz for the oscillators. They began exploring the use of that crystal in summer 1917. Crystals were grown and cut by General Electric laboratories, where Cady worked from July to October of that year, and Western Electric laboratories, the manufacturing arm of AT&T. By November Cady in New England experimented with Rochelle salt; he continued with its study and use until the end of the war. Anderson's group on the East Coast carried out parallel research during the first months of 1918.[23] The decision to examine Rochelle salt as an alternative to quartz followed its intense piezoelectric effect, more than two orders of magnitudes stronger than that of quartz, which suggested more efficient transducers. It was probably stimulated by difficulties in attaining large flawless quartz plates needed for Langevin's early 'one-crystal' resonator, and the high voltage that they required, which could be significantly reduced if one could use Rochelle salt plates of similar dimensions. Here again the size of the American research probably played an important role, as two and later one small group could pursue an alternative crystal, while other researchers worked on quartz.

Experimenting with Rochelle Salt

The use of Rochelle salt required more preparatory research than the use of quartz. Quartz had been the paradigmatic crystal in the piezoelectric research. The properties and performance of other crystals, including Rochelle salt, were less known. Previously, only one study, by Pockles, quantitatively examined the main piezoelectric properties of the crystal. Furthermore, it was more delicate than other crystals; it was prone to cracks and breaks, dissolving by humidity, and melting at 75°C. Moreover, Pockels left the largest piezoelectric coefficient of the crystal undetermined, since its value changed from one measurement to another. Early studies of the crystal by the American groups found that 'Rochelle salt well deserves the reputation it is gaining for erratic behavior.' With considerable efforts they managed to get quite stable results from

the non-piezoelectric ultra-sonic method see I. B. Crandall, report no. 16, 'High Frequency Directive Submarine Signalling', 8/8/18, in USNA, Washington office (box 50).

[23] Anderson reports, 'Behavior of Rochelle Salt Crystals', 13/3/1918, 22/4/18 (USNA, box 72); Cady, 'Outstanding Dates Relating to Work of W. G. Cady in Piezoelectricity', 2-page manuscript written probably in the 1960s, kept in AIP, unnumbered Notebook (NB), box 23, folder 2 in Cady's papers at ACNMAH; Walter Cady's diaries kept by the Rhode Island Historical Society, henceforth *Diaries*. I thank Chris McGahey for providing me with his research notes on them; Hackmann, *Seek and Strike*, 92.

the same plates, but marked peculiarities and differences between different plates, sometime even when they were cut from the same specimen. They, thus, examined the properties of Rochelle salt before testing a compound Rochelle salt–metal oscillator.[24]

Preliminary research of the crystal included basic questions about practical ways to handle and connect it with electrodes. Stabilizing the effect of electric tension and of mechanical stressed required time and much care. Cady, for example, found that backing plates at about 55°C improved the activity of problematic and fatigued plates. The researchers recorded electric properties like resistance, and elastic ones like their Young modulus. Voigt's piezoelectric theory taught that in order to have lengthwise vibration an oblique cut of the crystal is needed, since the crystal is electrically polarized only by shear stress. To enjoy the strong electric effect of sheer strain on the electric axis in the x direction, the Americans cut in a plane that included the x direction (c-cut 45° bar), measuring about $10 \times 10 \times 4$ mm. Their findings were surprising. In early 1918 Anderson's group examined the effect of a Rochelle salt plate placed between charged electrodes on the electric tension between them. The researchers found the effect to be asymmetric, depending on the sign of the electric charge. They thought that this might be the result of electrostriction (a quadratic mechanical deformation due to an electric field, which unlike piezoelectricity is dependent only on its absolute strength and not on its sign) or more exactly on the secondary effect of it. Although they were not sure how to explain the observations, they nevertheless saw an opportunity to enhance the sensitivity of Rochelle salt crystals used as receivers for ultrasonic waves. They suggested that the crystal might be more sensitive if put under constant electric field of about 100 volts/cm in the same direction that was expected to be electrically polarized by the returning ultrasonic wave.

Consequently, they examined the electric effect of pressure on plates that were also under electric field of various values. While the electric effect of pressure alone was quite linear, the addition of an electric field completely altered the results. The electric voltage in the detecting circuit was far from linear, dropping quite strongly at a particular range, depending on the applied voltage. In one case, it even changed its direction, first rising in value and then strongly dropping. Anderson candidly concluded: 'These results are surprising, to say the least, and the writer is sorry to admit that he has

[24] Anderson reports; Cady 'Report on Experiments with Rochelle Salt'; 8/5/18, NB 23; Shaul Katzir, *The Beginnings of Piezoelectricity: A Study in Mundane Physics* (Dordrecht: Springer, 2006), 203–4.

been unable to explain them. That electro-striction plays an important role is quite obvious . . . but the writer has not been able to combine electro-striction and piezo-electric effect in such a way as to get the results observed here.' Later research suggests that the results originated in a spontaneous polarization induced by the electric field in Rochelle salt, a property of ferroelectricity, a phenomenon unknown at the time. Yet, exploring the process and the laws that led to the unexplained findings was a low priority for Anderson's group. It concentrated on its technological aim—providing rules for designing ultrasonic transducers, and did not examine its results further.[25] Although he could not explain the observations, Anderson still inferred from the experiment a practical rule for using Rochelle salt as a receiver. Technological research had contrasting consequences for the discovery of ferroelectricity. On the one hand, observation of the effect originated in the group's research aimed at using Rochelle salt as an ultrasonic transducer and in particular in their hypothesis that an additional electric field would improve sensitivity. This setting was unlikely in a scientific exploration of the crystal's properties, without a particular practical aim. On the other hand, because of this same technical aim the group did not explore its surprising findings.

A month after Anderson, Cady reported on similar experiments, which he performed with Frank J. McGrath, a recent MA graduate and an assistant in physics at the University of Michigan, who served during WWI at the American Bureau of Standards. They also gained a few peculiar results, although not as bewildering as those of the California group. Among others they found that the effects of applying mechanical stress at once, in stages, and by removing weights produce different electric results, indicating a kind of hysteresis (a term which Cady did not use). High loads cause nonlinear electric effect, and after about 15 minutes also fatigue of the crystal (i.e. showing a weaker effect). Either following Anderson's report or their own observations, they also found that an electric field in the 'positive direction' enhances the electric effect of pressure. Yet, unlike the California team, Cady and McGrath examined the effect of pressure *after* the release of the electric tension, and 'found that the plate gave exceptionally large throws [in the galvanometer needle] for several hours later'. Yet, like their West Coast colleagues, they were not interested in exploring the phenomena but in exploiting it for their technical need for ultrasonic transducers. To their disappointment, however, at frequencies of ultrasonic waves

---

[25] Anderson, 'Behavior of Rochelle Salt Crystals', quote in the report of 22/4/18.

the exertion of voltage before the experiment had no effect. Still they assumed that Rochelle salt could be a good receiver, although it failed to show sharp resonance at any frequency. In this issue they disagreed with the California group, which by August 1918 discarded the use of the crystal for receivers.

Consequently, Cady and McGrath continued to study the properties of Rochelle salt as an oscillator among others in experiments in a water tank at Wesleyan University. They devised two kinds of Rochelle salt oscillators, and examined their performance and the effect of external conditions on them. Most of their research, however, was of an engineering kind. For example, they found that a sound screen behind the plates can increase the amplitude. In this case their research on the crystal and on the oscillator was very close since they did not form a crystal–metal sandwich for the oscillators, but crystals were put between two small tin plates. During this research, Cady observed a strong and unexpected electric reaction for Rochelle salt in resonance. After the war, this observation would become key to his discovery of sharp electric resonance in crystals, discussed in section 3.2. In August 1918 he and McGrath settled on a set of 12 to 18 Rochelle salt plates. For technical reasons, they found that is better to immerse the crystals in oil rather than cement them to a tin diaphragm.[26]

Quartz Transducers and the State of the Art at the End of the War

Although better known, the researchers examined also the piezoelectric and other physical properties of quartz. Often the study concentrated on quartz–steel sandwiches and their compound properties. Since piezoelectric vibrations had not been studied before, these studies also contributed to the general knowledge of the phenomenon. The quartz plates used in these sandwiches had been tested before they were insert into the compound to ensure that they do not have impurities and twists that would harm their performance. A few methods of testing based on static piezoelectricity were developed among others by Cady and Rutherford, who tested the plates used by the British.

---

[26] Cady, 'Report on Experiments with Rochelle Salt', 8/5/18, and 21/8/18, USNA box 50; Cady in 'Report on Conference of Physics and Engineering Divisions of the National Research Council', Washington, 18/7/18; Fuller and Ryan, 'Quartz Supersound Source Projector', both reports at the National Academies Archives (Washington), Central Policy Files, 1914–1918. On McGrath see *Wesleyan University Bulletin* (1918), 26 'Michigan Men in Service', *The Michigan Alumnus* 24 (March 1918): 357.

Cady also performed similar tests for Rochelle salt crystals.[27] In the early stages, researchers like Langevin and Cady studied also quartz plates without the metal coating under static and oscillating electric and elastic forces; they seemed to keep their findings to themselves, probably since they agreed with earlier published findings.[28]

Shirley L. Quimby carried out a more thorough study of quartz that included determining its major piezoelectric coefficient (and the only one relevant for the resonator). Quimby was a junior researcher at the largest group of researchers in the field, which was established at Columbia University and then moved to the naval research station in New London, Connecticut, under the actual directorship of Wills. His research was part of a larger theoretical and empirical study of the sandwich resonator, whose behaviour depends, among others, on the value of the coefficient. Beginning soon before the end WWI and continued until June 1919, this was probably the most comprehensive study of the issue. It included the first published (if only in a confidential report) theory of piezoelectric oscillations. Yet, earlier remarks of other researchers beginning with Langevin suggest that they also had made some, probably more basic, calculations of the interactions between electricity and elasticity in the metal–quartz sandwich. Based on theory of sound waves, piezoelectricity, elasticity, and electricity, Wills's group calculated the energy of ultrasonic waves produced by such a sandwich at various frequencies, and the electric voltage produced by a receiver, depending on the amplitude and frequency of the ultrasonic wave. While theory showed remarkable rise in the energy of the emitted wave at resonance frequency, the experiment showed much smoother peaks. Wills explained the gap by deviations of individual quartz plates from their assumed direction, and effects of low sound velocity at the ends of the plate, which were not considered in the theoretical account.[29]

---

[27] W. G. Cady, 20/6/18: 'Method of Testing Piezo Electric Crystals', USNA, Washington office (box 50, #9); On Rutherford's tests in Manchester, [Henry Andrews] Bumsead 24/4/18: 'Methods of Testing Quartz Crystals for Use in Supersonics'. USNA London files #57.

[28] Since I have learnt about Langevin's and Cady's studies only from their notebooks, I assume that a few other researchers have made similar examinations. Langevin, Notebook 'Servis US Cahier d'ordres du Directeur', undated by content, probably from 1917, L148/18 (discussed in Ch. 2), Cady NB 23.

[29] A. P. Wills, 'Asdic Laboratory Report' (held at Elias Klein papers in AIP): 'The following report covers the work done in the Asdic laboratory at the U.S. Naval Experimental Station, New London from the time of its establishment in November 1918 to June 1st 1919', 1.

In examining crystals under alternating voltage and pressure, Wills and his co-workers recognized the contribution of secondary effects of piezoelectricity, e.g. the effect of piezoelectricity on the dielectric constant. That meant that the crystal sandwich changes its electrical properties due to the effect, and that such a change should be observable when the inner motion of the crystal is relatively free, i.e. near resonance. The connection between their findings and the properties of piezoelectric resonance that Cady found (discussed in section 3.2) is clear in retrospect. Yet, although the report was written after Wills had learnt about Cady's findings (Cady was a visiting member of Wills's group and presented his findings to Wills), it does not account for the sharp electric resonance described by Cady.[30] Physicists did not predict the electric phenomena of resonance, since they required special considerations of the rather complex physical conditions, among others a separate treatment of resonance and non-resonance frequencies.

While they acquired further knowledge about physical phenomena, the goal of those researching the ultrasonic echo method was to produce a working practical technics. In that respect, the American groups made significant advances, building on the ideas and experience of the French and British. Beginning in autumn 1918 the New London group performed in situ experiments on the reception of echo from a ship and underwater objects, reaching a range of 400 to 600 m. A division of labour between the different American groups was planned to allow improvement of separate aspects (electronics, the crystals, and the transducers). While they adopted the quartz–steel sandwich as an emitter, the Americans continued to work with Rochelle salt receivers produced by Cady. They compared their reception in sea trials with that of the quartz sandwich. In December 1918, they decided to drop the research on Rochelle salt receivers, but half a year later, Wills still reported about their earlier use with approval. With the end of the war, however, the research became less intense, although Wills continued the research at New London with a smaller group.[31] Ultrasonic echo detection methods—ASDIC—did not reach operational use during the war.

Yet, its potential for commercial and military navies and its innovative character were clear. Its scientific origin was also known, and was emphasized by

---

[30]The report does suggest that the oscillator behaves electrically in a different way above and below resonance frequency, as established by Cady, but I suspect that this discussion followed Cady's findings. Ibid., 30–1.

[31]'New London reports', especially those of 29.10.18 (box 56, 12.12.18 (box 55), USNA papers at National Archive; Wills, 'Asdic Laboratory Report', Morecroft, 'Historical Survey'.

scientists like Langevin when they later discussed it in public. After the war it served as a leading example of the contribution of science and scientists to technology. It was, however, secondary to the success of scientists developing passive systems for locating submarines by the noises they produced (even if their tactical help is doubtful). In the United States the development of these methods for locating submarines was regarded as a triumph of the scientific method over that of the professional inventor, symbolized by Thomas Edison. It led to a wider appreciation of 'scientific research', i.e. a study of phenomena connected to the technics in use, by policy makers in the military and government.[32]

Research on ultrasound and methods of its production and reception continued during the interwar period. The improvement of sonar systems was a central interest in this research, even as it was much less intensive than during the war, and was developed at a much slower pace. Both military and civilian fleets adopted sonar technics; e.g. by the mid-1920s the British Admiralty had produced enough devices to equip up to a hundred vessels. Ultrasonic became also an interest for research, both for examining other effects and materials and for its own study, hitherto unexamined. Among other researchers, Langevin and Boyle carried out studies of the phenomena.[33] As will be discussed in Chapter 9, it also continued to affect the research of piezoelectricity, since piezoelectric transducers were employed in ultrasonic research. Despite these important development, and the extensive later use of ultrasonic scanning in medicine, industry, and the laboratory, the most valuable contribution of sonar research to technology was crystal frequency control technics. Based on the finding that followed the research on ultrasonic transducers it became and still is ubiquitous in electronics. How Cady discovered the peculiarities of piezoelectric resonance, on which the technics is based, is analysed in the second part of this chapter.

---

[32] Langevin, 'The Employment of Ultra-Sonic Waves'; Kevles, *The Physicists*, 119–38; Shaul Katzir, '"In War or in Peace": The Technological Promise of Science Following the First World War', *Centaurus* 59 (2017): 223–37.

[33] Hackmann, 'Sonar Research and Naval Warfare'; Lelong, 'Paul Langevin et la détection sous-marine', Zimmerman, '"A More Creditable Way"', 65. Roozbeh Arshadi and Richard S. C. Cobbold, 'A Pioneer in the Development of Modern Ultrasound: Robert William Boyle (1883–1955)', *Ultrasound in Medicine and Biology* 33 (2007): 8–11.

3.2 Cady's Discovery of the Electric Properties of Piezoelectric Resonance

Background and Research during the War

Until the events discussed here, Walter Cady had had a quiet and undistinguished career. Born in 1874 in Providence, Rhode Island, he 'decided on pure physics instead of engineering' during his college years at Brown University, where he stayed for his MA. Encouraged by one of his teachers, he continued his studies in Berlin, earning a PhD in 1900, on experimental research concerning the energy of cathode rays (electrons), on which he worked mostly with Walter Kaufmann, one of the world experts in the field. He returned for more pragmatic research at the magnetic observatory of the US Coast and Geodetic Survey, and in 1902 joined the faculty of Wesleyan University at Middletown Connecticut, where he taught physics until 1946. Cady recalled that:

> In one way, it was my good luck to be at a small college where I had no superiors to direct me, and where I had to choose my own subjects for research and devise my own equipment. For instance, my first patent was for an improved form of electrical connector.

This innovation, like a few other methods that he devised, differs from Cady's later invention of the piezoelectric method, since it followed his own particular need rather than a finding. Dividing his interest into a few areas, and assisted by master students and 'an expert mechanician', Cady continued also his study of magnetism. In 1908, following a finding by his student Harold Arnold (who will return in Chapter 4 due to his position at AT&T), Cady began investigating vibrations and rotations in electric discharges (such as the electric arc). Many years later he mentioned that as 'a good example of the fact that a chance observation can give rise to a long line of research'. Before beginning his study of piezoelectricity, he had investigated wireless telecommunication and related issues that included detectors of electromagnetic waves and high-frequency oscillations. His practice as a radio amateur probably stimulated his interest in electromagnetic waves. As already discussed, radio methods and devices (like triode circuits) were used in the research on ultrasonic transducers and vibrating piezoelectric crystals, and Cady employed his expertise in the field. Cady was also an amateur musician, and his lifelong interest in ornithology led to publications in this field. His students remembered him as 'a splendid, generous and kind man' with a wide education.[34]

---

[34] Cady, *Saving Ancestors* (1963), unpublished manuscript, a copy in AIP, 100–6, 116–18, 209–10, quotations on 100, 209, his papers in ACNMAH; *Diaries* (meetings about radio];

In spring 1917, i.e. before the Washington conference on antisubmarine measures, Cady 'did little thinking and small-scale experimenting, especially in the use of magnetic methods' for detecting submarines. In most of his trials he tried to detect the electromagnetic effects of a metal submarine in water (e.g. a change in the magnetic field, or in conductivity between two points in water). These attempts led to no practical results except to his invitation to the Washington conference. In this way, they were critical for his future career. Louis Bauer, his former supervisor at the Coast and Geodetic Survey, and then at the Carnegie Institution in Washington, DC, had just visited Cady, learnt about his attempts and therefore arranged his invitation. There he was exposed to Langevin's ideas. Its 'principle so novel and so suggestive', he wrote years later 'could not fail to excite the interest of many physicists'.[35] Cady clearly was among those excited. Soon after returning to Wesleyan he experimented with piezoelectric crystals, and two weeks later he became an employee of General Electric's (GE) research laboratory in Schenectady, NY, where he joined the group led by physicist Albert Hull and 'devised various forms of quartz and Rochelle salt [crystal] units for receiving ultrasonic signals'. 'The research on piezoelectricity', Cady recalled half a century later, 'was practically thrown at me. Fortunately the way had been somewhat paved by my previous work on vibrations, and it lay in my range of general interest. Anyway it has been my principal scientific concern ever since.'[36] Indeed, his experience with vibrations, including electronic circuits for their generation, was useful for his research on piezoelectric oscillations.

Cady left General Electrics in October and a month later 'began cooperation with Pupin, Wills and Morecroft at Columbia, though most of my [Cady's] share was done at Wesleyan'. There he got the wartime help of McGrath, who had just received his Master in Science. Cady continued to teach

personal communication with Gerald Holton (17/4/2007). For short descriptions of Cady's work see James E. Brittain, 'Walter G. Cady and Piezoelectric Resonators', *Proceedings of the IEEE* 80, no. 11 (1992); Sidney B. Lang, 'Walter Guyton Cady', *Ferroelectrics* 9, no. 1 (1975): 139–40; Sidney B. Lang, 'A Conversation with Professor W. G. Cady', *Ferroelectrics* 9, no. 1 (1975): 141–9; Hans Jaffe, 'Professor Cady's Work in Crystal Physics', in *18th Annual Frequency Control Symposium* (1964), 5–11, on his character and later work; on radio amateurs, Susan J. Douglas, *Inventing American Broadcasting, 1899–1922* (Baltimore: Johns Hopkins University Press, 1987), ch. 5, 144–86.

[35] Earlier non-piezoelectric attempts to detect submarines, in NB 18, pp. 170–1; NB 43, pp. 81–7. First quote Cady, *Saving Ancestors*, 210; second quote, Cady, *Piezoelectricity*, 676. On Bauer's visits, 'Transcription of Interview of Cady on September 28, 1972, by Sidney Lang' (an electronic copy is available from the author).

[36] Cady, 'Outstanding Dates'; Cady, *Saving Ancestors*, 211.

during the war.[37] In his research in Wesleyan Cady studied, among others, the general theory of piezoelectricity from Woldemar Voigt's *Lehrbuch der Kristallphysik*.[38] Among others, Cady examined empirically the piezoelectric and electric properties of static and oscillating crystals, measuring capacity, dielectric coefficients, inductance, and later resistance. By these studies Cady learnt general properties of piezoelectric crystals and acquired the skill to work with them in the laboratory. Like his colleagues', most of Cady's work concentrated on questions of design: efficient sizes and cuts of crystals, their mounting, the material to which they were mounted in the transducer (tin-foil, steel, aluminium, rubber, wax, castor oil),[39] and so forth. The inclusion of the crystals as transducers (mostly as receivers) in the electric circuit was another central question of design.[40] Cady experimented with the transducers, which usually consisted of one or a few crystal plates (dimensions of a few to a few dozen millimetres), and a thick metal in the case of quartz. Following Langevin and others, since February 1918 he had been connecting them to electronic circuits with a triode valve amplifier, inductance coil(s) and capacitor(s), sources of power (dc batteries), and a detector in various settings and branches, in some of them loosely coupled to the transducer's circuit.

Frequency was one among a few variables manipulated and measured in Cady's experiments. It was induced by electromagnetic means, like a known LC circuit. He recorded its value and its influence on other magnitudes. Near resonance variations in frequency led to strong changes in the inner motion of the crystal, lowering inner friction and allowing efficiency in energy transmission. As mentioned, most transmitters and receivers were designed to work near resonance frequency to minimize losses of energy (Rochelle salt receivers were an exception due to their high piezoelectric reaction). That was, for example, the advantage of the thick crystal first used by Langevin, and of his sandwich design. Therefore, since Langevin's early works on the subject, researchers had studied the behaviour of crystal transducers and circuits at

[37] Cady, 'Outstanding Dates'; McGrath is mentioned in a few reports from 1918.

[38] Cady first mentioned its study on 29/7/1917, *Diaries*. He probably used Woldemar Voigt's thorough *Lehrbuch der Kristallphysik* (Leipzig, 1910), to which he referred in his 1921 paper. Ironically, Cady used a German textbook in a way unexpected by its writer in an effort to fight Germany.

[39] Rochelle salt crystal had to be immersed in an isolating material since it is soluble.

[40] NB X, pp. 14–64; NB 20, pp. 53–65; NB 23, esp. pp. 1–32.

and near resonance frequencies.[41] No later than the beginning of 1918 Cady studied characteristics of resonance. For example, on February 12–15, he examined the resonance frequencies and its sharpness in rods of different lengths (from 14 to 28 mm), i.e. how sensitive is the intensity of sound to changes in frequencies.[42] Still, like his colleagues, Cady neither took up an intensive research on the shape of resonance nor studied closely its electric consequences, as resonance was basically an elastic effect. On August 15th while measuring capacity and resistance of a new Rochelle salt specimen, he calculated a negative capacity of plates at specific frequencies. Later he would associate the observation with the special electric properties of resonance, but in August he did not pursue the matter further.[43] As Cady was preoccupied with the technical research on submarines he did not follow what probably seemed to be an interesting observation but one that could not help improving ultrasonic detection.

In autumn 1918 Cady began paying more attention to the electrical properties of crystals near resonance. In October and November he employed a new, accurate measurement method that had been developed for capacitors in radio to probe the effect of the circuit frequency (among other variables) on the voltage, capacity, and resistance of the transducers. Apparently at that stage Cady suspected that interesting changes might occur to the electric properties near resonance. Still, the investigation was connected to his efforts to improve ultrasonic transducers, probably in an attempt to utilize properties of resonance for both senders and receivers (improving their energetic efficiency and, by using a narrow wavelength range, reducing their vulnerability to interference). Cady carried out important parts of this study in the New London navy station, performing underwater experiments from a ship, carrying out the

[41] See Paul Langevin, Procédé et appareils d'émission et de réception des ondes élastiques sous-marines à l'aide des propriétés piézo-électriques du quartz. FR 505703 (A), filed 17/9/1918 issued 5/8/20, and Langevin, 'Conférence Interalliée sur la recherché des sous-Marine par la méthode ultra-sonore: Historique des recherches effectuées en France' (19/10/18) L196/16. For other examples see the part of I. B. Crandall in 'Report on the Conference of Physics and Engineering Divisions of the National Research Council, Washington, July 18, 1918' and Fuller and Ryan's report 'Problem #324: Quartz Supersound Source Projector', from 1/8/1918, both in the National Academies Archives (Washington, DC), 1914–1918 CFP. I do not know of any examination of resonance electric properties before Cady's.

[42] NB 23, pp. 27–31. Cady mentioned 'dull resonance' of two new receivers in his reports to the NRC, (under the title 'Problem 190 Supersonic Vibrations', in AIP) from 29/8/1918; on these see NB 23 pp. 107–11 (10–15/8/1918) in which, however, he did not refer to resonance.

[43] NB 23, pp. 107–11.

above-mentioned comparison of Rochelle salt and quartz crystal transducers, including regarding their sensitivity to changes of frequency. He examined transducers usually consisting of a few crystals mounted together as 'senders' and 'receivers', as sandwich in the case of quartz. In December, he started calling them 'plates' (rather than receivers or transducers), while intensively experimenting with their electric properties (resistance, capacity) in different settings. These included variation in capacitors, crystal plates, arrangements, and frequencies.[44] Although, Cady's examination of the electrical behaviour of transducers at different settings and frequencies was not limited to the workings of the plates as ultrasonic transducers, apparently (at least through November) that was one of his central aims. Cady gathered knowledge that might be useful for designing more efficient submarine detectors, as he was still associate with Wills's group in New London.

Studying Crystals near Resonance Frequencies

Late December brought a break in Cady's research,[45] and probably helped him reorient his study from ultrasonic detection to the electric properties of crystals in resonance, which he suspected was responsible for peculiarities that he had observed in his circuits.[46] He began 1919 studying intensively the influence of frequency on the electric behaviour of quartz and Rochelle salt, including a reappraisal of earlier results. On January 4th Cady observed a sharp minimum in current in a circuit that contained a Rochelle salt crystal, at 'a critical frequency'.[47] The association between a minimum in current and resonance had been observed before. In his report for an interallied conference on

---

[44] An early graph of capacity versus frequency, which, however, did not show any peak, appeared on November 6th (NB 20, pp. 106–8). Other related entries: NB 20, esp. pp. 66–74 (1–2/10/1918), 114 (14/11/18), NB X, p. 68 (before 16/10/18—theoretical comparison of receivers at different frequencies, NB 12, pp. 76–81 (6–8/12/18), 82–7 (17–20/12/18). Cady's new method of measurement is first mentioned after 29/10/18 (NB 20, p. 100). It follows the one presented in *Circular of the Bureau of Standards, No. 74: Radio instruments and measurements* (Washington, DC, 1918), 180–2, 190–3.

[45] Cady went to two scholarly trips in December, one to New York (on the 9th) and the other to the American Association for the Advancement of Science meeting in Baltimore (on the 26th), *Diaries*.

[46] E.g. 20/12/1918, NB 12, p. 84.

[47] The term critical frequency appears in this context first on 20/12/1918 (NB 12, p. 84) and again on 9/1/1919. It appeared earlier (e.g. on 15/2/1918, NB 23, p. 31), but probably without the full meaning acquired by January 1919.

submarine detection in October 1918 Langevin described and explained the phenomenon:

> The intensity of the high frequency [current] shown by [a thermal am-meter at a secondary circuit] passes through a minimum at the moment of elastic resonance, due to the considerable deduction of the power that the ultrasonic emission exerts at this moment in the electric circuit, or, in other words, due to the increase produced at this moment in the emission resistance of the quartz condenser, comparable to an underwater antenna.[48]

Cady's observation was more significant than Langevin's in two aspects: he probably found a sharper minimum than those previously observed and certainly recognized its sharpness, and more importantly he followed this observation with an examination of the influence of frequency on current (in the circuits) and on the crystal's resistance and capacity.

Varying the settings by replacing Rochelle salt plates, capacitors, coils, and their arrangement, Cady obtained a clear change in resistance with frequency

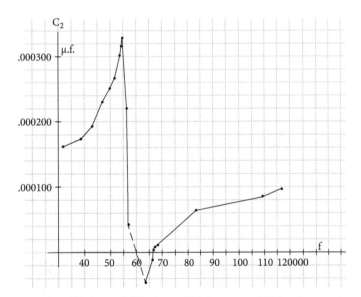

Figure 3.1 A graph drawn by Cady on 11 January 1919 of capacity as a function of frequency. Note the negative value around 65 KHz. *Source*: NB 25, p. 29. Figures from Cady's NB are courtesy of Smithsonian Institutions.

[48] Langevin, 'Conférence Interalliée', L(194)/16, p. 16.

and an even more conspicuous one for the capacity. Capacity near critical frequency had more than twice its 'normal value'. Moreover, its value dropped steeply to a negative value just near the maximum value, as Cady drew in a graph on January 11 (see Figure 3.1). As mentioned, he had observed a negative value five months earlier but did not make much of it then. Such a capacity now became meaningful as part of more general behaviour, and he verified theoretically that it has a physical meaning within the electric circuit. In analysing his findings about resonance Cady relied on a basic model of the crystal as a condenser and resistor in series (i.e. without considering inductance). In such a model the effective capacity and resistance vary with the frequency. At the end of the month and beginning of February he carried out a series of experiments on quartz plates like those that he had performed on Rochelle salt, finding even steeper curves near resonance. Subsequently, he investigated more closely the neighbourhood of resonance, narrowing the range of frequencies. In this research, Cady sought the best arrangement to enhance the effect, as is common in designing a technological device, yet, not less so, in experimental physics, where one looks for the clearest effects.[49] The discovery of a dramatic change in the electrical properties of crystals near its resonance frequency was very surprising as it did not have immediate analogy and was unforeseen by theory. Although it did not seem to contradict any basic theoretical tent, it took a few years until Cady and others accounted for it by the combination of elasticity, electricity, and piezoelectricity, with various assumptions and approximations (Ch. 9).

Cady's wartime research had been directed towards improving military technics; the armistice led to a reorientation. As a well-informed contemporary observed, '[w]ith the sudden termination of hostilities the problems confronting the scientific workers have to a large extent either suddenly changed their nature altogether or have been considerably modified'.[50] Yet part of Cady's research was still related to submarine detection, as he still worked with Wills's group at the New London naval research station. Thus, on Monday January 13th he began his day continuing the experiment on resonance from the previous week, but then changed to an examination of the sensitivity of Rochelle salt crystal as a receiver in an oil tank, apparently in

---

[49] NB 12, p. 89 (3/1/19); NB 25, pp. 1–47 (4/1–5/2/19).

[50] O. E. Jennings, 'Proceedings of the Baltimore Meeting of the American Association for the Advancement of Science', *Science* 49 (1919): 11. Jennings, general secretary of the American Association for the Advancement of Science, based his judgment on the December 1918 meeting of the society, which Cady attended. The society chose Baltimore 'partly because war conditions had brought together at Washington scientific men from all over the country' (11).

preparation for his visit in New London four days later. There he moved to salt water and compared the sensitivity of Rochelle salt and quartz to ultrasonic waves of different frequencies quantitatively and qualitatively by listening to the sound produced by a telephone receiver in their electric circuit, a common method used in radio and by Langevin's group. During these weeks he followed two distinct research paths: one on electric resonance and another on ultrasonic transducers. Within the latter, he attempted to exploit neither the specific changes in electrical properties nor the sharpness of resonance (or critical frequency, as seen in Figure 3.1). His work on the transducers did not require comprehensive research with crystals. Indeed, Wills's report on ASDIC research from June 1919, although mentioning Cady's work on resonance, concentrates on the energetic efficiency of the plates at different frequencies, and refers to the changes in capacity, inductance, and resistance only in passing. Coming back to Middletown, Cady returned to the electric study of quartz plates near resonance.[51] That he did not try utilizing his findings for sonar suggests that one should look at other factors that motivated his research, namely, the disciplinary logic of physics, and curiosity about natural phenomena. In other words, Cady's technological research would not have allowed him this comprehensive study of resonance. Moreover, the utilitarian regime of sonar development had probably hindered an earlier study of resonance by its researchers.

Cady's research during January and early February 1919 showed that piezoelectric crystals have a sharp resonance with clear electrical consequences. From the intensity of his work during these days and its focus on the behaviour near resonance, it seems that Cady realized that he had discovered an interesting phenomenon, although he still had much to learn about it. In later recollections he attributed the discovery to an earlier date:

> In the course of my tests of quartz and Rochelle Salt in 1917 and 1918, I had noticed that specimens cut from these crystals reacted in a peculiar way on the driving circuit when the frequency was close to that for a natural mode of vibration. In quartz the effect was extraordinary sharp. To me, this observation was like one of Sir Thomas More's 'Diamondes and Carbuncles upon certain rockes', discovered by the dwellers in Utopia: 'And yet they seke not for them: but by chaunce finding them they cutt and polish them.'[52]

[51] NB 25, pp. 32–9 (13–17/1/19), 40–1 (20/1/19), 42–6 (29/1–5/2/19 on electric resonance).

[52] Cady, 'Piezoelectricity and Ultrasonics', 49; the quotation from More appears without explanation, also as an epitaph of the chapter on the piezoelectric resonator in his textbook, Cady, *Piezoelectricity*, 284. Cady's choice of quotation is quite strange since More denies the value of

However, in this case the recognition that these are 'diamonds' was far from obvious. Although Cady had observed a few 'peculiar' behaviours of crystals during 1918 (such as negative capacity), at the time he had neither recognized the sharpness of resonance nor closely studied the phenomena before the beginning of the next year and the end of the war (for example, he didn't observe the quartz has sharper response until 1919). The previous August he had not cared enough to examine the resonance, nor did other researchers in the field (including Langevin), who most probably observed a few of these properties as they experimented with piezoelectric oscillators near resonance. Cady's discovery was accordingly a chance finding in a highly qualified sense. Although initially he had not looked for any special electric behaviour of piezoelectric resonators, a few occasional unconnected observations led him to undertake an investigation only by which he forged a discovery. He began the investigation in late 1918 but carried it through in a concentrated way only in the early part of 1919. Moreover, the stages described above suggest that 'the discovery' was the retrospective outcome of a gradual, uneven process that cannot be fully reconstructed. In a sense, it continued longer with studies that he carried out to elucidate further aspect of the phenomenon in the following months, which will be described in Chapter 4.[53]

3.3 Concluding Remarks

First World War research on submarine detection displays a concentrated effort of civilian scientists and engineers mobilized to the cause. Scientists worked alongside engineers, employing their knowledge and skills and dedicating their time to find practical solutions to critical war-related problems (e.g. submarine detection). Their achievements were impressive. Ultrasound detection technics could not have been developed in that period without the large-scale investment of effort, money, and scientific expertise. Indeed, measured by the number of scientists and engineers involved, the research

---

these findings, as *Utopia* continues: 'and with them [diamonds etc.] they adorn their children, who are delighted with them, and glory in them during their childhood; but when they grow to years, and see that none but children use such baubles, they of their own accord, without being bid by their parents, lay them aside.' Thomas More, *Utopia*, in a common, anonymous translation, book II (http://www.gutenberg.org/files/2130/2130-h/2130-h.htm) (Cady used a version with an older spelling).

[53] On the issue of a discovery as a process that takes time, I concur with Simon Schaffer, 'Making Up Discovery', in *Dimensions of Creativity*, ed. Margaret A. Boden (MIT Press, 1994), 13–51.

on submarine detection was one of the largest scientific-technical project conducted up to that point. It is difficult to imagine such an effort in another context. While known previously, almost all the researchers studied and experimented with piezoelectricity for the first time to help better design means of ultrasonic detection. Thereby, they made piezoelectricity better known in physics and engineering. Most of their findings were known to earlier students in the fields. A few experiments, however, indicated unknown effects, like that of electric resonance, or what seems as anomalies, like the behaviour of Rochelle salt. These results extended the knowledge of piezoelectricity, but the researchers neither explored them nor openly published them. Only afterwards did scientists examine surprising observations made during the war. Cady explored his earlier observations of resonance. Anderson did not pursue his observations of Rochelle salt, but one of the readers of his reports, W. F. G. Swann, suggested the subject to his PhD student Joseph Valasek, who found a permanent polarization in Rochelle salt analogous to permanent magnetism in iron. Consequently, this previously unknown state of matter, later known as ferroelectricity, became a dynamic field of investigation whose active research has continued unabated.[54]

Not all unexplained observations were later examined, however. In his tests of the suitability of quartz plates for transducers in early 1918 Rutherford sometimes observed that by a steady regular pressure 'that the electrometer went off scale and continued to do so over long intervals of time—almost as if the pressure had produced a steady E.M.F. in the circuit; he has, however, no time to investigate this', added the American scientific attaché in London. When he could return to 'pure science', Rutherford resumed his studies of the atom, and it seems that others either did not know it or did not find it interesting enough.[55]

Walter Cady liked to view his discoveries and inventions as results of chance and instantaneous insights. Indeed, from the perspective of Cady's scientific career, World War I and the quest for underwater detection methods were contingent events that led to his work on piezoelectricity. In the next stage, his

[54]Swann worked during the war on submarine detection with the National Bureau of Standards group, 'William Francis Gray Swann Papers, American Philosophical Society', http://www.amphilsoc.org/collections/view?docId=ead/Mss.B.Sw1-ead.xml (accessed 13/9/16). See also the short discussion and references in Ch. 9.

[55]Scientists might have thought that Rutherford's observation probably originated in some peculiarity of his experimental apparatus, rather than in a genuine phenomenon of quartz. Bumstead, 'Methods of Testing Quartz Crystals'; USNA, London files #57.

discovery of the sharp electric resonance was not only surprising, much like the discovery of any unexpected phenomenon or property (e.g. Geiger-Marsden experiment on atomic scattering); it was also accidental in the sense that it did not follow a programme aimed specifically at the study of resonance (unlike the scattering experiment).[56] Indeed, piezoelectric resonance was only a secondary subject in his research programme on underwater transducers, and Cady's first observations of a peculiar electric behaviour near resonance were made while examining other questions. Yet, these few initial observations were insufficient for him to recognize the phenomenon. Moreover, the discovery was not a singular event but a process in which Cady realized that there was a connection between the scattered data, and then recognized the electric properties connected to the resonance and its sharpness. This, however, required the deliberate experimental examination of crystal oscillations near resonance. Chance was perhaps important for getting the first hints (although other laboratories probably recorded similar results), but only Cady's subsequent research yielded anything like a true discovery.

This case shows that scientists' research programmes are often flexible, and that flexibility facilitates new and important findings. To follow the findings on resonance, Cady had to diverge from his research programme on ultrasonic transducers. Although the new interest was connected to the former, its further examination became a subject of novel research, which resulted in an altogether new programme that will be discussed in Chapter 4. Moreover, to make his discovery Cady had to leave his commitment to the technical aim of the utilitarian regime and return to a disciplinary research. Commitment to either scientific or technical aims was not merely an issue of rhetoric, but it directed researchers to study different sets of questions, and thus to different findings.

Cady's move followed his experimental findings and the scientific interest that they raised. Thus, the laboratory findings shaped not only empirical and theoretical conclusions but also the direction of research. Still, they did not determine the shift in Cady's research programme. His move seems instead to follow the normal path of disciplinary science: finding a subject of research that would promise potential new findings, publications, and recognition. His move (e.g. rather than to return to an earlier research programme) followed also his personal flexibility in choosing research topics, which characterized his work until 1917. A decade earlier he showed similar flexibility,

---

[56]The discoveries of X-rays and radioactivity are also examples of discoveries based on observations made within a different research programme.

with considerable but less impressive results, when he followed a student's findings to the field of vibration in electrical discharge. During the war Cady was
obliged to the research on ultrasonic transducers. With the end of the war, as
a master of its own time and means, Cady was free to pursue a new project.
Fortunately, the new project could be studied by his modest means. In piezoelectricity, he found a vein of rich ore, of which he became the most distinguish
researcher and which he had never left.

In the work on sonar, scientists like Langevin, Boyle, Cady, and Wills
worked like research engineers, or the rank-and-file industrial scientists (to
be discussed in the following chapters), in focusing their efforts on design
and questions related to its improvement. Their advantage, as clearly seen in
the case of Langevin, was in their ability to employ knowledge and experience from realms uncommon among engineers, and thus to complement the
latter's expertise. Another characteristic that helped them to attain better devices was their willingness to conduct a more general study of effects related
to the technics used.[57] This kind of research also led them to the observations
that would later be recognized as due to the peculiarities of piezoelectric resonance and ferroelectricity. They differed from their colleagues in industry
and government, however, by their ability to change their attitude towards
the research and return to open-ended disciplinary research. Here their experience with technology sometimes helped them, as they encountered new fields
of research. It revealed hitherto unknown findings that were unlikely to be
observed during that period outside the technological study, which they and
their colleagues could examine. Knowledge was not only transferred from science to sonar technics, as it clearly did, but also observations and findings, i.e.
knowledge, from technology were transferred back to physics.[58] Their wartime
research also made physicists aware of various understudied phenomena and
questions of potential scientific interest like those related to ultrasonics, which
were later studied by Boyle and Langevin.

[57]This tendency was common also among some engineering professors at the academy.
Vincenti, *What Engineers Know and How They Know It*.

[58]On a similar transfer of data on a larger scale see Yeang, *Probing the Sky with Radio Waves*.

4

THE APPLICATION OF PIEZOELECTRIC RESONANCE: FREQUENCY STANDARDS AND CONTROL

The sharp electric resonance of crystals is a general physical phenomenon dis-covered following a study of a particular technical artefact. Recognizing that this is a real phenomenon, rather than an artefact of particular experiments, and exploring its main properties required, as seen in Chapter 3, a turn from the study of ultrasonic transducers to resonance, and thus from technology to science. Ironically, the diversion from research related to ultrasonics toward questions remote from known practical aims led to technics that soon eclipsed sonar in societal importance. These technics, including frequency control and clocks, were based on Cady's recognition that the electric resonance of crys-tals is sharper and more stable than that of any known resonator of high frequency.[1] Cady himself was quick to utilize these properties for inventing methods for the exact measurement of high frequencies. Still, although this invention was used outside the laboratory, it fell within a traditional domain of physics. At least since the seventeenth century, physicists and astronomers had been developing and defining standards and measuring devices.[2] Like most highly accurate measuring methods, Cady's found only few practical employments. In this sense, they were very different from his later invention

---

[1] The resonator in use at that time was the tuning-fork, which is discussed in Chapter 5. In later years Cady made the comparison with the tuning-fork several times, e.g. transcript of oral interview by R. Bruce Lindsay and W. James King, 28–9 August 1963, AIP.

[2] One can think, for example, of the pendulum clock. From the 18th century, in France and later in Germany scientists were active in reforming weights and measurements; see Ken Alder, *The Measure of All Things: The Seven-Year Odyssey and Hidden Error That Transformed the World*, reprint edition (New York: Free Press, 2003); Klaus Hentschel, 'Gauss, Meyerstein and Hanoverian Metrology', *Annals of Science* 64 (2007): 41–75; Kathryn M. Olesko, 'The Measur-ing of Precision: The Exact Sensibility in Early Nineteenth-Century Germany', in *The Values of Precision*, ed. M. Norton Wise (Princeton, NJ.: Princeton University Press, 1995), 103–34. On electrical standards see Simon Schaffer, 'Accurate Measurements is an English Science', in ibid., 135–72.

Sonar to Quartz Clock. Shaul Katzir, Oxford University Press. © Oxford University Press (2023).
DOI: 10.1093/oso/9780198878735.003.0005

of a method to enforce an electronic circuit to oscillate at the frequency of the quartz resonator—crystal frequency control. Crystal frequency control enabled an unprecedented precision in determining the wavelength of electric oscillations, critical for the further development of the rapidly expanded network of electronic telecommunication, and therefore became widely used.

Cady's inventions followed his findings rather than a systematic attempt to answer specific technical or societal problems. In this sense, they differed from Chilowsky and Langevin's invention of the crystal ultrasonic echo method for detecting submarines, discussed in Chapter 2. The latter originated from an acute problem—how to locate enemy submarines—and was developed and improved in an organized project dedicated to solving the problem. Research and development in industrial laboratories also aimed at solving particular problems. The research arm of AT&T, for example, was established to provide the corporation with amplifiers for coast-to-coast telephony and control of wireless telephony.[3] The construction of a standard frequency and time system for the corporation, which implemented crystal frequency control as an alternative technics, was a similar project on a lower scale, which I describe in Chapters 5 and 6. Cady, on the other hand, undertook the investigation from his findings and asked how they could be applied. Like other 'technical solutions in search for problems', the initial design of frequency standard instruments resulted more from an interest in their working principles than from an interest in the ends for which they were used.[4] The development of ultrasonics is an example of a kind of technological change that stems from a particular societal need, although its implementation depended on the machinery of modern physics and engineering. This kind of technics can be called 'needs driven', as the effort for its development originated in needs recognized as such by the developers. It stands in contrast to 'knowledge-driven' technics, developed due to a wish to exploit familiarity with a particular phenomenon or behaviour, whether it originated in science or not. Needs is an inclusive

[3]Lillian Hoddeson, 'The Emergence of Basic Research in the Bell Telephone System, 1876–1915', Technology and Culture 22 (1981): 512–44; Leonard S. Reich, The Making of American Industrial Research: Science and Business at GE and Bell, 1876–1926 (Cambridge, UK: Cambridge University Press, 1985), 157–60.

[4]The inventors of such methods did envision some use for them before investing efforts in their development. For examples of other 'knowledge-driven' technics, see Richard H. Schallenberg, Bottled Energy: Electrical Engineering and the Evolution of Chemical Energy Storage (Philadelphia: American Philosophical Society, 1982) and Nicolas Rasmussen, Picture Control: The Electron Microscope and the Transformation of Biology in America, 1940–1960 (Stanford, CA: Stanford University Press, 1997), 25–35.

concept in this context that includes all kinds of societal requests and expectations from technics, including, but not restricted to, economic ones (other are e.g. military, ideological, political, and national).[5] Needs-driven solutions are often connected to a particular technical problem derived from a need, like the detection of submarines, and a subproblem, like the production of an ultrasonic transducer for echo-sounding. Piezoelectric frequency measurement standards and control methods form an example of a knowledge-driven technics, a type that became more common with the growth of scientific knowledge pertinent to technics since the late nineteenth century. The case of early piezoelectric technics, thus, provides clear examples of both the necessity- and knowledge-driven kinds. In other cases, the demarcation line between them is somewhat blurred.

With the invention of crystal frequency standards, research on piezoelectricity entered its fourth phase, that of exploring possible practical applications of the newly discovered phenomenon of sharp piezoelectric resonance, and their development into functioning technics. This chapter follows the application of the piezoelectric resonator from the invention of frequency measurement standards through the design crystal frequency control and its improvements to its early implementation in operating radio systems. It illuminates the role of knowledge of and research on piezoelectric resonance and of defining technical problems in this process of innovation. It examines how Cady recognized that piezoelectric resonance could be used as a practical meter for frequencies (section 4.1) and the way he managed to contrive methods of frequency control (section 4.2). At these stages of his work, Cady shared information with and consulted researchers at the research branch of AT&T. How and to what extent his exchange with them stimulated his invention of frequency control is examined within the discussion of the invention, shedding light on the relationships between corporative and academic researchers.

To his great disappointment, however, AT&T claimed legal rights on his inventions, relying on an earlier patent by one of its employees, Alexander M. Nicolson. Based on inner and outer correspondences of the corporation and

---

[5]Early proponents of what later came to be coined 'the demand-pull model' of innovation employed the term 'need', rather than 'demand', which restricts the factors that pull innovation to market-economic ones. They advocated 'a large or social conceptualization (need includes both societal needs and market demand)', of technological innovation. Although some doubt the analytical value of the term, since many things can be regarded as needs, I agree with Godin that 'needs have concrete manifestations'. Thus they can and should serve the study of innovation. Benoît Godin, *Models of Innovation: The History of an Idea* (Cambridge, MA: MIT Press, 2017), 120–5, quotes on 122 and 125.

Nicolson's laboratory notebooks, I examine in section 4.3 his research and the attitude of the corporation's researchers towards Cady's invention. In addition to interest in the actual accomplishments of the involved researchers and in the dealings between AT&T and Cady, analysing their interactions and comparing the work done in both places show mark differences in the approach to invention and research between a professor with his own small laboratory and a corporate laboratory. Although experts immediately recognized the potential of frequency control, its successful practical implementation in telecommunication systems required further non-trivial efforts. These included the invention of an improved circuit for frequency control by physicist George W. Pierce and its further development by John Milton Miller at the Naval Research Laboratory (NRL), and Cady's personal assistance in the first implementation of his frequency control method, which are examined in section 4.4. The chapter concludes with a discussion of the roles of research, scientific approach, and academic inventors in applying novel knowledge in new technics.

4.1 Frequency Standards and Study of the Resonator

Following his discovery of the sharp piezoelectric resonance of crystals, Cady looked for ways to exploit it scientifically and technically. One obvious direction for a scientist was to widen the knowledge of the phenomenon. He pursued that, and simultaneously examined paths for applications. Although not a professional inventor, Cady had enough interest in practical applications to seize the opportunities offered by the unique effect that he discovered. In this sense, he was an occasional inventor, seeking out inventions when an opportunity to contrive them came along. More than four decades after the invention of the piezoelectric frequency standard Cady recalled that the thought of it 'flashed on' him at one particular moment.

> I cut some thin bars from quartz crystals and set them into vibration by the current from a high-frequency vacuum tube oscillator. Each one vibrated strongly in resonance at a particular frequency, depending on its length. One night while I was getting ready for bed it suddenly flashed on me that these resonances were so sharp that such a rod could serve as a standard of frequency. It could be used to standardize, or calibrate, a frequency meter. At that time even the best frequency meters were not very accurate. It was in this way that the piezo resonator came to be invented.[6]

---

[6]Cady, *Saving Ancestors* (1963), unpublished manuscript a copy in AIP, 211–12. For a similar recollection with some variations, see Cady's interview by Lindsay and King, Thursday, 29 August 1963, session II: https://www.aip.org/history-programs/niels-bohr-library/oral-histories/4549-2 (accessed 3 Feb. 2020).

The frequency meter commonly in use was 'essentially a simple radio circuit, consisting of an inductance coil [L] and condenser [C] in series, with an ammeter or other device to indicate either the current flowing in the circuit or the voltage across the whole or a part of the capacity or inductance. Either the inductance or capacity [was] made variable and sometimes both.'[7] The resonance frequency (*f*) of the circuit is $2\pi/\sqrt{LC}$. Yet, as Cady indicated, these frequency values were physically unstable. Developed during WWI, tuning-fork-based meters, however, did suggest an accurate and reliable meter. In this device, an electromagnetically generated resonance of the tuning-fork enforced its frequency on a triode circuit to which it was coupled. Higher frequencies were generated by multiplying the frequencies by their 'harmonics'. It was a pioneering electronic device that considerably improved accuracy in measuring electromagnetic frequencies, and so became the basis for post-WWI frequency standards, shaping also the later piezoelectric standards. The tuning-fork frequency and time standards and their origins are therefore the subject of Chapter 5. In his experiments, Cady used LC wave meters, which he calibrated from time to time, perhaps using a tuning-fork standard (he could also compare them to radio wave transmission at a known standard frequency).

Cady's invention story, however, is not supported by his notebooks. They suggest a more complex and gradual process.[8] He began devising circuits that could be used both to study crystal resonance and to exploit these 'piezo resonators' for practical ends. These circuits were based on the discovery that the resonance of crystals is electrically sharp and the stability of its frequency, inferred from elasticity and supported by the results of his electromechanical experiments. Yet, it is less clear whether and in what sense Cady thought of resonating crystals as 'frequency standards'. Unlike professional inventors he did not indicate the goals of his experiments, and since his experiments could be used for more than one end, it is unclear if and when he had a particular application in mind. It is possible that during this research the specific idea of using the resonators as 'frequency standards' crystallized in Cady's mind in a way similar to his recollection. Even if so, Cady's account disregards or downplays

---

[7]On wave meters see D. B. Sullivan, 'Time and Frequency Measurement at NIST: The First 100 Years', in Frequency Control Symposium and PDA Exhibition, 2001. Proceedings of the 2001 IEEE International (2001), 4–17, 5; *Circular of the Bureau of Standards, No. 74: Radio instruments and measurements* (Washington, DC, 1918), 96–109, quotation on 98.

[8]The notebooks do not mention a list of such different bars in the relevant month of February–early March 1919, when Cady described his plans for a patent (see below). He continued working on condensers made of a few plates also after starting work on a patent.

the central accomplishments needed for the invention: the observation that the resonance is sharp and that its electrical consequences are profound (which he probably established by early February 1919), and the findings of proper means to detect it. The research on these questions was a more gradual process than Cady's recollections suggest.

Perhaps in testing the resonance of a four-plate quartz condenser during the first week of February, Cady attempted to examine its possible use for tuning wave meters. In these experiments, he employed a buzzer, and got a 'min[imum] of sound . . . not very sharp' at resonance frequency, when the crystal 'chokes'. This resembles a known calibration method of the commonly used wave meter, e.g. versus a tuning-fork standard. Cady's experiment, however, included an unsuccessful attempt to generate oscillations from the circuits (disconnected from the electric source), which is hardly consistent with the aim of designing frequency standards.[9]

A day after completing these experiments, Cady examined another use of crystal resonance, or perhaps another way of detecting it. That evening he attempted to receive a broadcast of electromagnetic waves (at 120 kHz) from the Navy station in Arlington Virginia (NAA) by coupling two circuits with Rochelle salt plates. A few days earlier he had received the station's signals by means of the known radio method of heterodyne detection, in use also in sonar.[10] However, the attempt with Rochelle salt failed, and Cady 'gave up' receiving the faint signals from NAA, but not the idea to employ piezoelectric plates to detect radio waves.[11]

---

[9]Cady used a buzzer coil whose sound would be 'chopped' in resonance frequency, NB 43, p. 31 (7/2/1919). Cady's notebooks are in ACNMAH. On the 'buzzer' method see *Circular No. 74*, pp. 107–8.

[10]As mentioned in Chapter 2, the heterodyne method is based on the beat heard by the difference in the close but unequal frequencies of two electromagnetic waves. Cady returned to the detection of NAA signals on March 21st. This time his detection method was not based on piezoelectricity, but he examined the influence of inserting crystals on the received signals. NB 25, p. 48 (8/2/1919), pp. 80–7 (21–5/3/1919).

[11]He might have made an earlier attempt to receive electromagnetic waves by the piezoelectric method on January 20th. On his way back from New London to Middletown that evening, he carried an electrometer on his car's backseat. 'A few miles later I ran machine off road while attending to electro[meter].' This suggests an attempt to detect electromagnetic waves. Yet, even in that case it is unclear whether he used a piezoelectric device for such an end. Cady's active interest in wireless communication was independent of and preceded his engagement with piezoelectricity; a week later he gave a talk to his local radio club on inducting coils. *Diaries*, 20/1/1919, 28/1/1919.

Two days later, he employed crystals to couple two new circuits, so that they would be electrically connected only at resonance frequency, enabling the detection of signals of that wavelength. In other words, if the arrangement allowed coupling only at a narrow range of frequencies, one had a means to compare frequencies to that of the piezoelectric plate. In this setting, an assistant sent waves from a nearby room, varying the frequency of the sender while keeping a constant current. The receiving antenna induced oscillation at the frequency of the electromagnetic waves in a 'primary' circuit consisting of a capacitor, a coil (this is an RC circuit), and a Rochelle salt plate (Figure 4.1). Metals coated the ends of the plate (near a, b and c, d in the figure), but not the middle. One coated end was in the primary circuit; the other side was

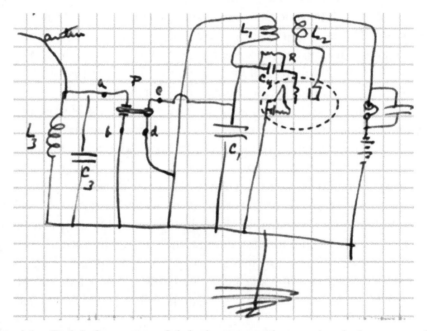

Figure 4.1 Cady's first successful device to couple two circuits by a crystal plate (P in the figure). An alternating voltage between a and b induced by electromagnetic waves received by the antenna on the left drives the plate to oscillate only near resonance. By the converese piezolectricity, the oscillation of the plates generates alternating voltage between c and d in the second circuit, which are amplified by the battery and audion (triode, marked by a dash oval added here) in the right-side circuit. As was common with a valve amplifier the plate (anode) branch feeds back to the grid through the coils L_2 and L_1. *Source*: NB 25, p. 48 (8.2.1919).

connected to a 'secondary' circuit, which included a triode valve amplifier, a battery, and a telephone receiver to detect the electric current in the circuit.

A year later in his patent application Cady explained the mechanism beyond a simpler version of this circuit: 'whenever an alternating current of the critical frequency flows in the circuit the [crystal] plate will be brought into energetic vibration through the agency of the alternating potential differences between the coatings [a, b in Figure 4.1]. These alternations will in turn generate potential differences in the coating [c, d], which will cause an alternating current of the same frequency to flow in the second circuit.'[12] In February 1919 he found that 'when λ [the wavelength] =3025 [metres] about, sound is max . . . and falls off in either dir[ection]. . . . [Yet] resonance [is] not at all sharp.' A week later, replacing Rochelle salt with quartz, he found that 'Results ca [circa] same as w. [with] RS, tuning somewhat sharper perhaps, but loudness at max. not as good.'[13] As Cady later remarked this device is a kind of a (bandpass) radio frequency filter, which transmits signals only at a particular range of wavelengths. However, at the time Cady did not use the term, and the electric circuits that he constructed were more complicated than those needed for a simple filter. Thus, he probably thought about the coupling of two circuits rather than a filter, although the difference is quite subtle.

Coupling of two electric circuits by the resonance of piezoelectric plates was Cady's immediate aim, but he did not indicate his further goals. His later use of 'the filter', his later recollection, and the similarity with the common practice of the day of using wave meters suggest that he tried to apply the resonance for the purpose of establishing a device that could set frequency standards.[14] Simultaneously, these tests examined the sharpness and electric effect of crystal resonance, which he continued studying in the following weeks. In the latter studies, which clearly did not aim at a practical device, he varied not only the frequency but also the arrangement of the circuits and their components to enhance the effect and found sharper and narrower resonances. At the end of the month he found 'extremely sharp' resonance in a quartz crystal (now part

---

[12]Walter G. Cady, The piezo-electric resonator, US1,450,246, filed 28 February 1920, and issued 3 April 1923.

[13]NB 25, pp. 48–9 (8/2, 10/2, 17/2/1919).

[14]One might suggest that Cady considered the use of the filter for radio receivers. However, it does not seem to have any clear advantages on contemporary devices. That it is limited to a narrow frequency range, would be regarded as a disadvantage for a useful receiver since transmission frequency was unstable. Only frequency control allowed a stable enough transmission that opened the possibility of a narrow receiver.

of a 'sandwich' transducer). Cady, thus, could focus on narrower ranges of frequencies, e.g. between 71,500 and 71,606, which he extended to 73,600 Hz and between 57,288 and 57,373 Hz (less than 0.15%) with another crystal. In comparison, in the previous month he examined frequencies between 43,500 and 76,000 Hz in one experiment. The numbers that Cady recorded for the frequencies in the experiment exceeded the accuracy of current wave meters; in 1919 their accuracy was estimated not to exceed 1/100. Yet, absolute values were not important for these experiments, and differences between frequencies could be read to a higher precision.[15]

A few days earlier, on February 22nd, Cady 'showed resonators to [Harold] Arnold of Bell Labs, [George V.] Wendell of Columbia, and [Karl] Van Dyke of Wes[lean] [then a PhD student in Chicago]. Told them of the coupling device, also my observation of the negative capacitance of Rochelle salt plates at certain frequencies.'[16] Cady's display of his results reveals a confidence in his findings and their significance. His openness differs from the behaviour typical of professional inventors or commercial company employees. As a student, Arnold 'came very close to [him] in the laboratory', and they had maintained a scientific connection over the intervening decade.[17] But Arnold was also a high-ranking employee of AT&T. He directed the corporation's research branch, then under Western Electric, the manufacturing arm of AT&T,

[15] NB 25, pp. 52–9 (21–8/2/1919). The experiment also revealed in such cases a small plateau for the value of the electric magnitudes in resonance, which could be helpful for filters. For an earlier experiment see NB 25, pp. 20–1 (10/1/1919). On precision of wave meters Henri Abraham and Eugène Bloch, 'Amplificateurs pour courants continus et pour courants de très basse fréquence', *Comptes rendus* 168 (1919): 1105–8, 1106; see also Sullivan, 'Time and Frequency Measurement at NIST', 5.

[16] Since Cady's memories of more than 40 years after the events are the only source about this meeting, it is impossible to know exactly what Cady told his colleagues and former students and what their reaction was. W. G. Cady, 'Outstanding Dates Relating to Work of W. G. Cady in Piezoelectricity', 2-page manuscript written probably at the 1960s, kept in AIP. Cady consulted his notes in writing this report probably in the early 1960s. He mentioned the same occasion in the 1963 oral interview with Lindsay and King (AIP). There, he recalled that he asked Arnold 'if there was anything his company might be interested in'. However, he was not so clear about the year of the meeting (since it was Washington's birthday, the date was easier to remember) and he did not mention the other visitors, so his written testimony should receive more weight.

[17] Cady interview by Lindsay and King (AIP), p. 19. Cady's meeting with Arnold was not in itself unusual. Among other former students who maintained contact after leaving Wesleyan were Karl Van Dyke, who returned to the university as a professor and helped Cady build a strong centre for the study of piezoelectricity, and Gerlad Holton, who dedicated his second book to him. Cady's diaries mention meetings with Arnold in December 1918 and October 1919, as well as several later meetings that appear to be related to professional concerns.

which was the kernel for Bell Telephone Laboratories (Bell Labs) incorporated in 1925.[18] Cady may have informed him not only as a colleague but also to learn whether his findings might interest that company. Unlike academic physicists free to share their ideas within their community, industrial scientists and engineers were often restricted by corporate policy, and Arnold did not inform Cady of associated research done under his supervision (see below). Still, Cady did not dismiss the possibility of financial gain from his invention. On 2 March 1919, he discussed his plans to patent a wavelength standard with Commander de Frees of the New London research station. That he could file a private patent indicates that his research on piezoelectric resonance was conceived as separate and independent from the work on submarine detection (as discussed in Ch. 3). That Cady was not an employee of the Navy probably also helped.[19] Cady further mentioned his plans for a patent in January of the following year before publishing his findings. A year later, he publicly presented his method of crystal frequency control a month *before* filing a patent on the subject.[20]

Patents and possible applications were one goal of Cady's research; scientific knowledge and publications were another. To a large extent, these aims went hand in hand because more thorough knowledge and understanding of the resonator contributed to its technical design. In particular, research that Cady carried out on variant cuttings and the mounting of crystal plates and on new directions of oscillations helped improve the sharpness of the resonance, and in adjusting it to needed frequencies. Such research on the relation between shape and crystalline vibrations was, however, less relevant for designing circuits for detecting and exploiting resonance frequency. 'The filter' that Cady constructed in February was his only device for that purpose when he announced his plans for a patent. Eight months later, when he filed the patent, he suggested five additional circuits (or methods) to detect and use resonance.

---

[18] James E. Brittain, 'Harold D. Arnold: A Pioneer in Vacuum-Tube Electronics', *Proceedings of the IEEE* 86 (1998): 1895–6.

[19] Cady, *Diaries*, 2/3/19. In his 'Outstanding Dates' Cady mentions that he talked with de Frees on March 30th. However, the contemporary evidence is more reliable.

[20] Cady, The piezo-electric resonator. He first published on these methods a year later in Walter G. Cady, 'The Piezo-Electric Resonator', *Physical Review* 17 (1921): 531. Yet the following April he presented technics of frequency control in Walter G. Cady, 'New Methods for Maintaining Constant Frequency in High-Frequency Circuit', *Physical Review* 18 (1921): 142–3, which he patented in Walter G. Cady, 'Methods of Maintaining Electric Currents of Constant Frequency', 1,472,583, filed 28 May 1921, and issued 30 October 1923.

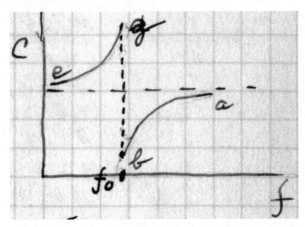

Figure 4.2 Cady's schematic drawing of the discontinuity in the value of the capacity, which becomes very high when approaching resonance frequency (f_0) from below and tends to zero when approaching from above. *Source*: NB 25, p. 74 (8.3.1919).

Cady continued examining the electric behaviour of crystals in resonance, varying crystal plates, frequencies, settings, and additional components of the circuits such as capacitors and coils. Among other results, close examination of quartz plates under forced oscillations clarified that the value of current and capacity near resonance is discontinuous. If resonance is approached from low frequencies the capacity rises, while approaching from higher frequencies causes it to decrease (see Figure 4.2). A change in the frequency, thus, induces a discontinuous change in the apparent capacity of the plate, completely changing its electric behaviour. This explains how the actual capacity of the resonating plates can acquire a value that puts the whole circuit at its resonant frequency. Apparently, for Cady, this observation made such behaviour intelligible; i.e. it provided a reasonable explanation for the phenomenon.[21] Yet, he still wondered why the capacity is changing in this way. Whatever was the cause, the observation led to the conclusion that the resonant frequency is insensitive to a change in the capacity and inductance of auxiliary coils and capacitors, a conclusion of important practical consequences, which Cady exploited for frequency control two years later. In March 1919, however, he

---

[21] I agree with Dear that scientists often strive to make the natural world intelligible, i.e. to give an account that would explain the causes of the observed phenomena and regularities in manner that gives practitioners a sense of understanding, albeit according to changing standards. Peter Dear, *The Intelligibility of Nature: How Science Makes Sense of the World* (Chicago: University of Chicago Press, 2006).

did not make this conclusion explicit, although he implicitly assumed it in explaining a few results to himself.

During 1919, his research included general questions regarding piezoelectric behaviour, like possible variations in the value of piezoelectric coefficients (especially problematic with Rochelle salt, because of what was later known as its ferroelectricity) and the influence of other variables such as temperature. He began developing a theory of the piezoelectric resonator and rules and methods for computation. This theory was based, on the one hand, on his empirical findings and, on the other hand, on piezoelectric and elastic theories. To understand better the behaviour of the crystal rods, he worked also on the laws of their (damped) mechanical vibrations, which involve a complicated mathematical theory for which he provided a solution in autumn 1919. He continued elaborating his experiments and ideas and sent for publication two brief and one detailed exposition of the behaviour and theory of piezoelectric crystals in 1921 (discussed in Ch. 9). Among others, he continued his attempt to model the resonator to a capacitor and a resistor in series, which he first tried in January 1919. Yet, he failed to contrive such an 'equivalent circuit' that would mimic the behaviour of the piezoelectric crystal near resonance. In retrospect one can observer that his model was too close to the phenomenon. As will be discussed in Chapter 9, an equivalent circuit in this case also requires inductance, although it is not manifested by the crystal. In the tradition of Voigt's theory (Ch. 1), Cady's theories of mechanical vibration and piezoelectric oscillations were phenomenological accounts that did not suggest a process or mechanism that underlies the phenomena.[22]

Cady's studies of piezoelectricity, resonance, and rod oscillations were connected to questions of design. More thorough understanding of the behaviour near resonance was likely to help improve frequency-standards devices (although it happened to be more important for Cady's later invention of frequency-controlled circuits than to this one); articulated theory of rod oscillations was likely to help design the crystal–metal resonating plates.

[22]Cady, NB 25, pp. 68–110, esp. 68–70, 72–5; NB P, p. 1–1a (analogy with synchronous motor) from March–May 1919. *Diaries*, 1919, Walter G. Cady, 'Note on the Theory of Longitudinal Vibrations of Viscous Rods Having Internal Losses', *Physical Review* 15 (1920): 146–7. Cady's research on the piezoelectric resonator resembles in character that of Bell scientists after the discovery of the point contact transistor in 1947. In both cases, scientists developed novel theoretical understanding of an unpredicted phenomenon through an intensive experimental study: Michael Riordan and Lillian Hoddeson, *Crystal Fire: The Invention of the Transistor and the Birth of the Information Age* (New York: Norton, 1997), Chs. 7–9.

Notwithstanding, Cady's research went beyond the direct needs of design, suggesting that the latter was not his sole motivation. His work on the vibration of rods provides a good example of Cady's quest for knowledge that went beyond direct technological ends. For the design of plates a direct study of their behaviour would have been sufficient. Moreover, even if one chooses a mathematical–scientific approach one does not have to suggest a general solution, whereas Cady did. The disciplinary logic of physics, however, did call for such contributions.

4.2 Crystal Frequency Control

Cady believed that the novel stable frequency standards for high frequency could be valuable for the expanding number of researchers and engineers in electric communication. Wireless communication rapidly expanded following the successful implementation of continuous wave transmissions based on triode oscillators during the First World War. Continuous waves allowed the transmission of voice over the ether, instead of communicating via Morse code common before the war, and made wireless telephony possible whether as an independent system or as complementary to a wired system. The formation of the Radio Corporation of America (RCA) between October 1919 and March 1921, owned by the major American corporations manufacturing telecommunication and electronic devices GE, AT&T, and Westinghouse, marked the recognition of the commercial and military strategic importance of electromagnetic waves. Among the lessons one could learn from the creation of RCA was the crucial role of patents in radio. A central justification for its creation as a monopoly was the need to employ patents that were under the rights of each of its owner corporations for an efficient wireless transmission. While managers and lawyers in these companies negotiated the division of the markets between them, wireless evolved toward an unexpected direction. Radio amateurs exploited continuous waves also for sending music on the ether. A few of these 'broadcasters' became very popular, inspiring the creation of the first 'radio stations' and leading to a 'broadcasting boom' that swept the USA from late 1920. Broadcasting opened two new markets, one for simple-to-use receivers and the other for advertisements by broadcasters, expanding the investment in wireless technics.[23]

---

[23] Susan J. Douglas, *Inventing American Broadcasting, 1899–1922* (Baltimore: Johns Hopkins University Press, 1987), 285–91, 298–303.

Cady engaged indirectly with these changes. He presented his findings to physicists and engineers and sent crystals resonators for examination and calibration to the radio section of the US National Bureau of Standards (BoS and later NBS) and to other relevant organizations. Notwithstanding, sharp, stable electric resonance remained a solution in search of a problem. Frequency standards had a limited market—primarily researchers and regulatory agencies.[24] After the emergence of broadcasting, the BoS, for example, used them to measure wavelengths of radio transmitters, helping to enforce commercial broadcasting at narrower and steadier frequencies, thereby opening a small, fixed market for such devices.[25] Furthermore, the new method also suggested a means for an international comparison of local standards. From April to July 1923, Cady travelled to Italy, France, and Britain with his resonators and compared measurements of their frequency at the BoS with those done at local laboratories.[26] Yet, controlling the frequency of electric circuits was the major problem piezoelectric resonance eventually found.

The practical advantages of enforcing oscillations to particular frequencies, rather than merely measuring them, were obvious already in 1919. The immediate beneficiaries of narrower and more precise wave bands were radio and multiplex telephony and telegraphy. The former was arguably the most exciting technology of the time; the latter a highly valuable method that allowed carrying a few conversations, each in its own frequency band, on the same wire (more on multiplex in Ch. 6). Maintaining the frequencies of the oscillations at particular values would prevent the problem of interference between the different transmission channels in bidirectional wireless communication,

---

[24] In the 1930s the piezoelectric filter, which Cady suggested in his 1920 patent, found a problem in telephony, where it enabled the transmission of more than 480 conversations over one pair of conductors, and in radio. As will be discussed in chapter 8, however, the piezoelectric filter was based on further developments in piezoelectricity and electronics; see W. P. Mason, 'Quartz Crystal Applications', in *Quartz Crystals for Electrical Circuits, Their Design and Manufacture* (New York: D. Van Nostrand, 1946), 11–56, esp. 14–15.

[25] Christopher Shawn McGahey, 'Harnessing Nature's Timekeeper: A History of the Piezoelectric Quartz Crystal Technological Community(1880–1959)' (Ph.D. thesis, Atlanta, Georgia Institute of Technology, 2009), 131–73.

[26] W.G. Cady, 'An International Comparison of Radio Wavelength Standards by Means of Piezo-Electric Resonators', *IRE Proceedings* 12, (1924): 805–16. That Cady omitted Germany, where he himself had studied, suggests that he joined the post-WWI boycott. See Cady, *Saving Ancestors*, 220; 'The Use of the Piezo-electric Effect for Establishing Fixed Frequency Standards', Radio Laboratory, Bureau of Standards, 11/9/1920, AIP; on the boycott see, Brigitte Schroeder-Gudehus, 'Challenge to Transnational Loyalties: International Scientific Organizations after the First World War', *Science Studies* 3 (1973): 93–118.

multiplex and later with broadcasting. Although the financial benefits of a practical method for controlling frequencies were obvious, its invention had to wait more than two years after the discovery of electric resonance. Moreover, it was not developed by researchers from the telecommunication industry, but by Cady.

Still Arnold from AT&T, the larger company in the field, encouraged Cady to invent the method. Years later Cady recalled:

> [Arnold] was present at New York when I gave a talk for the Physical Society on the resonator [on 26 February 1921]. . . in the conversation that Arnold and I had afterwards, he said, 'it would be awfully nice if you could find some way to make the crystals not only resonate like this, but also control frequency'. Well, I hadn't thought of that and on the train going back to Middletown, I thought the thing over and instead of going home and going to bed as I should have done, I went right up to the laboratory and started setting things up and in a few days I began to get definite results in the stabilizing of the current by means of the crystals and was working out the theory of it.[27]

Cady's notebook supports the main story. On Monday, February 28th he contrived the first electric circuit stabilized (or controlled) by the resonance frequency of a piezoelectric crystal (Figure 4.3). Relying on his previous studies, for example on his explanation of electric behaviour near resonance, he began intensive research on the theoretical and experimental aspects of circuits to stabilize frequency, and filed a patent three months later.[28] The patent described and explained the working principles of a refinement of the invention from February 28th along with three other methods, which were variations on the original circuit.

Cady's idea was to enforce an electronic RLC circuit to oscillate at or close to a natural frequency of a piezoelectric crystal. To this end he inserted a crystal into a circuit so that it strongly coupled two branches of the circuit only when at resonance. Due to its strong electric effect, at resonance the current generated by the crystal was stronger than that of any other oscillator and therefore enforced its frequency. The circuit that he contrived was based on a

---

[27] Walter G. Cady, interview by Lindsay and King, p. 18.

[28] Cady's notebook entry on Monday 28/2/1921 is titled 'Control of Oscilla[tion]s by Quartz Resonator'. There he summarized his attempt on Saturday evening, and his experiments on crystal plates on Sunday. Cady, NB 25, pp. 237–87; Cady, 'Methods of Maintaining Electric Currents'. This was probably not the first circuit to be crystal controlled, but it was the first output of a deliberate attempt to contrive such a circuit; see the discussion in section 4.3.

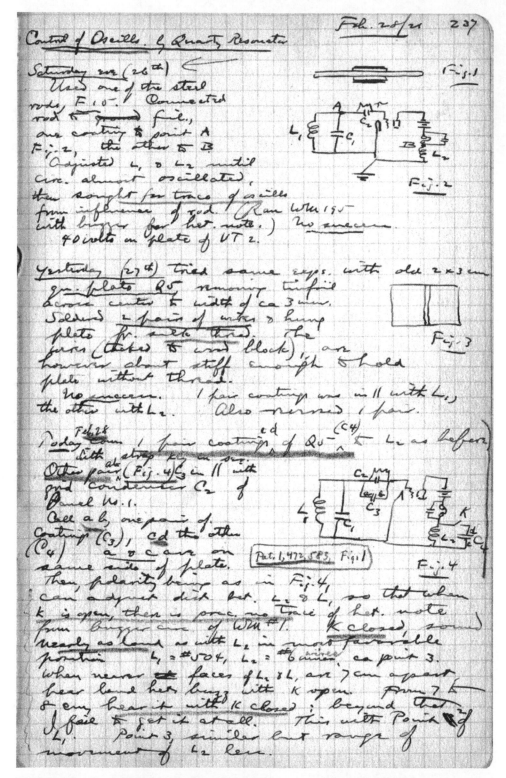

Figure 4.3 Cady's notes on his attempts for 'control of oscillations by quartz resonator' from 28 February 1921. *Source*: NB 25, p. 237.

common triode oscillator in which the grid and anode branches are coupled by coils (7 and 8 in Figure 4.4).[29] Without a piezoelectric crystal, the circuit would oscillate at a frequency determined by the value of the capacitors in the grid branch and the inductance of its coils. A piezoelectric crystal electrically connected by two pairs of small metallic coating (13–14 and 15–16) also couples the two branches. Using a variable capacitor (9) one varies the frequency of the circuit oscillator until it approaches a natural frequency of the crystal. Then the alternating electric field between the coating 13 and 14 would cause strong elastic vibrations, which would induce strong electric oscillations in coating 15–16, stronger than those produced in coil 7, and would determine the oscillator's frequency in that branch. Once oscillating, Cady could manually loosen the coupling between the two coils, so the frequency is determined solely by the resonator. He found that the frequency is stable under a normally disturbing effect, like changes of voltage.

The piezoelectric crystal in this circuit enforces its resonance frequency on an RLC oscillator, which without a crystal, or if the crystal is far from resonance, oscillates at its own unstable frequency. Such a 'stabilizer', to use a later term, is less effective and more precarious for most needs than a 'crystal-controlled oscillator', in which the crystal resonator alone induces oscillations in the circuit. Unlike a stabilizer, an oscillator generates oscillations only when the crystal resonator is very close to its resonance frequency.

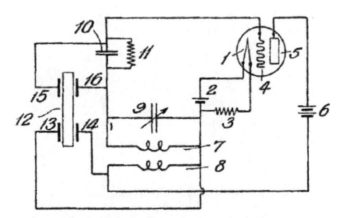

Figure 4.4 A schematic of Cady's first circuit for enforcing the frequency of a crystal on a circuit; as depicted in his patent, 4 is the grid and 5 the anode. *Source*: Cady, 'Methods of Maintaining Electric Currents'.

[29] The figure in the later patent presents the circuit better than the original figure in Cady's notebook, since it shows clearly that the same crystal plate is connected to two pairs of electrodes. See Cady, 'Methods of Maintaining Electric Currents'.

Thus, the resonator did not have to enforce its own frequency on other components of the circuits. The two kinds can be technically very similar. The first crystal-controlled oscillator circuit originated in the first stabilizer. While modifying the stabilizer circuit, Cady found that '[t]he greater the amplification constant of the vacuum tube, the more widely may the electrical constants of the circuit be altered without affecting the frequency.' By using a cascade of triode amplifiers (in a commonly used 'resistance amplifier') he could dispense with the components of the original LC oscillator and couple the grid and anode branches solely by the piezoelectric resonator and not by coils as in a common oscillator. He thereby got a crystal-controlled oscillator. In this circuit, first constructed on May 3rd, the piezoelectric crystal is connected with one coating pair (plates 13, 14 in Figure 4.5) to the anode of the third triode (5), and the other pair (16, 15) to the grid of the first triode (4). A small alternating potential on the anode 5 sets the crystal into vibrations (through coatings 13, 14) at its natural (resonance) frequency. The mechanical vibrations induce alternating potential on the grid, which is amplified and leads to higher potential difference in the anode, now at the resonance frequency, which induces further vibrations of the resonator.[30]

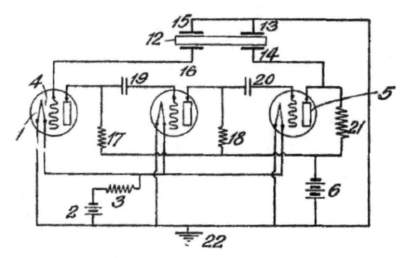

Figure 4.5 Cady's 3 May 2021 method for frequency control as depicted in his later patent. *Source*: Cady, 'Methods of Maintaining Electric Currents'.

---

[30]Cady, 'Methods of Maintaining Electric Currents'. Cady, NB 25, pp. 272–3 (3/5/1921), 276–7 (5/5/1921).

These two methods were based on the ability of the resonator 'to transmit energy in the form of an alternating current into another circuit', only when its frequency is very close to resonance. In two additional stabilizing circuits, Cady employed his finding of the discontinuous and asymmetrical change of the resonator capacity near resonance frequency. In these circuits, the resonator did not couple the grid and anode branches, but enforced the frequency of the current in the grid branch to be a little below its resonance frequency; as he observed in early 1919; at this state the crystal's capacity is maximal and therefore with proper setting the current could be maximal. In these circuits the crystal was connected to only one pair of coatings. All the methods relied on Cady's understanding of the abrupt changes in the piezoelectric crystal's electrical behaviour near resonance frequency, which he had acquired in his two years of research. He considered this theoretical understanding crucial also for his claim for invention, since it demonstrated that the circuits indeed function as claimed, and explained why they do so. His patent application therefore included a discussion of the means by which the resonance properties secure stable oscillations.[31]

From Cady's recollections, it seems as though he had only needed the right question to put him on track 'to make the crystals . . . control frequency', at least with the knowledge that he had acquired by early 1921. Yet, the question was not new for him. The tuning-fork standards, which he knew, provided an example of a resonator that enforces an electronic circuit to oscillate at its resonance frequency, and thus a possible inspiration to do the same with piezo-resonators.[32] Moreover, Cady was not oblivious to the potential practical advantages of controlling frequencies. His meeting with Arnold was needed neither to learn about them nor to attempt to contrive an appropriate circuit. Three months earlier, he had already considered 'a notion that the resonators can perhaps be used in connection with a regulating device to keep the speed of high-frequency alternator, or any sort of high-frequency device, very rigidly constant.'[33]

His notebook entry from January 22nd (i.e. a month before the encounter with Arnold) reveals an attempt to construct a circuit that would oscillate at

[31] Cady, 'Methods of Maintaining Electric Currents', quotation on p. 1.

[32] Cady's piezoelectric frequency standards, on the other hand, only made it possible to observe the resonance frequency, e.g. by detecting sudden decreases in a current at a particular circuit branch. See Chapter 5 on the tuning-fork standards.

[33] Cady to Arnold, 9/11/1920, AT&T Archives Loc 79 10 01 03.

the frequency of piezoelectric resonator. In this RLC circuit he used a crystal with one pair of coatings, one of its electrodes connected to the grid branch in point *h* and the other to the anode of the triode below *a* in Figure 4.6. As in his crystal-controlled oscillator of May 3rd, the two branches of the circuit were connected only through the resonator; i.e. there was no coils coupling. Yet, in this earlier method Cady did not separate between one pair of electrodes connected to one branch that would have an electric oscillation that induces a mechanical vibration (like plates 13, 14 in Figure 4.5) and another pair in which the mechanical vibration of the resonator induces alternate electric voltage (like plates 15, 16 in Figure 4.5). Instead, he assumed, according to theory, that a change in voltage on one electrode should suffice to induce mechanical change and that its vibration (induced by a coil and capacitor) would induce an alternating electric voltage on the other electrode (*b*), leading to a positive feedback; later frequency-controlled oscillators, like the popular Pierce circuit, used one pair (see below). In this arrangement, however, the effect was not strong enough, and he abandoned the circuit, as well as attempts to generate oscillations in resonance frequency for a month.[34]

When he returned to the question with Arnold's encouragement, Cady made two changes that helped him achieve a functioning method: he

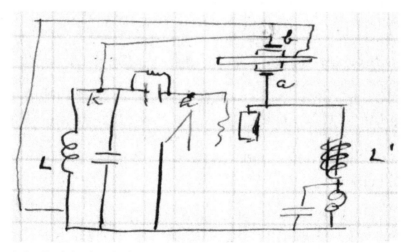

Figure 4.6 Cady's circuit from 22 January 21. *Source*: NB 25, p. 225.

[34]Cady, NB 25, p. 225, 22/1/1921. I discuss (in section 4.4) the differences between the methods of Cady and Pierce and plausible reasons for Pierce's success.

connected two rather than one pair of metal plates to the crystal, and he used the crystal to stabilize an independent oscillator, rather than making it the generator of the oscillations (he had hinted at a separation between the control and oscillations already in November 1920).[35] Neither of the changes was necessary. As mentioned, during March–May 1921, Cady found that each of these changes alone sufficed for his aim. Pierce later constructed a piezoelectric oscillator with one pair of plates (discussed in section 4.4). Yet historically Cady managed to contrive the first crystal frequency control circuit by making these two changes. Through studies of and variations on the first successful method of late February, which in retrospect can be seen as unnecessarily complicated, he achieved simpler methods, which opened the door for even simpler methods suggested by others. It is probably impossible to tell what exactly Cady gained from Arnold in their talk in New York, aside from an encouragement to re-examine the construction of a frequency control circuit. Perhaps one of Arnold's remarks steered Cady toward one of the changes that he ended up making, or he just needed to describe his attempts to someone in order to rethink them himself.

The interaction between Cady and Arnold began before the New York meeting and continued after it was over. Maintaining a good personal and professional relationship, Arnold continued to meet Cady and made sure that Western Electric maintained a professional connection with the Wesleyan professor by freely loaning amplifying tubes, resonators, measuring instruments, Rochelle salt crystals, and other laboratory devices that Cady needed for his research. Loaning apparatuses to a university laboratory was a general policy of AT&T's at the time.[36] At professional meetings and personal visits to Cady's university laboratory in Middletown, CT, and at the Western Electric research laboratory in Manhattan during 1919–21, Cady informed Arnold, and members of his staff, about his progress with the piezo-resonators and at the same time learned about potential uses for his discovery and inventions.

[35] Cady to Arnold, 9/11/1920, AT&T Archives Loc 79 10 01 03.

[36] For AT&T's policy of loaning see Arnold to Cady, 29/1/1920; for mention of devices requested, sent, and used, see e.g., Arnold to Kendall, 13/2/1918 (tubes), Cady to Arnold, 14/12/1918 (resistors); Arnold to Cady, 10/4/1920 (a 'special high frequency oscillators' built for Cady); Cady to Arnold, 30/4/1920 ('thermo-elements' for measuring low current), 6/11/1920 (a vacuum thermo-couple), all in AT&T Archives Loc 79 10 01 03. And a late reference to Rochelle salt crystals, Arnold to Cady, 5/2/1925. Walter G. Cady, 'The Piezo-electric Resonator', *IRE Proceedings* 10 (1922): 83–114, esp. 91.

Arnold was 'much interested in [his] work with quartz crystals' and pointed out that 'real practical use for a standard of frequency' could be found for particular wavelengths. He encouraged Cady to extend his research on resonance and its practical usage. Cady responded, among others, by developing resonators that resonate at the frequencies of interest for industry. In January 1921, he began sending the corporation (and a few other companies) piezoelectric resonators with detailed instructions for their use. In the following October, he visited Western Electric's laboratory and laid out the details of his methods.[37]

Arnold's comment that it would be 'awfully nice' if Cady could make crystals control frequency as well as resonate implies a belief that his former teacher's experience with the piezo-resonator made him more qualified to solve the technical problems than were AT&T's own staff. In addition to the acknowledged advantage of special expertise, Arnold trusted Cady due to their work in a small college laboratory some ten years earlier. More broadly, the personal connection between professor and former student signals an obvious truth: personal relationships are important to the flow of knowledge about science and the technological needs of industry.[38] In this initial stage both parties enjoyed the personal relationship: Cady was informed about industrial needs, Arnold could use Cady's research as a kind of outsourcing. The corporation enjoyed the fruits of his research, without the need to invest much resources in a field whose practical value for its interests was still doubtful. Such exchanges between industrial and academic researchers formed connections between the academy and industrial laboratory. The interaction between Arnold and Cady also shows the understandable value industry placed on products, and that even a scientifically trained industrial researcher was more inclined to concrete technological goals than was a technically oriented college professor.

[37] Arnold to Cady, 8/11/1920 ('much interested'); Cady to Arnold, 9/11/1920 (large wavelength range); Cady to Arnold, 26/1/1921 (sending a resonator), AT&T Archives Loc 79 10 01 03. Cady, *Diaries*; frequent visits are mentioned on 17 December 1920. Cady recalled that he had asked Arnold 'if there was anything his company might be interested in' (Cady, interview by Lindsay and King, p. 18). Cady to Arnold, 17/2/1924 (a visit in October 1921), AT&T Archives Loc 79 10 01 05.

[38] Arnold's Western Electric laboratory had a close connection with Robert Millikan, who sent a few graduates from Chicago University's physics department, including Arnold, to AT&T; see Hoddeson, 'The Emergence of Basic Research Bell', 526, 532–3.

4.3 Interference Procedure

The collaboration with AT&T left Cady bitter. In April 1923, the corporation filed a division application for a patent on 'piezophony' filed by its employee Alexander M. Nicolson in 1918. The division included 'claims corresponding to the broader claims in Cady's applications' on frequency standards and frequency control, which the corporation's patent department had received already in November 1921, accompanied by a memorandum written by Arnold. When Cady's patents were issued (respectively, on 13 April 1923 and on 30 October 1923), 'the claims were revised to correspond in exact wording to Cady's claims'. On this division, the corporation applied for the rights for most of Cady's claims in his two patents.[39] The exchange with Cady, thus, not only enabled the technical personnel of the corporation to direct the professor to study questions of their interest, and to examine possible applications of his findings before he made them public, but it also enabled its legal expert to assert rights on his inventions. Cady, on his part, should have been aware of the technical advantage that he provided the corporation. He was not oblivious to the financial value of his inventions. On the contrary, he informed the corporation about the content of his applications, with the hope of gaining royalties on them. Apparently, however, he was unaware that this move exposed him to a financial risk. He did not consider that the corporation could assert priority on the employment of the resonator to stabilize frequency. Arnold strengthened that impression by hinforming him that if the inventions proved useful for AT&T it would be interested 'to at least have rights under his patent'.[40]

Indeed, AT&T did not claim that Nicolson had foreseen the application of piezoelectric resonance suggested by Cady. Instead, its patent department relied on a recent court decision according to which 'a party can make claims when the function claimed is inherent, even though it was not originally described'. Following Cady's invention, it found out that a figure of circuit in Nicolson's original application (Figure 4.7) can be reinterpreted to suggest frequency control; i.e. the described circuit can be made to oscillate at the

---

[39] Quote from J. G. Roberts (from AT&T's patent department) to Arnold, 17/12/1923. Alexander M. Nicolson, 'Generating and Transmitting Electric Currents', US2212845 (A), a division filed 13 April 1923, and issued 27 August 1940.

[40] Arnold, an inner memo, 4/3/1920, quotation on p. 3, reporting on his discussions with Cady on 21/2/1920 (Wesleyan) and 27/2. (Western Electric laboratory), AT&T archives loc. 79 10 01 03. Note that he had made this statement already regarding Cady's first patent. On Cady's hopes to get revenues from AT&T, see e.g. Cady, 'Problems Confronting the Independent Inventor', a manuscript of a talk, 6/8/1963, in AIP archives, p. 2.

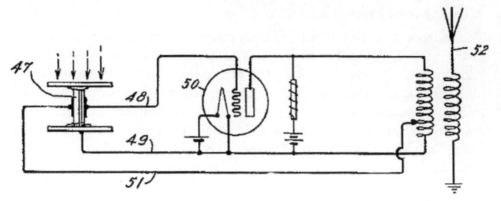

Figure 4.7 The figure from Nicolson's patent application used by AT&T to claim the rights for piezoelectric control. The piezoelectric crystal is 47. *Source*: Nicolson, *Piezophony*.

crystal's resonance frequency. In this arrangement, it explained, 'oscillations are obtained by feeding output energy [through conductor 51 in Figure 4.7] from an amplifier [50] to its input circuit [48] through a piezo-electric crystal [47] without electrical tuning. The mechanical resonance of the crystal determines the frequency of oscillation.'[41] This resembles Cady's idea of connecting the anode to the grid through the crystal, in a way that would allow feedback only in resonance. Nicolson, however, used only one pair of electrodes, while Cady employed two pairs on the same crystal slab that would be coupled electrically only in resonance. Indeed, Cady later claimed that he had not been able to make Nicolson's circuit oscillate at resonance frequency.[42] In his original application, Nicolson did not refer to the crystal's ability to control the frequency of the circuit. His goal was 'modulation of high frequency [electric] waves in accordance with [a sound] signal'. Accordingly, 'the piezo device 47 is actuated by both the signal [sound waves] and by high frequency [electric] impulses and effects a modulation of the high frequency waves in accordance with the signal'. He connected the anode and grid through the crystal to allow high-frequency electric oscillations generated by the triode and the coils needed for wireless and telephony. The sound (shown by arrows on

[41] Roberts to Arnold, 24/4/1924 and 17/12/1923.

[42] 'Transcription of Interview of Cady on September 28, 1972, by Sidney Lang' (an electronic copy is available from the author).

47) changes the amplitude of the high-frequency oscillations.[43] Thus, Nicolson was not interested in exploiting the inherent frequency of the crystal. It is highly unlikely that he recognized that this circuit could, in principle, be made to oscillate at the resonance frequency of the crystal. Restricting the oscillations to resonance frequencies would have heavily limited the usefulness of the method for use with electric oscillations in all the useful spectrum.

In his research and original patent application, Nicolson did not show much interest in the resonance properties of crystals but in their ability to convert sound, especially speech, to electric signals for use in telephony, exploring possible applications of Rochelle salt transducers as microphones (receiver) and speakers of sound, and as amplifiers for long-distance calls. He had begun his study of piezoelectric transducers immediately after the June 1917 Washington conference on submarines. While the other American groups worked on submarine detection, Nicolson's small group was investigating possible applications of piezoelectric oscillators to telephony, the main business of AT&T. Although Nicolson's group contributed to the war effort by growing Rochelle salt crystals, in the case of this group it was the corporation rather than the state that benefited from the information received from the French and British delegations.[44] Apparently, the company did not see a problem in exploiting knowledge that its researchers received in these circumstances. On behalf of Nicolson, it even claimed rights on the use of piezoelectric crystals for detecting submarines (in this same patent on 'piezophony'), although it is clear from the records that its researchers began studying the effect only after they had learnt about the method from the allied at the Washington meeting.[45]

[43] Alexander Mclean Nicolson, 'Piezophony', US1495429, filed 10 April 1918, and issued 27 May 1924. p. 3.

[44] In an inner report on the research during 1918, Arnold stated that '[i]n connection with some of our war work it became know that certain crystals which showed the Piezo-electric effect ... could be made to act as transmitters [for voice]'. Arnold to Colpitts, 5/2/1919, p. 7, AT&T Archives: Loc. 79 10 01 13. Nicolson's first entry on piezoelectricity is from 29/6/1917, two weeks after the Washington conference, attended by representatives of his laboratory. Nicolson NB 141, pp. 105–13, AT&T Archives (in microfilm).

[45] Nicolson, 'Piezophony'. Some authors from AT&T even claimed that 'Mr. A. M. Nicolson, then in the Research Department of the Western Electric, independently initiated work on the utilization of piezoelectric devices for picking up compressional waves from sea.'Raymond A. Heising, 'Introduction', in *Quartz Crystals for Electrical Circuits, Their Design and Manufacture* (New York: D. Van Nostrand, 1946), 1, repeated in S. Millman, ed., *A History of Engineering*

It is interesting that only following the practical use of piezoelectricity by scientists like Langevin and the personal briefing about the piezoelectric transducer did the research arm of AT&T began examining the possible applications of the effect for its own needs. The task was given to Nicolson, a well-educated and experienced researcher. Born in 1880, he studied and instructed at the Finsbury (city and guilds) technical college in London, and then carried out research at Siemens and Halske in London and Berlin, before completing two-year post graduate studies of applied science at Harvard. In 1912, he joined Western Electric.[46] He began by studying the literature on the phenomenon, which he had hardly known previously. Within two weeks he started exploring applications for telephony. He then carried out his own research on the piezoelectric crystals that he used. Similarly to the researchers working on ultrasonic detection, discussed in Chapter 3, his research was closely connected to the expected usage of the crystals. He studied somewhat general questions, like the strength of the piezoelectric effect in Rochelle salt and its elongation, and the dependence of capacity and resistance of crystals on their frequency. He also took preliminary examination of specimens that he planned to use, commenting, for example, that a particular specimen which he connected to a button microphone is 'very efficient at speech frequency'.[47]

As with its use for ultrasonics, the company's choice of Rochelle salt crystals for speech followed its strong piezoelectric effect. This choice led Nicolson to observe erratic behaviour, as found also by Anderson and Cady. He had not dwell about it, however, before he studied Cady's report of 8 May 1918 on the issue (Ch. 3). In October 1919, he reported on this behaviour and on hysteresis in public. By this date, he had also learnt about Cady's experiments with resonance, which he shared with members of Nicolson's laboratory. Nicolson

and Science in the Bell System: Physical Sciences (1925–1980) (New York: Bell Telephone Laboratories, 1983), 529. AT&T enjoyed the fact that Langevin had not taken the adequate steps according to US law, and claimed to be the first to file the method in the USA. It did not need to claim priority over the work done in Europe. Frederick V. Hunt, *Electroacoustics: The Analysis of Transduction, and Its Historical Background* (Cambridge, MA: Harvard University Press, 1954), 56.

[46] Anon., 'A. M'L. Nicolson, Video Pioneer, 69', New York Times, 4 February 1950, 11, and Raymond A. Heising, 'Alexander Mclean Nicolson 1880–1950', *Bell Laboratories Record* 28 (May 1950): 221.

[47] Study from publications (mostly in German), NB 141, pp. 105–13 (beginning on 29/6/1917); a 'report on some piezo-electric applications (made between July 13th and October 1st 1917)', NB 180, pp. 5–6 (9/10/1917); research on length, p. 101 (17/12/1917), on capacity and resistance, pp. 116–20 (21/12/1917–2/1/1918); quotation on p. 128 (29/1/1918) and passim. Notebooks are in AT&T Archives.

replicated the observed dramatic change in the electric properties of the crystal, including negative capacity, near resonance. Yet, even then he did not elaborate on this behaviour and recorded less steep resonance curves than Cady. Instead he focused on strengthening the electromechanical effect of the crystals by paying attention to growth methods, desiccation (regarding the crystal's structure), and the way the pressure was exercised.[48] Probably restricted by his corporation, Nicolson's research stayed close to the practical uses of Rochelle salts for transducing and modulating voice for telephony, where the electromechanical effect is important.[49] He did not invent frequency control, despite some claims to the contrary by authors connected to the Bell System.[50] This should not be a surprise; frequency measurement and control were not useful for the technics that he was developing.

That Nicholson's research was directed towards the immediate applications of transducers can explain why he did not pursue Cady's and his own findings about the peculiar electric properties of resonance. Unlike Cady, Nicolson, a rank-and-file researcher in an industrial laboratory, was obligated to a research programme that fitted his company's commercial goals. As historian Leonard Reich observed, before 1925 'the AT&T laboratory initiated very little research that was not concerned with some type of technology already under development. This greatly limited the scope of its research effort and often stood in the way of those conceptual leaps that led to major breakthroughs in technology.'[51] Nicolson's case suggests that Cady's academic status freed

[48]Nicolson referred to the erratic behaviour of a specific specimen on 4/2/1918 (NB 180, pp. 130–1); he copied in full Cady's report, NB 180, pp. 138–65; see also NB 141. Alexander Mclean Nicolson, 'The Piezo Electric Effect in the Composite Rochelle Salt Crystal', *IEE Proceedings* 38 (1919): 1315–33. On resonance see fig. 17, p. 1329; it is discussed briefly in relation to impedance, pp. 1331–2, on the preparation of the crystals and the manners by which pressure was exerted, pp. 1319–28.

[49]'The commercial exploitation of' the electroacoustic devices suggested by Nicolson 'did not get under way for nearly a decade, however, owing to delays in devising suitable mounting and discovering suitable crystal "cuts", not to mention the difficulty of securing pure and homogenous crystal specimen... Rochelle salt finally did come into its own as an electromechanical transducer after C. B. Sawyer and C. H. Tower had solved the twin problems.' Hunt, *Electroacoustics*, 52.

[50]The official history of Bell Labs for this period credits Nicolson for building 'crystal controlled oscillator which was operated successfully in 1917'. M. D. Fagen, ed., *A History of Engineering and Science in the Bell System—The Early Years* (Murray Hill, NJ: Bell Laboratories, 1975), 318. This false claim is repeated in many other places.

[51]Reich, *The Making of American Industrial Research*, 215.

him to follow his own findings and that his 'scientific' attitude (i.e. independence of a particular technical objective) encouraged his exploration of the
properties of the piezo-resonator. Cady defined the research that led him to
the discovery of piezoelectric resonance as 'pure research in a new branch
of applied physics'. By that, he implied that unlike what applied research (or
engineering science, to use a later term) might suggest, he did not aim at providing tools for improving useful devices, but at producing new knowledge, as
expected by the disciplinary logic.[52]

Furthermore, even after Cady had discovered the sharp piezoelectric resonance and the ability to use it as a frequency standard, AT&T did not use its
rich resources to initiate its own research on the effect. Instead, it preferred
to encourage Cady to continue with his own efforts for contriving methods
of frequency control. It was probably only after it began using Cady's devices during 1921, or even later, that the company recognized the value of the
methods for its system and extended its research on resonators.[53] The early un-
derestimation of Cady's findings for the interests of AT&T has a parallel in an
earlier neglect by Arnold and his colleagues of the value of the triode amplifier,
the most important component of pre-transistor electronics. Fritz Lowenstein
demonstrated an amplifier based on the triode ('audion') in January 1912 and
showed its details to the company's engineers in June, but although they were
looking for such a device that could be used as a 'repeater' for telephony, they
missed its worth for the company. Only two months later, Arnold and his
colleagues became enthusiastic with the roughly equivalent invention of de
Forest's.[54]

[52]'Pure' does not usually go with 'applied'. Yet, Cady's words can be interpreted as consistent (and in agreement with my finding here). By this interpretation, his research was connected
through its subject (through questions and possible goals) to applications, but, unlike engineering research, did not aim directly at a better design and departed from questions of design. Cady,
'Problems Confronting the Independent Inventor', AIP, p. 1.

[53]AT&T probably began inner considerations of the value of Cady's patents in 1921: 'It
will be remembered that this matter [of Cady's patents] was first given consideration some
three years ago when the value of the patent was probably considerably less than it is today
[in 1924].' An inner memorandum regarding the material value of Cady's patents by E. L. Nelson, 31/7/1924, p. 2 (AT&T archives Loc: 70 10 01 05). On 26/12/1923 Arnold asserted that 'the
use of such resonators for wave-length standards and for maintaining the frequency of electric
circuits, particularly radio circuits, looks fully as hopeful as it has done in the past' (a letter to
O. B. Blackwell, Loc: 70 10 01 05).

[54]Hugh G. J. Aitken, *The Continuous Wave: Technology and American Radio, 1900–1932*
(Princeton, NJ: Princeton University Press, 1985), 245–6. De Forest's invention did not have

Edward L. Nelson, an engineer with Arnold's research branch responsible for developing radio apparatus, used the analogy to the early audion circuit to emphasize that the crystal frequency control device was still not robust enough for practical usage, and that it needed further development. 'Whether subsequent progress will carry the analogy further is difficult to predict,' he added. Nelson also worried about the variation of the resonance frequency with temperature, which could make the resonator impractical in the short-wave range. As shown in Chapters 6 and 9, the problem of variation with temperature continued to worry engineers and physicists during the following decade, yet they were more confident that their work would help diminish the thermal variations. Already in mid-1924, when Nelson wrote his report, one of them, his colleague Warren Marrison at the research branch, had developed a crystal cut that resonated at more stable frequencies under temperature variations for use in the corporation's radio transmitters.[55] Thus, experts at the research branch, who looked on more long-term research and development, were more confident that crystal frequency control would prove itself practical as the audion had done.

Arnold's staff began early checking with the resonators sent by Cady in early 1921. Among others its frequency standard group considered the use of crystals for that end already before May 1922, although at that time it preferred using a tuning-fork. Research on piezoelectric resonators became a major occupation of the group only by late 1923, when the corporation realized that it would need crystal frequency control for broadcasting and for secondary frequency standards to be used in the field (on these needs of AT&T, see Chs. 5 and 6). By mid-1924 the field became an important subject of study for the research branch.[56]

During 1924, in response to the request of the patent department, researchers at Western Electric tried to estimate the value of Cady's patents for

any clear technical advantage over Lowenstein's. According to Hong it was based on Lowenstein: Sungook Hong, *Wireless: From Marconi's Black Box to the Audion* (Cambridge, MA: MIT Press, 2001), 181–3.

[55] An inner memorandum regarding the practical value of Cady's patents by E. L. Nelson, 31/7/1924, AT&T archives, loc: 70 10 01 05; Marrison to William O. Baker, 12/7/1977, states his work is for broadcasting transmitters (loc: 86 08 03 02); it is hinted in Horton to Hartley, 3/7/1924 (loc: 79 10 01 02), and the potential of his work in Arnold to Beatty, 21/7/1924 (loc: 70 10 01 05).

[56] Marrison, NB 1112–2, p. 38 (29/5/1922), a list of 'suggestions' of Marrison, NB 2161, pp. 1–4 (4/5/1926). The research notebooks of the group from the period were not preserved. In 21/7/1924, Arnold wrote (to Betty): 'We have, however, already spent in the order of $5,000, in laboratory experimentation on devices using these crystals.' AT&T archives, loc: 70 10 01 05.

the company. '[T]he engineers who consulted on this matter were much impressed with the evidence of utility which grew as they talked over the various applications,' Arnold wrote. They pointed out many possible uses. Beyond their application for primary and secondary standards three uses seemed central: in carrier telephony, i.e. telephone conversation that are carried at a different frequency than that of the original audial speech (including multiplex); for controlling frequency in broadcasting stations (which had already been in use, see section 6.1); and for control frequency in a sender–receiver radio apparatus in point-to-point communication, used also to complement its wired telephone network. The first two were largely developed by the researchers of AT&T to answer the company's special needs. The latter was initiated by researchers at the NRL in late 1923, in search for a transmitter and receiver with predetermined settings at the short-wave range. Although unforeseen by Arnold and his staff, they immediately considered entering into the new market and assumed that other governmental agencies would request similar devices.[57]

'On general principles we [at the research branch, wrote Arnold in an internal memo] believe that the next few years will see very severe demands on the radio art for holding carrier frequencies with great accuracy. We know of nothing so accurate as these crystals for holding the frequencies used in short wave length broadcasting. It seems reasonable therefore that such crystals will find a use of large value in the radio art.' Yet, for the question of the patent department about its monetary worth, he answered that '[t]he value of Cady patents still remains shrouded in a cloud of mystery'. The question became more complicated since Arnold did not know the extent of Cady's patent vis-à-vis AT&T's own claims and a patent of George W. Pierce on a more efficient method for controlling frequencies by a piezo-electric resonator. Still, Arnold wrote that its 'value might readily be of the order of $50000'. The same figure appears also in a report of Hartley from his staff. It might have been suggested by the patent department, following its negotiations with Cady and other radio companies.[58] By April 1924 Cady

[57]Arnold to Beatty, 21/7/1924, p. 1; Hartley, a memo on the value of Cady's patent, 13/9/1924; on learning about the Navy's plans, see a memo by Lewis M. Clement, 21/3/1924. Cady mentioned his work at the NRL in his letter to Arnold on 3/11/1924. See also the other reports in AT&T archives, loc: 70 10 01 05. On the work at the NRL: Linwood S. Howeth, *History of Communications Electronics in the United States Navy* (Washington, DC: Government Printing Office, 1963), 526–30; McGahey, 'Harnessing Nature's Timekeeper', 107–9.

[58]Arnold to Beatty, 21/7/1924, pp. 2, 1, 2; Beatty to Craft, 21/6/1924; Hartley to Arnold, 15/9/1924, loc: 70 10 01 05.

granted the rights to make the device and to license them to others, to the General Radio Company, a small company of Cambridge, MA specialized in instruments for radio apparatus. Like Arnold, Melville Eastham, its founder and manager, had visited Cady's laboratory and observed the working of the resonators already in 1921. The company continued to manufacture resonators and later engaged in the construction of an early quartz clock (Ch. 7).[59] With a new attorney, David Rines, who also represented Pierce, Cady negotiated in the second half of 1924 with AT&T. Rines saw that 'Cady was still hoping to find some way to get out from under that litigation,' and thus offered a competitor, Westinghouse, the full rights for Cady's patent for the said amount of $50,000. In January 1925, the negotiations led to the sale of the patent rights to RCA, partly owned by Westinghouse (although also to AT&T).[60] Although this was a handsome sum, about ten times his annual salary, Cady felt deceived by AT&T. He had good reason to believe that the economic value of his invention was much higher; and thought that the company abused the information that he had shared with it in order to claim his rights.[61] The legal proceedings between AT&T and RCA continued for three decades.[62]

[59] Beatty to Arnold, 24/4/1924; Cady's diary entries from 31/5/1921 and 30/8/1921; Arthur E. Thiessen, *A History of the General Radio Company* (West Concord, MA: General Radio Co., 1965), and pp. 225-8 and 240 below.

[60] *Memoir of G. W. Pierce by David Rines in the form of a letter addressed to Prof. F. V. Hunt, December 15, 1956*, from the *Papers of George Washington Pierce*, Harvard University Archives (HUG 4693.54F), 6–9, quotation on 8.

[61] Arnold to Cady, 5/2/1925; Cady to Arnold, 11/2/1925: 'My chief aim in life from now on will be to try to keep the cat in the bag in Middletown'. The value of US$50,000 in the late 1920s is about $720,000 in 2018 in real wealth, and about $4,000,000 in relative income. Incidentally, the amount is close to the reward that Langevin and Chilowsky received from the French Navy for the invention of sonar. In addition, Langevin had initially asked for 50,000 GBP from the British Admiralty and eventually received 2,000, which were about US$10,000. Regarding salaries: David B. Potts, *Wesleyan University, 1910–1970: Academic Ambition and Middle-Class America* (Wesleyan University Press, 2015), 136, 190. For Cady's view that the corporation abused its power, see e.g. the interview with Lindsay and King (session I), and a text of talk 'Problems Confronting the Independent Inventor', AIP Cady's dossier. On Langevin see David Zimmerman, '"A More Creditable Way": The Discovery of Active Sonar, the Langevin–Chilowsky Patent Dispute and the Royal Commission on Awards to Inventors', *War in History* 25 (2018): 48–68.

[62] The court also discussed the rights of AT&T versus Langevin and Pierce; for a summary of the events from Cady's application to the court's decision of 1953, see Hunt, Electroacoustics, 54–7.

4.4 Frequency Control beyond Cady's Circuit

Cady's invention of crystal frequency control stimulated work on its improvement by inventors and engineers who hitherto had not worked on piezoelectricity. In this section I discuss the origins of the two oscillators that became most popular for crystal frequency control, those of Pierce and John M. Miller and explore the reasons for their success. The entrance of new researchers to the field and the effort directed on the improvement of existing technics and their implementation in practical telecommunication systems, rather than on new technics, mark a change in the character of the research; among others it did not require expertise with the physical phenomenon. This effort can be seen as a move towards the enterprise of the fifth phase of the evolution of piezoelectricity, that of a field known to be technically useful.

In summer 1923 when he invented a new oscillator that vibrates at a resonance frequency of a quartz crystal, Pierce probably did not know about the emerging legal and financial dispute between AT&T and Cady.[63] His main novelty was in feeding back the voltage from the anode to the grid by a piezoelectric resonator with a single couple of electrodes (Figure 4.8). In this design he used only one triode and most importantly sustained highly stable oscillations 'determined by the frequency of the crystal alone'.[64] The design was considerably simpler than Cady's. In Cady's 'oscillator' (more effective for frequency control than a 'stabilizer'), the crystal coupled two branches of the electric circuit (the anode of one triode and the grid of another), by two pairs of electrodes (see Figure 4.5) using three triodes in cascade (although already in 1922 Van Dyke had showed that a single tube is quite sufficient). Indeed, in two of his 'stabilizers' Cady connected the crystal resonator with a single pair of electrodes and inserted it between gird and cathode (rather than to couple two branches). Yet, the crystal did not connect the two electrodes of the triode directly but through additional capacitors and coils in parallel and series (depending on the particular arrangement). Furthermore, stabilizers would find limited use in comparison with the very popular oscillators.[65]

[63] Cady's first letter to Arnold that reveals his knowledge of AT&T's claims for his patent is from 17/2/1924 (AT&T Archives, loc:79 10 01 05). From its content, it seems that Cady had not known about that much before.

[64] George W. Pierce, 'Piezoelectric Crystal Resonators and Crystal Oscillators', *Proceedings of the American Academy of Arts and Sciences* 59 (1923): 81–106, on 81.

[65] Cady, 'The Piezo-Electric Resonator'; Walter G. Cady, *Piezoelectricity: An Introduction to the Theory and Applications of Electromechanical Phenomena in Crystals* (New York: McGraw-Hill, 1946), 493–4.

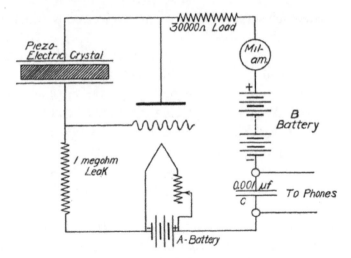

Figure 4.8 Pierce's improved piezoelectric oscillator. *Source:* Pierce, 'Piezoelectric Crystal Resonators'.

In Pierce's oscillator, unlike the case with a stabilizer, unless the piezoelectric resonator vibrates there are no electric oscillations in the circuit. Due to a potential difference between the tube's grid and plate, the crystal would slightly deform through the converse piezoelectric effect. Any small variation in current would generate a mechanical shock in the crystal and make it vibrate at one of its natural frequencies. By the direct piezoelectric effect, the vibration would induce an alternating voltage on the grid, which causes a variation in the anode current at the same frequency. Under proper impedance values, it would strengthen the crystal vibration and lead to a positive feedback until a balance between the energy given by the electric battery, friction, and loss of heat to impedance is reached. In this way, the method ensures, as Pierce asserted, that the circuit would oscillate only at a natural frequency of the resonator.[66] He further claimed that his method allowed for a more constant source of electric oscillations than previously known and that it was practical over a wide range of frequencies, including high frequencies, which had become very useful for radio telecommunication. This improvement, Pierce

---

[66] Similar to Cady, Pierce detected the oscillatory current by a drop of current in an ammeter or by the heterodyne method using earphones. Pierce, 'Piezoelectric Crystal Resonators and Crystal Oscillators'.

claimed, originated in his use of only one pair of electrodes, which allowed for a better mounting of the crystal.[67]

Pierce's circuit, which became very popular, is very similar to Cady's above-mentioned unsuccessful attempt to construct a piezoelectric oscillator in January 1921, mentioned earlier. In both cases the crystal was connected between the plate and the grid, replacing the coils of common triode oscillators. Cady's original explanation of his attempts fits the later successful circuit: 'Theory: vibr[ation] of rod maintained by changes in plate poT [potential]. Charges induced on *b* [the electrode connected to the grid branch] furnish feed-back [through the grid].' Probably the most significant difference was that Cady induced vibrations in the crystal's voltage through a capacitor and a coil in parallel. In other words, far from resonance his circuit allowed for alternating current between grid and cathode, and for varying direct current in the anode branch. Perhaps he regarded that as necessary for initiating vibrations in the crystal, and perhaps also in order to keep the crystal at one rather than another of its resonance frequencies.[68] Pierce's success, however, showed that there was no need for such electric oscillations. These oscillations might have masked the effect of the crystal. It is also possible that components used by Cady (e.g. the crystal, tube, connectors) were less suitable for enhancing the effect. Apparently, he did not pursue the attempt for long. He did not even try a variation that he suggested in his notebook, namely connecting the resonator directly to the grid rather than through a resistor.[69] In other words, it seems that the difference between Cady and Pierce was more in the practical-technical performance than in a general insight about the possibilities offered by piezoelectricity and electronics. One may speculate that Pierce's broader experience with electronics helped him construct the circuit and realize that normal variations within the tube and batteries would suffice to initiate a feedback process in the crystal.

[67]George W. Pierce, Electrical System', US2133642 (A), filed 18 April 1930, and issued 18 October 1938, pp. 1–2; this is a renewed application from 1930 of an original application of 1924. On the working principles of the circuit see also Cady, *Piezoelectricity*, 495–7.

[68]Crystals can vibrate at a few different resonance frequencies (and in their multiplication), depending on their dimensions.

[69]Cady, NB 25, p. 225, 22/1/1921. In his notes, Cady suggested connecting the crystal electrode *b* directly to the grid at point *h* rather than through a resistor in parallel to a capacitator (connection at point *k*, Fig. 4.6) as he had tried in the laboratory.

Most of Pierce's earlier research was closely connected to wireless technics, from studies of electric oscillations and electromagnetic waves (on which he wrote textbooks), through research on crystal rectifiers (no connection to piezoelectricity) and on the use of gases for wave detection, to the invention of a three-electrode mercury-vapor-discharge tube (1913–14). He gained further expertise with cutting edge electronics in the First World War, when he developed systems to combine signals arriving from several hydrophones compensating for the distances between them. Although he made almost his entire career from a graduate student to a full professor at Harvard Physics Department, Pierce regarded himself as positioned midway between the Physics Department and the School of Electrical Engineering. His work combined research on physics mostly on topics of technological interests with research and development of technics, leading to a dozen patents before he embarked on the piezoelectric oscillator.[70]

Pierce considered the use of crystal frequency control following a calibration of his university wavemeter standards that he undertook. Regarding his expertise in the measurement of wavelengths since his 1900 dissertation, it is no wonder that he took over the task. Higher requirements from frequency meters in terms of range and accuracy, on the one hand, and better performance of new tuning-fork-based standards that employed the recent multivibrator, on the other (both will be explained in Ch. 5), created a need to calibrate older standards in use. Pierce conceived that the new piezoelectric-based circuits, recently presented by Cady, could serve as an even better foundation for calibrating frequencies especially in the high range as their resonance frequency was a few orders of magnitudes higher than that of tuning-forks.[71] Pierce enjoyed knowledge of electric oscillations and radio and a direct contact with Cady, who brought him, as well as a few other colleagues, piezoelectric crystals. Moreover, on January 1923 Pierce compared the frequency of a quartz resonator with his laboratory standard, gaining experience with the piezoelectric resonator and with frequency control.[72] He could quickly recognize the potential of Cady's invention for radio. His inventor's attitude

[70] Frederick A. Saunders and Frederick V. Hunt, 'George Washington Pierce', *Biographical Memoirs. National Academy of Sciences* 33 (1959): 351–80.

[71] Pierce, 'Piezoelectric Crystal Resonators and Crystal Oscillators', 81–5.

[72] Cady, 'International Comparison of Wavelength Standards', 814; 'Transcription of Interview of Cady on September 28, 1972, by Sidney Lang', p. 24. Pierce and Cady had met at least a few times before.

and practical expertise in electronics allowed him to assume that he might improve on Cady's design, as he would actually do.

While Cady followed his scientific findings and looked for practical ways to exploit them, Pierce followed his immediate needs for better frequency standards and the larger socioeconomic interest in controlling frequencies. The development of his method did not require previous expertise with piezoelectricity but he probably enjoyed the encounter with Cady. His invention was not only very successful practically, becoming widely used, but also financially, yielding royalties of about a million dollars (recall that Cady received $50,000). Apparently, Pierce's experience with patenting, the advice of his attorney, and, according to his attorney, his talent in the legal field prepared him to secure his patent rights. He was willing to fight for them against the pressures of the large corporations. Fortunately for him, the working principles of his oscillator were different from those of Cady, and those that could be attributed to Nicolson's circuit. Thus, although Pierce was not original in contriving a frequency control oscillator, he could protect his claim for an original design of such an oscillator.[73]

Pierce protected his patent rights not only against large corporations but also against other inventors. The most important among them was John Milton Miller, a physicist working as a radio engineer at the US Navy's Radio Laboratory, which was incorporated into the newly established NRL in April 1923.[74] Miller suggested two main innovations to Pierce's circuit: (1) adding a tuneable LC element (consist of variable coil and capacitor in parallel) to the plate branch (right side of Figure 4.9), and (2) connecting the crystal between grid and filament (cathode) instead of between grid and plate. Miller worked on improving crystal oscillators for a particular technical end: the high-power, high-frequency (short-wave) transmitters of the Navy to be used for long-distance communication. He enjoyed direct contact with Cady and his colleague and former student Van Dyke who served as consultants for the NRL in 1924. Miller also learnt about Pierce's practice by visiting the latter's laboratory.[75] Working on the implementation of Pierce's circuit, Miller

[73] Memoir of G. W. Pierce, esp. 9–10; Saunders and Hunt, 'George Washington Pierce'; Charles Süsskind, 'Pierce, George Washington', in *Dictionary of Scientific Biography*, vol. 10 (Detroit: Charles Scribner's Sons, 1975), 604–5.

[74] Louis A. Gebhard, *Evolution of Naval Radio-Electronics and Contributions of the Naval Research Laboratory* (Naval Research Lab Washington DC, 1979), 33–4.

[75] John M. Miller, 'Quartz Crystal Oscillators', Monthly Radio and Sound Report (US Navy), June 1925, 53–64; John M. Miller, 'Piezo-electric Oscillation Generator', US1756000

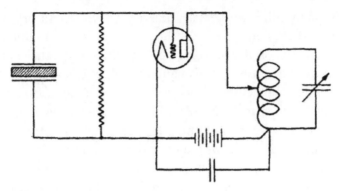

Figure 4.9 Miller's piezoelectric oscillating circuit. *Source*: Crossley, 'Piezo-electric Crystal-Controlled Transmitters'.

and his colleagues noticed two problems. First, the power output was considerably lower than the normal power of the tube, hindering their aim of employing the oscillator in a high-power transmitter. Second, the circuit 'permits the generation of a number of oscillations at one time should the crystal be so constructed that there are two possible oscillations . . . close to the coupling frequency'.[76] To prevent these oscillations, Miller added the tuning circuit, which by generating electric oscillations at a frequency close to the desired resonance frequency (as crystal might have a few) made it possible to choose a particular vibration. This is in contrast to Pierce's idea to allow oscillations only when the crystal is in resonance. Miller's addition induces oscillations also when there is no crystal. Thus, his innovation was in an opposite direction to that from Cady's circuits to Pierce's circuit. Yet, the oscillator was not less stable. So distinct designs possessed different properties that were advantageous under particular conditions and for different aims. Pierce tacitly acknowledged that in his patent application, suggesting a few versions of his circuit that included also a tuneable part. Still, since he renewed the application in 1930, it is not clear when he conceived them.[77]

(A), filed 10 September 1925, and issued 22 April 1930; Howeth, *History of Communications-Electronics in the United States Navy*, 326–30 (on Cady's consulting); *Memoir of G. W. Pierce*, 21 (on Miller's visit to Pierce's laboratory).

[76] A. Crossley, 'Piezo-Electric Crystal-Controlled Transmitters', IRE Proceedings 15 (1927): 18.

[77] Pierce, 'Electrical System'. Cady, who was in contact with Pierce and with Miller at that time, claimed that Miller 'introduced the tuneable element in the anode circuit'. Cady, *Piezoelectricity*, 495.

From reports by other workers at the NRL radio laboratory it seems that Miller considered changing the location of the resonator following his addition of a tuneable RC oscillator to Pierce's circuit. Placing the crystal between grid and filament made it behave like an inductance element, while in Pierce's design it acted as a capacitor. Miller saw that for some uses, an inductance element would secure more stability than a capacitance element, yet both designs have advantages for some settings. In making variations on his invention Pierce also contrived a circuit with a crystal between filament and grid probably independently from Miller; thus, it is sometime called the Pierce–Miller circuit.[78] The NRL researchers found the Pierce–Miller circuit more reliable 'because in such a circuit there is no tendency for short-circuiting the high-voltage plate circuit should the crystal crack or slide out from between the contact plates'. Useful at high and low frequencies, Pierce–Miller's circuit was employed by the Navy in its new crystal-controlled transmitters, probably the first to implement crystal frequency control in late 1924. Still, the production of a practical device useful for high-power transmission that the Navy needed required further electronic adjustments, which researchers of the NRL radio division carried out during the mid-1920s.[79]

Trained in physics, Miller worked on questions related to technology and on technical development from an early stage. In 1915, he completed his dissertation as an employee of the National Bureau of Standards. It is a good example of research shaped by technological interests. He examined the effect of frequency on the effective resistance of iron and bimetallic wires, a complicated question from a theoretical point of view (due to their changing permeability) but important for practice since these wires were becoming useful in some telephone networks. He confined the study to the case of low magnetic field, theoretically simple and also useful for the telephone systems. Four years later he became assistant to the director, and was responsible for precision instruments in the NRL radio division. His piezoelectric oscillator

---

[78] The court granted the invention to Pierce, but Cady claimed that Miller reached the design independently (Cady, *Piezoelectricity*). His claim is supported by Pierce's willingness to form a partnership with Miller regarding these patents, although in general he was ready to fight for his rights. Still, Rines, who reports about that, thought that both innovations are of Pierce rather than of Miller; '*Memoir of G. W. Pierce*, 21.

[79] Crossley, 'Piezo-Electric Crystal-Controlled Transmitters'; Gebhard, 'Evolution of Naval Radio-Electronics', 49–51. Miller, 'Quartz Crystal Oscillators'.

was probably inspired by a popular triode oscillator (of Hartley's).[80] Expertise in electronic was useful also for Pierce in realizing more efficient ways of harnessing the piezoelectric resonator. Of course, Cady also needed to know the field in order to experiment and manipulate with the piezoelectric crystals.

One can regard the move from Cady to Pierce, Miller, and the team working at the radio division as a transition from physicist's expertise in crystals to radio engineers, experts in electronic technics. This was also a move from the phase of research and development to applying piezoelectric resonance (the forth phase in the scheme) to the enterprise of improving the basic technics contrived during the forth phase, as part of the fifth phase of research. Then electronics became at least as important for further developments than knowledge of piezoelectricity, even if the properties of the crystalline vibrations remained central for their various uses by the Navy and by others (they still called for Cady's expertise). This was also the case with the development of the quartz clock, described in Chapters 5–7.

With the Pierce–Miller oscillators and the growing demand for frequency control, the method became a domain of radio engineers also beyond AT&T. Most major telecommunication corporations in the USA and Western Europe such as the German Telefunken, the Dutch Philips, and the American GE, Westinghouse, and RCA dedicated research and development efforts for improving crystal frequency control methods and for their implementation in various communication systems. By 1924 radio experts regarded the technics especially important for short-wave radio, where a change of a tiny portion in the frequency has a decisive impact on reception and interference. Following the experience of radio amateurs, including their success in two-way transmission across the Atlantic in winter 1923–4, researchers in corporations and government laboratories recognized that these wavelengths are useful for long-range communication.[81] The usefulness of frequency control

[80] Miller received the Institute of Radio Engineers Medal of Honor mainly for his works in the first half of the 1920s. His professional activity as an engineer was manifested in his 1925 move to research and development of receivers at a commercial company. John M. Miller, 'Effective Resistance and Inductance of Iron and Bimetallic Wires', *Bulletin of the Bureau of Standards* 12 (November 1915): 207–67; Howeth, *History of Communications-Electronics in the United States Navy*, 326–30. 'John M. Miller', Engineering and Technology History Wiki, https://ethw.org/John_M._Miller (accessed 2/12/2018).

[81] Chen-Pang Yeang, *Probing the Sky with Radio Waves: From Wireless Technology to the Development of Atmospheric Science* (Chicago: University of Chicago press, 2013), 111–43.

for short waves was the major reason for its study at the NRL. Westing-house and RCA pursued similar research and development projects, reaching first experimentally functioning crystal-controlled transmitters probably in the winter of 1924–5. This led to further research and developments of the technics to attain higher precision and stability and to be applied to a wider range of frequencies and ends, as is common with engineering development.[82] The process of innovation and invention did not end with the Pierce–Miller circuit.

4.5 Concluding Remarks

'World War I led to a widespread quickening of interest in and enthusiasm for industrial R&D in the United States' and in Europe. Yet, although indus-trial and state laboratories became central for technical innovation, the early applications of piezoelectricity beyond sonar shows the importance of inven-tors working alone—or with a few assistants, like Cady and Pierce.[83] Cady reached his most important findings and inventions alone, choosing his own research questions. Pierce developed his invention working at his personal uni-versity laboratory. This suggests that independent inventors, especially from the academy, continued to be crucial for new fields and new methods. The case of AT&T indicates that corporations were reluctant to embark on the development of entirely novel technics. That Arnold encouraged Cady to de-velop crystal frequency control shows that the reluctance did not stem from

[82] McGahey, 'Harnessing Nature's Timekeeper', 100–3, 111–15; Kenyon Kilbon, 'Pioneer-ing in Electronics: A Short History of the Origins and Growth of RCA Laboratories, Radio Corporation of America, 1919 to 1964' (1964), 21, manuscript at http://www.davidsarnoff. org/kil.html, accessed 10/2/2020. Balthasar Van der Pol, 'Het gebruik van piëzo-electrische kwarts-kristallen in de draadlooze telegrafie en telefonie', in *Gedenkboek ter herinnering aan het tienjarig bestaan van de Nederlandsche vereeniging voor radiotelegrafie, 1916–1926* (Nauta, 1926); A. Meissner, 'Piezo-Electric Crystals at Radio Frequencies', *IRE Proceedings* 15 (1927): 281–96.

[83] David A. Hounshell, 'The Evolution of Industrial Research in the United States', in *En-gines of Innovation: U.S. Industrial Research at the End of an Era*, ed. Richard S. Rosenbloom and William J Spencer (Boston: Harvard Business School Press, 1996), 13–85, quotation on p. 34, for the decline of the independent inventor. See also Thomas P. Hughes, *American Genesis: A Century of Invention and Technological Enthusiasm, 1870–1970* (New York: Viking, 1989), 138–9.

an overall failure to recognize the potential value of the new methods, but probably from an effort to avert the risk in exploring unknown territory.[84]

Yet, Cady was not an Edison. When he was described as an inventor he commented: 'I would rather have "inventor" omitted, and just say "physicist." I am not a professional inventor.' He did not practice like a professional inventor; for example, he neither kept a patent notebook nor recorded his intentions.[85] Cady's self-image thus reflects a significant difference with the 'professional'; he sought neither devices to invent nor areas requiring technical solution. His investigations were sparked by his findings, inventions sometimes followed, but they were not his primary goal. This attitude probably describes many twentieth-century scientists who may be described as occasional inventors. Pierce became more of a professional inventor than Cady, actively seeking areas in which he could exploit his expertise, although he reached frequency control through his laboratory work. Langevin presents another kind of an occasional inventor.[86] He became an inventor for the occasion of the war. Like Cady, he worked in the transitory science and technology regime, leaving the disciplinary for the utilitarian regime for a short period.[87] Like Cady, he kept an interest in his invention and its development also after the war, returning from time to time to the transitory regime, although working mostly in the disciplinary one.[88] Yet, Langevin's inventions of the dielectric (with Chilowsky) and piezoelectric sonar stemmed from technical-social needs rather than from novel findings, and in that sense better resemble the purposeful inventions of

[84] Apparently by 1936 Bell Labs had changed its strategy as it began investing in research and development towards a solid-state amplifier, a technics that did not exist at the time, leading to their in-house invention of the first transistor in late 1947.

[85] Cady's letter to president of the Academy of Applied Science Robert H. Rines, 25 October 1963, ACNMAH. The patent notebook opened by Charles Steinmetz in early 1891—before he had filed any patent whatsoever—both symbolized and established his new professional identity as an engineer; see Ronald R. Kline, *Steinmetz: Engineer and Socialist* (Baltimore: Johns Hopkins University Press, 1992), 37.

[86] Shaul Katzir, 'Technological Entrepreneurship from Patenting to Commercializing: A Survey of Late Nineteenth and Early Twentieth Century Physics Lecturers', *History and Technology* 33 (2017): 109–25.

[87] Terry Shinn, *Research-Technology and Cultural Change: Instrumentation, Genericity, Transversality* (Oxford: The Bardwell Press, 2008), 171–82; Marcovich and Shinn, 'Regimes of Science Production'.

[88] On Langevin's interwar involvement in the development of sonar see Benoit Lelong, 'Paul Langevin et la détection sous-marine, 1914–1929. Un physicien acteur de l'innovation industrielle et militaire', *Épistémologiques* 2 (2002): 205–32.

engineers than Cady's type of occasional invention. With the maturation of the technical field and recognition of its possible relevance to various technical problems, its study followed needs and specific technical problems rather than new phenomena. The utilitarian logic became dominant. The improvement of Cady's original frequency control circuit, and the development of the filter are examples in point. An independent academic well connected to industry (Pierce) and a researcher at a governmental laboratory (Miller) were central in the former enterprise. Employees of AT&T and RCA's research laboratories developed the latter methods.

Cady's success in inventing a method of frequency control, based on new physical principles previously employed only by him for measuring instrument points, was an advantage that he as a physicist had over engineers. He mastered the phenomena of piezoelectric resonance better than anyone, and used his experimental and theoretical understanding in contriving a circuit that oscillates at the crystal's frequency, for example in comprehending the behaviour of the crystal near resonance (Figure 4.2). He and Pierce showed also more flexibility than their colleagues at the industrial laboratories. The role of thoroughly rounded knowledge of the phenomena is suggested also by the power of direct connection to Cady. Pierce enjoyed it as did AT&T's research arm (at the early stage); The researchers at the NRL asked for the physicist's help in implementing his own invention, notwithstanding their expertise in electronics and their own distinct contribution. The role of physicists in these inventions and their improvements suggests that in addition to the above-mentioned reasons, Arnold might have preferred leaving the task of invention to Cady because he assumed that the expertise of academic researchers would be more useful than the experience gained at the industrial research laboratory. Scientists, thus, played a crucial role in the fourth phase of research, harnessing the findings of the previous phase (piezoelectric resonance) to novel technics, employing the tools and expertise that they cultivated in their study of the phenomena. In this aspect, their role resembles scientists' role in the second phase, notwithstanding that at the later phase the phenomena were considerably better known in the technical community.

5

THE TUNING-FORK ROOTS OF THE QUARTZ CLOCK

Among the technics based on crystal frequency control the quartz clock is clearly the most famous. By 1929 the accuracy and precision of the new quartz-based timekeeper competed with state-of-the-art pendulum clocks, a technology that had been improved and refined during more than 250 years of horology. Within a decade, the new electronic quartz clock surpassed its mechanical predecessor, arguably becoming the world's most precise and steady measuring device. While first a laboratory device, later with miniaturization, the quartz clock revolutionized personal timekeeping. It put a highly precise and cheap timekeeper in the hands of virtually everyone in the affluent world. Its precision accompanies modern life with clocks of many shapes, wristwatches, mobile phones, and virtually any electronic gadget.[1] This combination of accuracy and price, made possible by quartz-controlled electronic circuits, could not have been achieved with earlier mechanical timekeepers. Moreover, the quartz clock technology allowed for expanded broadcasting and therefore distribution of precise time signals by radio as well as by the new clocks. Consequently, the knowledge of precise time became central to an increasing number of activities, leading to a further spread and adoption of the notion and discipline of clock time.[2]

Although it revolutionized timekeeping, the quartz clock originated in research and development efforts on crystal frequency control, which followed the needs and interests of telecommunication rather than those of horology. Improving crystal frequency control and implementing it in

---

[1]Federation of the Swiss Watch Industry, 'The Swiss and World Watchmaking Industry in 2012', https://web.archive.org/web/20130512140245/http://www.fhs.ch/statistics/watchmaking_2012.pdf; for telephones, 'Global Mobile Phone Sales to End Users from 2009 to 2014, by Vendor', https://www.statista.com/statistics/270243/global-mobile-phone-sales-by-vendor-since-2009/.

[2]On the spread of time discipline see Shaul Katzir, 'Time Standards for the Twentieth Century: Telecommunication, Physics, and the Quartz Clock', *Journal of Modern History* 89 (2017): 119–50.

Sonar to Quartz Clock. Shaul Katzir, Oxford University Press. © Oxford University Press (2023).
DOI: 10.1093/oso/9780198878735.003.0006

telecommunication devices was a large-scale endeavour taken up by many
engineers working in industrial, governmental, and academic organizations.
This further research and development on the technics invented by Cady and
Peirce was one of three enterprises that characterize the fifth phase of research
on piezoelectricity, that of a field recognized to be technically useful. One can
see the NRL's efforts to implement quartz oscillators in their transmission
(mentioned in Chapter 4) as an early manifestation of this endeavour. Within
this work on crystal frequency control the development of a quartz clock fulfilled specific interests, and was therefore pursued only by a minority of the
researchers in the field.

The researchers at Western Electric and its successor Bell Labs, the first
to contrive and operate a quartz clock, searched for a highly accurate and
stable method for measuring frequency. Following its long-standing monopolistic business strategy AT&T, the parent company of Bell Labs, sought some
means to integrate its complex telecommunication system (the Bell System),
the largest in the world. A prerequisite for its integration was the establishment
of an accurate method for measuring the frequencies of the various oscillations
used in the system. In other words, to achieve its business goals the corporation needed a frequency-measuring standard.[3] National agencies needed
such a standard to coordinate the communication networks under their jurisdiction. The expansion of radio traffic, and the birth of broadcasting in the
early 1920s, led regulating agencies to seek a stricter allocation of the useful
wavelengths between their various users. These agencies needed to provide
precise and practical means for their measurement, a task given to national
laboratories. Two such laboratories, the British National Physical Laboratory
(NPL) and the Italian Navy's *Istituto Elettrotecnico e Radiotelegrafico*, devised
quartz-based systems.

Other, more specific interests of AT&T in radio and telephony defined a
need not only for measuring frequencies but also for their control. Controlling frequencies became essential for its plans for a national broadcasting
network and for multiplying the number of signals delivered in its wire and

[3] 'Standard' in this context is a device that keeps a known particular value of a physical magnitude (in this case frequency) allowing for the comparison and measurements of other quantities
of the same magnitude (e.g. frequency) in its own terms. Although its existence is a prerequisite
for a regulating standard, normally a 'measuring standard' does not set any regulations. E.g.,
the frequency standard does not determine the breadth of the waveband allocated to each radio
station, but to enforce such a breadth one needs a reliable means for measuring frequency. See
also the discussion later in this section.

wireless lines. The unique combination of interests of the telecommunication giant made its research arm the only laboratory to carry out serious research into the two distinct but connected goals of frequency standard and frequency control. As I show in the current chapter and Chapters 6 and 7, this combination provided the Bell group specific knowledge and experience that enabled them to construct the first quartz clock. Notwithstanding, other organizations also explored ways of controlling, measuring, and manipulating radio frequencies, if for different specific aims and socioeconomic interests within electronic telecommunication. In three cases, those of MIT's short-wave laboratory in the USA, the industrial research laboratory of Philips in the Netherlands, and Japan's radio research committee, this research led to independent inventions of different technics for deriving lower known frequencies from radio-frequencies, technologies which enabled the construction of a quartz clock.

Interest in manipulating frequencies of electromagnetic oscillations and in their exact measurement had begun neither with broadcasting nor with the invention of crystal frequency control. It aroused from the advancement of wireless technics that followed the introduction of the triode, its intensive research and development during the First World War. Due to these developments, the exact value of the frequencies and the ability to manipulate them became technically, and thus economically, socially, and militarily significant. Physicists and engineers working for the war effort, for national laboratories, and for commercial companies contrived methods for measuring and manipulating frequencies based on the recently invented electronically maintained tuning-fork, discussed in section 5.1. Moreover, the NPL and Western Electric contrived and constructed frequency standard systems on these tuning-forks, which suggested previously unknown accuracy in measuring high frequencies. Furthermore, to ensure the accuracy of these measuring standards they designed another novelty—the first electronic clock. Although the mechanical vibrations of the tuning-fork controller continued to determine its pace, designed for and based on electronic technics the new clock symbolized and helped the turn from mechanical to electronic technology.

A groundbreaking achievement in itself, the tuning-fork frequency and time standard systems are central to this history of piezoelectricity due to their seminal role in the development of the quartz clock. In contriving technics based on crystal frequency control, researchers relied on devices, techniques, and methodologies developed for the tuning-fork standard. For the group at AT&T, the first to construct operating clocks on both principles, the tuning-fork system also served as a starting point and a model for its later quartz clock. As shown in this chapter and Chapters 6 and 7, research on the latter was a continuation of research on the former. This chapter, therefore, examines the history of the turning-fork standards as one of the two main sources of the quartz clock, along with crystal frequency control technics, a major outcome of the history described in Chapters 1 to 4. Among others, the history of the nineteenth-century roots of the tuning-fork standards reveals the connection of the quartz clock to the tradition of exact measurements. Chapters 6 and 7 discuss the development of the quartz clock itself. Chapter 6 closely examines the innovation process within Bell Labs, building on an analysis of their tuning-fork system, presented in the last part of this chapter. Chapter 7, on the other hand, suggests a comparative history of the simultaneous invention of technics that enabled the construction of a quartz clock in the few laboratories mentioned above.

Historians have shown that standards were often developed to enable the transfer and communication of results, knowledge, and artefacts from one location to another, for the needs of science, technology, and commerce. To a large extent this was also the case with the frequency standards. In this case precise knowledge of the frequencies used in the corporate and national networks enabled the efficient operation of these systems. Yet, this frequency standard occupied a different level from the standards that have received the most historical attention, helping to make it much less contested than were standards in other examined cases. On the one hand, this was not a standard that defines a magnitude, since frequency was completely defined by time, one of the four 'basic' magnitudes (along with mass and distance to which electric charge was added). Frequency is simply the inverse of time, and thus another aspect of the same physical phenomenon of change. In this sense, it differed from most 'derivative' magnitudes, like force and electric resistance, which depended on more than one basic magnitude, which therefore could not be compared directly to any of them alone and required its own definition. A comparison of frequency to a clock, however, suggested a way for monitoring its value in 'absolute units', i.e. in terms of the basic magnitudes. This simple dependence of frequency on time standard removed many of the potential methodological, ideological, and political overtones that sometimes accompany the continuous process of defining standards.[4] On the other hand, the

---

[4]The definition of and methods for measuring both basic magnitudes like length and derived magnitudes like electric resistance involved controversies with such overtones. Simon Schaffer, 'Metrology, Metrication, and Victorian Values', in Bernard Lightman (ed.), *Victorian Science in Context* (Chicago: University of Chicago Press, 1997), 438–74. Joseph O'Connell identified

accepted stable and known vibrating device, as developed by AT&T and the NPL, did serve as a primary frequency standard for all the system, i.e. the authoritative reference for any frequency measurement in the system. It thus differed from a secondary standard used to facilitate the distribution of a magnitude determined by a primary standard, but often contestable by itself, as illustrated for example by the electric 'resistance box'.[5] Clearly, the tuning-fork and quartz-crystal frequency standards were very far from regulating standards. This kind of standard—on one hand, completely defined by one basic standard, on the other, stable and useful for allowing a very high precision competitive with those of basic standards—has not received adequate discussion.[6] Due to this connection to the basic time standard, much of the research on the frequency standard focused on methods for comparing it to a clock. This resulted in new electronic timekeepers based on the basic vibrators of the frequency standard—the tuning-fork and the quartz crystal, respectively. In addition to a comparison with a highly precise basic standard, connecting a frequency standard to a clock suggests a highly sensitive method for discerning small systematic errors as they accumulate and are displayed by its mechanism.

The tuning-fork measuring systems and standards originated in the same research sites that contrived and developed the piezoelectric technics: civilian scientists and engineers mobilized to WWI research efforts, and industrial and national research laboratories, in France, Britain, and the USA. Their study, however, was technological in character and did not include wider exploration of the phenomena, as was the case with the study of piezoelectric resonance. In this chapter I follow the research at the various sites, examining the immediate goals of the researchers within the larger technical and social (military, governmental, economic) objectives of their research, and the resources useful for their innovations. Two institutions receive special attention here: the

a 'Calvinist' ideology in modern standards, in 'Metrology: The Creation of Universality by the Circulation of Particulars', *Social Studies of Science* 23 (1993): 153–7. On these and other aspects of standardization see also the articles in M. Norton Wise, ed., *The Values of Precision* (Princeton, NJ: Princeton University Press, 1995), and literature cited in this chapter, esp. footnotes 5 and 37.

[5] Bruce J. Hunt, 'The Ohm Is Where the Art Is: British Telegraph Engineers and the Development of Electrical Standards', *Osiris* 9 (1994): 48–63.

[6] An important exception is David Pantalony, *Altered Sensations: Rudolph Koenig's Acoustical Workshop in Nineteenth-Century Paris* (Dordrecht: Springer, 2009), 99–105, which discusses Koenig's determination of his tuning-forks as a standard of acoustic frequency by a chronometer in 1877–9.

national laboratory and the industrial research laboratory, as the NPL and AT&T were the two main sites for the development of the frequency standard systems. As discussed in the Introduction to this book, these became central sites for research and innovation in the early twentieth century. By examining the characteristics of the research on tuning-fork standards in these laboratories, this chapter contributes to our understanding of the practice of research and development in these sites. In particular, analysing the developments of the tuning-fork (in the last part of this chapter) and quartz standards (in Chapter 6) make it possible to examine the process of innovation within the corporate laboratories at the time they became a major source for technical innovation. As seen in Chapter 4, the industrial laboratory confined its researchers mainly to assignments deemed relevant to the needs of the corporations. Yet at the same time, it put richer intellectual and material resources at their disposal, which led to modified methods of invention. As with other technics, researchers at AT&T relied on earlier external innovations. In this case, as in the case of sonar, they originated in French World War I research and developments. Moreover, in both cases, scientists who mobilized to the effort invented the groundbreaking technics.

5.1 Abraham and Bloch

On the eve of the First World War, the tuning-fork was the central device in exact measurement of frequencies for music and scientific ends alike. Since 1834, the tuning-fork suggested a portable and thus useful 'measuring stick' for frequencies, calibrated by a few audial and visual methods.[7] Moreover, tuning-forks were incorporated in a few mechanical and electromagnetic instruments that allowed for maintaining their oscillations and their use for measuring electromagnetic oscillations. These instruments, however, could not be directly used for wireless since tuning-forks could not vibrate at the high frequencies commonly used in radio communication. A way to apply the tuning-fork for measurements in the new realm was suggested by a new device—the multivibrator.

The 'Multivibrator' originated in the French military research on radio communication. Henri Abraham, a physics professor at the *École normale supérieure*, and Eugène Bloch, an active physicist and a teacher at a lycée, invented the device. Like Langevin, Abraham and Bloch were mobilized to

---

[7] M. W. Jackson, *Harmonious Triads: Physicists, Musicians, and Instrument Makers in Ninteenth-Century Germany* (Cambridge, MA: MIT, 2008).

help the war-related research effort, in their case on radiotelegraphy. They examined some anomalies with the behaviour of triode-valve amplifiers in use. As mentioned in Chapter 2, French researchers were latecomers to the study of the triode, but after they had received American and German knowledge and actual prototypes, they compensated for that with intensive research. With the help of Abraham and other civil scientists and engineers, the French military radiotelegraphy devised and produced state-of-the-art valves and multivalve amplifiers.[8]

Examining such multivalves amplifiers during 1916–17, Abraham and Bloch 'noticed irregular discharges in these devices'. On further investigation, they identified the cause of the discharge in the way the triodes were connected to each other. At this point they realized that they could employ this inter-fering discharge, which they originally tried to eliminate, to determine high frequencies used in radio, a basic step for further measurements in radio re-search. Their measurement posed a challenge, which the war sharpened as it brought both significant advancements in radio and scientists with an interest in exact measurements to its study. As mentioned in Chapter 4, contemporary wave meters, however, were inadequate at this range; they hardly exceeded an accuracy of 1 per cent. Abraham and Bloch, therefore, seized the new effect they found to improve the accuracy of frequency measurements.[9]

Consequently, the two physicists changed the goal of their research. Instead of using the triodes for amplifying alternate current without changing its frequency as they had initially tried, they used triodes to multiply known frequencies. Rather than eliminate the discharge as needed for their original goal, they augmented it by coupling the triodes in a new manner: connecting the grid of each triode to the anode of the other triode through a capac-itor (see Figure 5.1). Thus, instead of coupling the grid and anode of the same triode as was common in valve oscillators they crossed between the two triodes. This connection generated periodic discharges between the valves. The new device could generate electric oscillations in many frequencies that are exact integer multiplications of the input frequency. Following how the

[8]Michel Amoudry, *Le général Ferrié et la naissance des transmissions et de la radiodiffusion* (Grenoble: Presses universitaires de Grenoble, 1993),162–95.

[9]H. Abraham and E. Bloch, 'Mesure en valeur absolue des périodes des oscillations électriques de haute fréquence', *Journal de physique théorique et appliquée* 9 (1919), 211–22, on 212–13; H. Abraham and E. Bloch, 'Amplificateurs pour courants continus et pour courants de très basse fréquence', *Comptes rendus* 168 (1919), 1105–8, on 1106. They described their re-sults in confidential reports in 1916 and 1917. On earlier measuring devices, see *Circular of the Bureau of Standards, No. 74: Radio Instruments and Measurements* (Washington, 1918), 96–109.

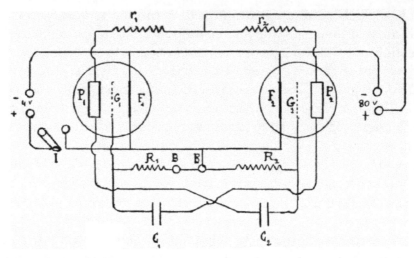

Figure 5.1 The multivibrator: the two circles are the triodes, where F is the cathode (filament), P the anode (plate), and G the grid. Note that the grid of the left triode (G_1) is connected to the anode of the right triode (P_2) through a capacitor (C_1) and vice versa. The figure also includes resistors and direct current power sources. *Source*: Abraham and Bloch, 'Measure en valeur absolue', 214.

oscillations appeared (and were heard) in acoustics, students of periodic phenomena dubbed such oscillations 'harmonics'. According to Abraham and Bloch their new device 'is truly extraordinary rich in harmonics reaching an order of 200 or 300. We gave this device the name *multivibrator*, which reminds this remarkable property.' Due to its production of high harmonics, the multivibrator made it possible to generate high-frequency oscillations from low-frequency ones. Since precise and stable low-frequency vibrations were easier to produce, the method offered high-frequency oscillations of higher stability than could be produced by direct means. These oscillations could be the basis for a new frequency meter for radio frequencies.[10]

Abraham and Bloch chose the tuning-fork as the reference for the low frequency multiplied by the multivibrator. Thereby, they employed its mechanical precision for the novel electronics. During the nineteenth century, scientists like Ernst Chladni, Jules Lissajous, Herman von Helmholtz, and Lord Rayleigh and instrument makers like Rudolph Koenig studied and improved the stability, purity, and exactness of the tone produced by the U-shape tuning-fork in common use by musicians. In their hands it became the most precise device to measure frequencies. They also employed it to regulate

[10]Abraham and Bloch, 'Mesure en valeur absolue', 212–13.

the production of sound pitch and as a timer for short intervals.[11] By employing the tuning-fork, Abraham and Bloch connected the new field of electronic oscillations with the established tradition of research on mechanical vibrators.

Yet, the multivibrator could not be directly connected to the mechanical vibrations of the tuning-fork, but only to electric oscillations. Thus, Abraham and Bloch connected the multivibrator to a circuit whose frequency was equal to that of the tuning-fork. To ensure the equality, they transduced the electromagnetic oscillations of the electric circuit into mechanical waves, i.e. sound, through a telephone receiver, a common tool in early electronic laboratories. Comparing the sound heard through the telephone with the tuning-fork's pitch, they reached a deviation of no more than one thousandth between the two, as higher differences produced audible beats. This electrical–mechanical transduction reduced the accuracy by one order of magnitude, since the frequency of their 1000-Hz (near an operatic soprano's do) tuning-fork was stable to within one in ten thousand. As common in the acoustic tradition, the latter frequency was determined by comparison to a clock. Using a known method, Abraham and Bloch recorded the tuning-fork's vibrations on photographic film and compared them to the second beats of an astronomical clock. To reduce the rate of its marks on the film, they connected the tuning-fork to a small synchronous motor. The harmonics produced by the multivibrator-based tuning-fork could be served as a high-frequency wave meter, of about one order of magnitude more precise than those in common use.[12]

Within two years, researchers dispensed with the manual audial comparison, which complicated the use of the two devices and reduced the precision. Instead they incorporated the tuning-fork and the multivibrator into the same electronic circuit. Curiously, at about the time they developed the multivibrator, Abraham and Bloch invented a triode circuit that oscillates at the frequency of an embedded tuning-fork. Still they did not connect this circuit to the multivibrator. In their method, a magnetized tuning-fork (or a pendulum)[13] was put between two electromagnetic coils, which were connected to two sides of a triode amplifier (of two or three valves). (Figure 5.2 shows a circuit of a later simplified version with one valve amplifier.) Produced by the amplifier, the alternate electric current in the coils induced a changing magnetic field, which exerted a force on the magnetic tuning-fork. The force put

[11] Pantalony, *Altered Sensations*, 22–5, and *passim*; J. H. Ku, 'Uses and Forms of Instruments: Resonator and tuning-fork in Rayleigh's Acoustical Experiments', *Annals of Science*, 66 (2009): 371–95.

[12] Abraham and Bloch, 'Mesure en valeur absolue'.

[13] The circuit of the pendulum was somewhat different; the description here is of the tuning-fork circuit.

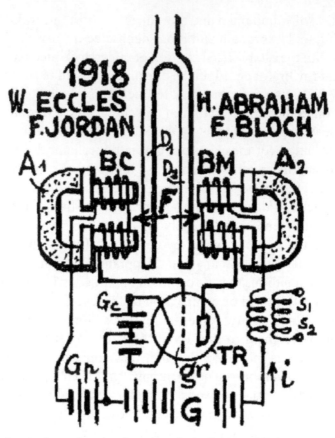

Figure 5.2 A triode-maintained tuning-fork: BM and BC are two induction coils on the magnet, connected to the triode, one to its grid (gr) and the other to its anode (TR). The coil on the right side is used to transmit the oscillation to another circuit. *Source*: M. Lavet, 'Propriétés des organes électromagnétiques', 184.

the tuning-fork in motion, and thus made it vibrate at its resonance. Consequently, the tuning-fork exerted an alternating magnetic force on the coils. Since the triode circuits were flexible enough to oscillate at a range of frequencies (not too far from the tuning-fork resonance), this alternating magnetic force enforced the frequency of the electric current in the circuit to its own period. The magnetic force of the coils kept the mechanical vibration of the tuning-fork from damping.[14]

---

[14] Henri Abraham and Eugène Bloch, 'Entretien des oscillations méchaniques au moyen des lampes à trois électrodes', *Comptes rendus* 168 (1919): 1197–8; Henri Abraham and Eugène

Originally, Abraham and Bloch developed pendulum-controlled triode circuit to maintain steady alternating currents of a few cycles per second, in order to amplify low-frequency oscillations for unspecified needs of the military radiotelegraphy.[15] Coupling a pendulum to an electromagnet was common in electric pendulum clocks.[16] Abraham and Bloch's novelty lay in coupling the mechanical vibrator to an electronic circuit based on the new triode valves. Once they had a low-frequency circuit based on the pendulum, extending their method to tuning-fork was a simple step. Beginning with Lissajous and Helmholtz in the 1850s, scientists and instrument makers had suggested magnetic coupling of vibrating tuning-forks to electric circuits. The connection of the tuning-fork to electromagnetic systems was, thus, common within a scientific tradition of exact measurement. Moreover, in 1915, Amédée Guillet, a lecturer of physics at the Sorbonne devised the first electric tuning-fork chronometer, which, like Abraham and Bloch's device, was based on magnetic coupling. Yet, unlike Abraham and Bloch, Guillet did not use electronic valves, but employed, instead, a carbon microphone as a current rectifier. His device was exact only for short intervals, but probably satisfied its role for laboratory measurements. As a student of Gabriel Lippmann, Guillet had a strong interest in metrology, especially of electromagnetic units but also of time. His chronometer incorporated a method for maintaining the vibrations of the tuning-fork, which he had previously developed for exact measurement, with a novel method for counting the electric oscillations. Since Guillet published his suggestion, and moved in the same circles as Abraham and Bloch they most probably had heard about his method.[17] Yet, they did not need Guillet's

Bloch, 'Entretien des oscillations d'un pendule ou d'un diapason avec un amplificateur à lampes', *Journal de physique théorique et appliquée* 9 (1919): 225–33.

[15] Normally, triode valve circuits oscillated at much higher frequencies.

[16] On electric clocks, e.g., M. Viredaz, 'Horloges électriques', at http://www.viredaz.name/Horloges/francais.htm [accessed 28 February 2014]; Randall Stevenson, 'Mechanical and Electrical Clocks', in Derek Howse (ed.), *Greenwich Time and the Discovery of the Longitude* (Oxford: Oxford University Press, 1980), on 213–19. Ku, 'Uses and Form', 376.

[17] Marius Lavet, 'Propriétés des organes électromagnétiques convenant aux petits moteurs chronométriques a diapason', *Annales Françaises de chronométrie* 15 (1961): 183–96; N. Hulin, 'Un mutualisme pédagogique au tournant des XIXe et XXe siècles. Informer, échanger, centraliser', at http://www.aseiste.org/documents/53f9941495beb59b3ecf87631053f420.pdf [accessed 26 June 2012]; Charles Maurain and A. Pacaud, *La Faculté des sciences de l'Université de Paris de 1906 à 1940* (Paris: Presses universitaires de France, 1940), 53. A. Guillet and V. Guillet, 'Nouveaux modes d'entretien des diapasons', *Comptes rendus* 130 (1900): 1002–4; A. Guillet, 'Roue à denture harmonique, application à la construction d'un chronomètre de laboratoire à mouvement uniforme et continu', *Comptes rendus* 160 (1915): 235–7; see also the list of Guillet's

particular suggestion to know that one can couple tuning-forks to electric circuits. Like Guillet, Abraham and Bloch were physicists; apparently, they maintained an interest in precise measurements also in their technical research on practical devices.

5.2 Eccles, Jordan, and Smith

British researchers, mobilized to the war effort, independently invented a method of 'sustaining the vibration of a tuning-fork by a triode valve', a case of simultaneous invention. William Eccles and Frank Jordan developed the method to measure the magnifying power of valve amplifiers. Interestingly, that was the goal of Abraham and Bloch's research that led to the multivibrator rather than to the triode-maintained tuning-fork. Later, however, Eccles traced the origins of the device to commercial motivations. Namely, in 1914, he suggested the use of a tuning-fork to circumvent a patent of the German firm Telefunken, which made it possible to employ only one amplifier for both sound and radio waves regardless of the differences in their frequencies. It reduced, thereby, electric consumption and expenses on additional amplifiers and thus threatened to dominate radio. Eccles tried unsuccessfully to utilize the harmonics of a tuning-fork to reach both audial and radio frequencies from one vibrator. At that time, he had been a notable expert on radio communication both in its theoretical understanding and in its practical use in industry and the academy. A reader at University College London, he had carried out many studies on a range of radio topics, from wave detectors to the ionosphere.[18]

During the war, Eccles consulted a few military arms on radio. For the needs of his new clients, he modified the tuning-fork circuit, most importantly by incorporating the newly introduced triode. With the help of Jordan, an 'electrician' and a 'lecturer of physics' at the City and Guilds College where Eccles became a professor of applied physics and electrical engineering in 1916, he 'found that a generator of remarkable constancy [in voltage and

publications in Web of Science. On Lippmann's metrology see D. J. Mitchell, 'Measurement in French Experimental Physics from Regnault to Lippmann: Rhetoric and Theoretical Practice', *Annals of Science*, 69 (2012): 453–82.

[18] J. A. Ratcliffe, 'William Henry Eccles. 1875–1966', *Biographical Memoirs of Fellows of the Royal Society* 17 (1971): 195–214, quotations, p. 198; W. H. Eccles, 'The Use of the Triode Valve in Maintaining the Vibration of a tuning-fork', *Proceedings of the Physical Society of London* 31 (1919): 269.

frequency] had arisen'.[19] While the constancy of the frequency stood at the centre of their design, the precise value of the frequency was less important for their measurements. Yet, they soon found another military application for the triode-maintained tuning-fork—secret transmission of pictures, which hinged on precise synchronization of frequencies.

A few methods for transmitting pictures by electromagnetic signal were known at the time. Eccles and Jordan based their system on the 1904 method of the German physicist Arthur Korn. As with most facsimile methods, in Korn's method an electric 'eye' moved in front of the picture, scanned it and translated its luminosity into electric signals. A light beam repeated the motion of the eye in the receiver, producing a copy of the picture on chemical paper. Korn suggested a mechanism by which the motion of the source controlled the motion in the receiver to ensure synchronization. Instead, Eccles and Jordan controlled the period at each end through a separate tuning-fork, dispensing with the need to send signals about their motion. Since one could not reproduce the picture without knowledge of the eye's period, the system could be used for secret signalling.[20] The system required a high degree of agreement in the frequencies of the two tuning-fork circuits. Since the mechanical period of the electric eye was much lower than that of a tuning-fork, Eccles and Jordan needed a mechanism to reach lower frequencies. This was easy to find; a means to couple tuning-fork to low-frequency electric oscillators was well known. In 1875, a Danish inventor, Poul la Cour, had invented the phonic wheel—a kind of electric motor that turns at a known fraction of a tuning-fork's frequency—for synchronization in telegraphy. Rayleigh claimed to invent the device independently for his needs in acoustical research, three years later.[21] The phonic wheel consists of a cogwheel whose centre is an electromagnet and whose circumference is made from soft iron. When the iron circumference

[19] Jordan's affiliation is mentioned in his patents, e.g. 'Improvements in Application of Thermionic Valves to Production of Alternating Currents and in Relaying', GB 155854, filed 17 Apr. 1918. 'A List of the Principal Reports of Experiment and Investigation Received by the Board of Invention and Research from August 1915 to February 1918', ADM 293/21 also in 212/159.

[20] Arthur Korn, *Elektrische Fernphotographie und Ähnliches*, 2nd edn. (Leipzig: S. Hirzel, 1907), 66–9. Eccles's recollection is quoted in Ratcliffe, 'William Henry Eccles, 1875–1966', 198; 'Fourth Periodical Schedule of Reports concerning Experiment and Research', ADM/159. Unfortunately, I could not locate Eccles and Jordan's report itself. Therefore, the description of the system is my reconstruction.

[21] J. W. Strutt Baron Rayleigh, *The Theory of Sound*, 2nd edn., vol. 1 (New York: Dover, 1945), 65–70.

moves near a tuning-fork, itself made from a magnetized metal, it becomes magnetic. Consequently, magnetic forces on the teeth, from the electromagnet and the tuning-fork, adjust the wheel to rotate in phase with the tuning-fork. As with Abraham and Bloch, the novelty of Eccles and Jordan's design was in connecting known mechanic and electric methods to the new triode valve.

Soon Eccles and Jordan's triode-maintained tuning-fork found another usage in measuring very short time intervals. Frank E. Smith, the head of the NPL division of electrical standards and measurements, did not try to improve telecommunication, but ballistics, and needed some means to determine projectile velocities. With David Dye of his laboratory he recorded the electric current induced by projectiles passing through large coils on running cinematograph paper and compared their marks with steady marks of time. For the latter end, Smith 'used first a purely electrical arrangement, and found that a very constant frequency (about 1000 per second) could be obtained . . . For certain reasons, however, it was thought desirable to employ a tuning-fork, and one of those had been borrowed from Prof. Eccles.' The marks made by the tuning-fork attained 'an error not greater than one fifty-thousandth of a second'.[22] This was probably the first application of an electronic circuit for a precise time measurement, although the tuning-fork had been used previously for timing in mechanic and electromagnetic instruments. Smith's transformation of the electronic tuning-fork method from frequency to time measurement displays the close connection between standards of time and frequency.

5.3 Dye's Tuning-Fork Frequency Standard

Ballistics was not part of the regular expertise of the NPL's division of electrical standards. Normally, the division engaged with studies related to electromagnetism and its commercial use. With the development of radio communication, frequency measurements became a central concern of the division, answering a growing interest from the government, the military, and private companies. With the end of hostilities the military remained the major user of wireless communication with high stakes in measuring and controlling frequencies. High frequency (by contemporary standards), stability, and

---

[22] First quotation by Smith in Eccles, 'The Use of the Triode Valve', and second quotation from *Annual Report of the National Physical Laboratory for the Year 1919* (Teddington: National Physical Laboratory, 1920), 50 (the same publisher for consequent reports mentioned in the following). This part refers to a work done in 1918.

accuracy of senders remained a central concern of the admiralty, which requested apparatuses for measuring wavelengths from the NPL throughout the 1920s. With the rapid development of civilian wireless broadcasting, the British General Post Office, responsible for civilian communication, joined the request for high-frequency standards in the middle of the decade; towards the decade's end a commercial company such as Marconi's Wireless Telegraph Co. deemed the field useful enough to fund related research carried out at the NPL.[23]

Until 1921, wireless was used overwhelmingly for point-to-point (or to a few points) two-way communication, by the military arms, a few commercial companies, and amateur operators. The rapid emergence of public broadcasting, i.e. transmitting signals from one broadcaster to a large number of listeners who are unable to transmit signals back, posed new challenges for governments. Along with the growth of two-way transmission, the emergence of broadcasting stressed government interest in dividing the useful electromagnetic spectrum into many communication channels. To this end governments had to allocate relatively a small bandwidth for each station. Interference between transmissions at overlapping, or even nearby, wavelengths posed another problem for the allocation and regulation of the electromagnetic spectrum.[24] Such regulations meant standards in two senses of the term: as stipulating technical requirements, like the deviation of the actual frequency from the allocated one, and as providing means to check whether these rules are maintained. The latter required refinement in the precision and accuracy of measuring methods.

Within the NPL, frequency standards became the expertise of David Dye. Dye joined NPL's electrical measurement division in 1910, after studying engineering at the London City and Guilds Technical College. 'A brilliant but rather irascible scientist,' 'he showed . . . a wonderful instinct for measurements of the very highest accuracy; and especially for the attainment of this accuracy by means of perfection of the mechanical construction of his

---

[23]On the request from these bodies see NPL *Annual Report* for the years 1920 (p. 63), 1923 (p. 84), 1924 (p. 77), 1926 (p. 11), 1928 (p. 13), and D. W. Dye, 'A Self-Contained Standard Harmonic Wave-Meter', *Philosophical Transactions of the Royal Society of London. Series A*, 224 (1924): 259–301, esp. 300.

[24]E.g. Susan J. Douglas, *Inventing American Broadcasting, 1899–1922* (Baltimore: Johns Hopkins University Press, 1987), Ch. 9, pp. 292–314; C. P. Yeang, 'Characterizing Radio Channels: The Science and Technology of Propagation and Interference, 1900–1935' (PhD dissertation, MIT, 2004), 327–56.

instruments.' Thus, upon Smith's departure in 1919, the 32 years-old Dye succeeded him as head of the division. While heading the division Dye returned to formal studies, attaining a Doctor of Science degree in 1926.[25]

The war-related research provided Dye with new techniques for measuring high frequency. In 1919 he adopted Abraham and Bloch's multivibrator and their methods in examining extant standards of radio frequency. Improving on the inventors he combined, probably for the first time, the multivibrator with the triode-maintained tuning-fork, of which he had learnt from Eccles and Jordan. Well familiar with the use of the latter from his research on ballistics, Dye saw its combination with the multivibrator as a simple step.[26] He regarded the combined system as the future standard for high frequency and continued improving it in the following years. Dye directed his efforts toward two main goals: increasing the precision of the system and extending its use for higher frequencies. By 1922 he could measure frequencies of 10^7 Hz (10 MHz). To this end he connected two multivibrators in cascade.[27] Already in 1921 the NPL was satisfied enough with the accuracy to adopt the tuning-fork–multivibrator device (with a single multivibrator) as its standard for radio frequency in its reliable range of up to 150 KHz (useful for radio, but below broadcasting range). Yet, as common in the work on standards, Dye continued to refine the precision of the apparatus, and especially that of the electronically maintained tuning-fork, on which it relied. He experimented with variations in a tuning-fork's frequency under changing physical conditions like temperature, magnetic field, and modifications in the triode circuit, comparing the affected tuning-fork to one that was 'kept invariable as possible', under constant temperature and pressure.[28]

In order to ensure the accuracy and stability of its standard, Dye contrived a method of enabling a direct comparison with a standard astronomical clock. He was not satisfied with the indirect comparison offered by Abraham and

[25] E.V.A. [Appleton], 'David William Dye. 1887–1932', *Obituary Notices of Fellows of the Royal Society (1932–1954)* 31, (1932): 75–8; quotations from Louis Essen, *Time for Reflection*, Chapter 2, unpublished memoirs (1996), in Ray Essen's personal library; L. Hartshorn, 'D. W. Dye, D.Sc., F.R.S', *Proceedings of the Physical Society* 44 (1932): 608–10.

[26] Dye casually mentioned the use of the tuning-fork circuit in the 1919 NPL report (p. 51). Only in the 1921 report did he add that its combination with the multivibrator was 'an improvement first introduced at the Laboratory' (p. 75).

[27] NPL Reports for 1922 (p. 83), 1923 (p. 85).

[28] David Dye, 'The Valve-Maintained Tuning-Fork as a Precision Time-Standard', *Proceedings of the Royal Society of London. Series A, Containing Papers of a Mathematical and Physical Character* 103 (1923): 240.

Bloch, who marked the vibrations of the tuning-fork on a tape. Forty years earlier, the instrument maker Rudolph Koenig had designed a mechanical clock controlled by a tuning-fork. Like Dye, Koenig had not had a direct interest in horology but devised a clock to determine his tuning-fork's frequencies by comparing the rate of the clock to that of a standard pendulum clock.[29] Dye followed a similar track, yet he relied on the new triode technology. With that technology, Dye constructed what was arguably the first electronic timekeeper.[30]

Following Eccles and Jordan's design for a secret facsimile, Dye employed a 20-teeth phonic wheel to reduce the 1000-Hz vibration of the tuning-fork. The phonic wheel drove a 50:1 worm wheel, which closed an electric circuit 'once each 1000 alternations', marking thereby a dot on a chronograph tape, in a manner similar to that suggested by Smith to record the motion of projectiles. Moreover, by adjusting the electromagnetic properties of the circuit, the motor that drove the chronograph was put in synchronization with the tuning-fork. 'In this way the tuning-fork records its own frequency directly on the chronograph without any attention and with extreme accuracy.' Comparing the 'second' dots made by the tuning-fork with those marked by a standard second pendulum clock, Dye observed the accumulated error in the period of the former. With assistants he continued to carry out such comparisons for longer intervals of up to a week on the same tape. By 1932 '[t]he frequency stability over hourly periods [was] of the order of 5 parts in 10^8, and over weekly periods, 3 parts in 10^7'; the latter was on the same order of magnitude as the best mechanical pendulum clocks of the time, and the short-range accuracy approached that of the most exact electro-pendulum clocks.[31]

Beyond the purposes of wireless communication, for which he had begun the research, Dye regarded the tuning-fork circuit as a precision time standard. As such he deemed it useful for measuring relatively short intervals of time (as he had done with Smith), possibly even 'to observe variations in the hourly

[29] The idea of a mechanical tuning-fork clock preceded its use for calibration, and it continued to be used also for other ends. Pantalony, *Altered Sensations*, 100–5.

[30] By electronic I mean circuits whose mechanism rely, to a significant extent, on the properties of discrete electrons, as distinct from other electromagnetic properties, as is the case with valves and transistors.

[31] Dye, 'The Valve-Maintained Tuning-Fork', on 257; D. W. Dye and L. Essen, 'The Valve Maintained tuning-fork as a Primary Standard of Frequency', *Proceedings of the Royal Society of London. Series A, Containing Papers of a Mathematical and Physical Character* 143 (1934): 285–306, on 306.

rate of standard clocks'.[32] He had not, however, connected the tuning-fork-controlled phonic wheel to a clock mechanism that would allow continuous reckoning of time and its continuous display. That Dye did not take this step does not seem to originate in technical difficulties. It rather reveals his lack of interest in making such a continuous display tuning-fork clock, as it did not seem to serve any concrete aim, and its construction did require meticulous work. In addition, from 1924 Dye turned his attention to the new frequency and time standard based on piezoelectric frequency control.

5.4 Bell's Tuning-Fork Frequency Standard and Clock

While state agencies sought exact standards of frequency for coordinating wireless communication under their jurisdiction, AT&T needed them to integrate its extensive telecommunication system, known as the 'Bell System'. In 1920 or early 1921 managers in the corporation concluded that '[r]efinements of [the] methods [of electrical communication] have reached a point where it is imperative that determinations of the frequency of any . . . alternating currents may be made with an accuracy considerably higher than has been possible hitherto.'[33] Consequently they assigned the development of a new system for accurate measurement of frequencies to a group at the research branch of Western Electric.[34] AT&T found a special need for a higher accuracy of frequencies in two of its central areas of interest: wireless communication and telephony. Exact frequencies became crucial for assigning each wireless transmitter a narrow wave band, enabling higher traffic in the bands allocated to the corporation, which it used to complement its wired telephone lines, especially in intercity lines. It also became crucial for multiplex telephony and telegraphy. The financial benefits were obvious, as these novel methods would obviate the need for multiple expensive copper wires. The saving would be most dramatic for long-distance lines, where the corporation enjoyed a monopoly. Western Electric, thus, developed and produced systems for multiplex telephony for use in the Bell System. In its version of multiplex telephony, the regular electric current produced by a telephone transmitter was impressed upon an electric oscillation of a higher frequency through a vacuum tube in

---

[32] Dye, 'The Valve-Maintained Tuning-Fork', 259–60.

[33] Joseph W. Horton, Norman H. Ricker, and Warren A. Marrison, 'Frequency Measurement in Electrical Communication', *AIEE Transactions* 42 (1923): 730–41, quote on 730; Marrison, NB 1112–2, p. 38 (29.5.1922), which mentions also earlier work; Riker, NB 1097–2 (June 21–Dec. 22). All notebooks of AT&T's researchers are from AT&T's Archive.

[34] The division was also called the Research Department.

a process called modulation. At the receiving end of the call, a similar circuit 'demodulated' the current and yielded the original wave. It was necessary that both ends would agree about the frequency of the oscillator, which set the frequency range of the transmission. Higher accuracy with the frequency could also allow for smaller 'safety ranges' between the frequency bands used for each call, and thus increase the number of calls transmitted on the wire. This was also true for wireless, allowing for increased traffic in a given breadth of wavelengths. More accurate standards would thus enable the corporation to hold its electronic oscillators to narrower frequency bands for both radio communication and multiplex telephony.[35]

Beyond its usefulness for radio communication and multiplex telephony, accurate determination of frequency was imperative for integrating these two technics along with others used by the Bell System into one intercommunicating system. AT&T regarded its 'interdependent, intercommunicating, universal system' as a strategic advantage over smaller, local competitors, a justification for its monopoly and a way to maintain it. Its slogan, 'One System, One Policy, Universal Service', reflected its business strategy, namely constructing an integrated system that would allow each user to connect to any other subscriber to the system in the USA (hence 'universal'). None of its competitors could have offered a similar service. Universal service, the heads of the corporation claimed, required a central coordination of the subsystems in use. '[N]o aggregation of isolated independent systems not under common control, however well built or equipped, could give the public the service that the interdependent, intercommunicating, universal system could give,' the corporation pronounced.[36] Central control required coordinating and integrating the various wire and wireless technics used in the world's largest and most complex communication network. A common method for measuring the frequencies of the various signals in use was a prerequisite for such coordination. That, in turn, required an agreed standard frequency to serve as a reference for measuring all the oscillations used by the corporation, and methods for comparing them to this reference. Historians have pointed out

---

[35] E. H. Colpitts and O. B. Blackwell, 'Carrier Current Telephony and Telegraphy', *AIEE Transactions* 40 (1921): 205–300.

[36] Robert MacDougall, 'Long Lines: AT&T's Long-Distance Network as an Organizational and Political Strategy', *Business History Review* 80 (2006): 297–327, the first quote is from AT&T president Theodore Vail in 1910, on p. 303. Milton Mueller, *Universal Service: Competition, Interconnection, and Monopoly in the Making of the American Telephone System* (Cambridge MA: American Enterprise Institute, 1997), 92–103; second quote (also from 1910) is on p. 98.

that standardization was often established to make it possible to coordinate work done in different locations by providing a common framework and thus a means of comparing and transferring results from one location to another.[37] Usually standards answered the interests of a few independent laboratories, or companies and the state, which was concerned about their coordination. In the case of AT&T, however, the diversity and size of its own system were large enough to warrant the establishment of a frequency standard to address its interest in integrating the Bell System.

The size of the Bell System was nothing new, even if it continued growing. Its use of different frequencies for transmitting signals, however, was rather novel. During the first forty years of its existence, AT&T had hardly exploited frequencies beyond the audio-frequency range up to approximately 10 kHz, as the electric signals in the wire followed the original voice frequency. Yet, the rapid development of triode-based radio and telephone technologies during the 1910s changed that. Researchers in AT&T had a central role in the improvement and early applications of the triode for amplifying currents and for emitting and receiving electromagnetic waves, as both were deemed important for the interests of the telecommunication giant. The corporation adopted these new radio technics, which made it possible to transit human voice over distance, to complement its extensive wired network and the array of services it provided. Circa 1920 wireless communication used a spectrum ranging from a few ten thousands to a few millions hertz. With the voice and a multiplex range meant to measure all alternating currents used in the system, the group had to provide a standard that would 'cover the entire range between a few cycles per second and several million'.

Still, the novel goal of attaining high accuracy originated not only from new needs, but also from new opportunities. The electronically maintained tuning-fork made the construction of a reliable primary frequency standard whose 'absolute value . . . is known to be one part in 100,000' a feasible enterprise.[38] Such high accuracy of the primary standard was calculated by allowing the secondary calibrating devices to be in error higher by one order of magnitude, while allowing for still one order of magnitude higher error for

[37]Simon Schaffer, 'Rayleigh and the Establishment of Electrical Standards', *European Journal of Physics* 15 (1994): 278; Hunt, 'The Ohm Is Where the Art Is'; Arne Hessenbruch, 'Calibration and Work in the X-Ray Economy, 1896–1928', *Social Studies of Science* 30 (2000): 397–420.

[38]Horton, Ricker, and Marrison, 'Frequency Measurement in Electrical Communication', 730.

the frequency of the circuits practically used in the system. To make such a central standard useful, the corporation needed some means to transmit the basic frequencies to the dispersed locations within the company. To that end, the research branch developed a way to send the frequency signals through telephone or radio lines, and portable secondary standards that maintain a stable frequency as they were physically carried from one location to another. Often, the researchers examined the use of the same method both as primary and as secondary standards. This would assist in the development of the quartz standard.

These tasks were assigned to a group headed by Joseph Warren Horton. Born in 1889, Horton joined Western Electric in 1916, after studying physics and chemistry at MIT, where he also 'spent a year as an assistant in the Physics Laboratory, and a second year in the Senior Electrochemical Laboratory, in Dr. [Harry] M. Goodwin's [physics] department'. When the USA entered the First World War, Horton left for the naval underwater laboratory in Nahant, MA. He regarded that period as part of his education gained from experience working with notable scientists from General Electric and his own company. 'Almost the only problem which was wholly under [his] control while at Bell Labs had to do with the measurement of frequency.'[39] In later years Horton would return to MIT as a PhD student and later joined the faculty, studying the application of physics and electrical engineering to biology and medicine. From this perspective he recalled that 'In general the problems [with which I dealt at Bell] called for the utilization of existing knowledge rather than for the acquisition of new knowledge. Although required to do many things that had never been done before I would have said, at the time, that I was an engineer rather than a scientist.'[40] Still part of the knowledge utilized by Horton and his group, especially regarding the quartz clock, had been recently acquired in scientific research, albeit outside the Bell System.

[39] Joseph Warren Horton, *Excursions in the Domain of Physics*, a typed manuscript, 1965, held by the American Institute of Physics library, quotations on pp. 2,4; obituary in *IEEE Spectrum* 4 (1967): 38–9; James E. Brittain, 'Joseph Warren Horton', *Proceedings of the IEEE* 82 (1994):1470. On his teachers and their teaching and work, see John W. Servos, *Physical Chemistry from Ostwald to Pauling: The Making of a Science in America* (Princeton, NJ: Princeton University Press, 1990), 103–6. Despite Brittain's claim (which is repeated by others), Horton probably did not have prior experience with piezoelectricity. Although he did research ultrasonic detection, he worked in Nahant on passive hydrophones, which unlike sonar-like detectors did not employ the effect.

[40] Horton, *Excursions in Physics*, 3–4.

By December 1922 Horton headed a group of 25 employees, most of them probably with a degree in physics or engineering, divided into four subgroups, one of them under his direct responsibility. Only one or two of the smaller groups under Horton worked on frequency standards. Arriving with a fresh PhD in physics in summer 1921, Norman H. Ricker headed one of these groups until his departure from AT&T at the beginning of 1923. A young subordinate who had already worked on tuning-fork standards, Warren Marrison succeeded Ricker. Born in 1896, Marrison earned a BSc in engineering and physics, taking advance courses in the latter subject from Queen's University of Ontario in 1920, after serving part of the war period in the air force of his native Canada, where he employed his experience as a radio amateur. With some practical and laboratory experience with 'high frequency work' he arrived at Western Electric where he spent the summer of 1920. In September of that year he registered for graduate studies of physics and mathematics at Harvard, where he chose courses on a subject directly related to his work at the company: electric oscillations and their application in radio. After attaining an A.M. degree, he returned to the company in September 1921.[41] Eight months later, he joined the study of tuning-forks, continuing the experimental work of one J. C. Davidson, who had left the group. The quite easy replacement of one researcher by another is a mark of the team approach of the group and AT&T's laboratory in general. Although individuals often worked alone, or with one assistant or co-worker, the research was ultimately a group effort. Problems and questions were discussed in both official and informal meetings, and questions were allocated from one individual to another and even from one unit to another.[42]

Returning from Nahant, Horton was assigned to work on multiplex telephony. One of the major technical difficulties that hampered its implementation was the production of many known frequencies from a basic oscillation

[41] 'Chart of Western Electric Company Research Department, Dec. 1922', manuscript at 'Norman Hurd Ricker Papers', in Woodson Research Center, Foundren Library, Rice University (14/142); W. R. Topham, 'Warren Marrison—Pioneer of the Quartz Revolution', *National Association of Watch and Clock Collectors Bulletin* 31 (April 1989): 126–34. Marrison's first dated notebook entry in AT&T is from June 1920; Marrison's 'record card' in Harvard University Archives and his application in UAV161.201.10, Box 71. I thank the staff of both archives for providing me the relevant material.

[42] NB 111–2 began by Davidsson (10.21–3.22), continued by Marrison from 29.5.22. Leonard S. Reich, *The Making of American Industrial Research: Science and Business at GE and Bell, 1876–1926* (Cambridge, UK: Cambridge University Press, 1985), 202–4. Horton, Ricker, and Marrison, 'Frequency Measurement in Electrical Communication'.

frequency. In 1920, Horton suggested a way to produce 'any required harmonic frequency wave, or a series of such waves from a given sine wave of fundamental frequency'. His scheme enabled the combination of circuits of different frequencies that are integer multiples of a basic frequency to reach any integer multiplication of the original. This work provided him resources for developing Western Electric's frequency standard, and probably contributed to making him responsible for its development.[43]

During 1921–3, Horton's group developed a compound system that would serve as a frequency standard for any radio wavelength used by AT&T. The system resembled earlier systems for measuring radio frequencies like that of Dye with a few significant changes and additions. It included a valve-maintained tuning-fork that provided a fundamental exact frequency, a 'harmonic generator' for multiplying this frequency, and modulators to reach intermediate frequencies between the values produced by the harmonic generators. Since AT&T required a means for measuring frequencies to within 100 Hz, the group chose a tuning-fork with that natural resonance instead of the tuning-forks used by other groups, which resonated at approximately 1,000 Hz. It also decided to design a new more complex harmonic generator to multiply the basic tuning-fork frequency, instead of Abraham and Bloch's multivibrator. This harmonic generator (also called a 'harmonic producer') multiplied the basic standard frequency to its tenth harmonic, which went into a second generator and into a third reaching 100,000 Hz. Electronic circuits oscillated at each of these frequencies (in the later working system its exact frequency could be tuned by a variable circuit). 'Balance modulators' produced oscillations whose frequencies were the additions and subtractions of the incoming frequencies. The scheme was similar to the one suggested by Horton in 1920 for multiplex telephony, showing the similarity between the projects on multiplexing and frequency standards. Individual balance modulators had already been used for telephony in the Bell System. The modulators allowed the system to generate any intermediate frequency that is an integer multiple of the fundamental 100 Hz.[44] The resulting alternating wave could be compared

[43] Joseph W. Horton, 'Harmonic Generator System', US patent 1519619, filed 23 December 1920, patented 1924, quotations on 1; Colpitts and Blackwell, 'Carrier Current Telephony and Telegraphy'.

[44] Marrison, NB 1112–2, 38–9 (29.5.1922); Horton, Ricker, and Marrison, 'Frequency Measurement in Electrical Communication'. One could attain a specific frequency also directly by multiplying the original oscillation by the desired number. This was the strategy of David Dye at the NPL, using one or two multivibrators. Horton's group divided the process into two separate

to and thus could measure the frequencies of other waves. The system also included an integral quality-control ingredient, a method for comparing the basic frequency standard to a more reliable standard, i.e. a pendulum clock.

Horton's group developed the individual components of its system of frequency standards mostly in parallel to each other. Apparently the electronically driven tuning-fork was the earliest piece of the circuit to work; the group had used it already before June 1921. A few of the group's members continued to improve it until the spring of 1922. In its design they relied directly on the device of Eccles and its improvement by E. Eckhardt et al. at the American National Bureau of Standards. To prevent variations in its frequency, the Bell group kept the tuning-fork under constant temperature and pressure. The sensitivity of the tuning-fork's frequency to temperature variations followed from elastic theory and had been established experimentally in the nineteenth century.[45] The group at Bell thus used extant scientific knowledge and did not need to carry out its own research on the effect of temperature to recognize the need to prevent temperature variations. As Horton later remarked, his group did not engage in discovering new knowledge. It did examine the performance of its own tuning-forks, including measuring the effect of temperature on their specific frequencies to suggest improvements in their design. For example, it modified their shape and the location of the electromagnets that induce their vibration.[46]

In working on Western Electric's system, Horton probably knew that Dye controlled the basic frequency of his multivibrator by a tuning-fork circuit, although the latter had not published the details of his apparatus at the time.[47] Still, he chose a somewhat different approach to the problem of harmonic generation, an approach that would shape also its research on the frequency divider for the quartz clock. The group looked for a

steps, first generating known high harmonics and then combining them to reach any specific frequency. Horton, Ricker, and Marrison claimed that this division leads to higher accuracy and fewer mistakes with higher frequencies since direct multiplication of high order is prone to lead to a non-intended harmonics (ibid., 741). Dye, 'The Valve-Maintained Tuning-Fork'.

[45] E.g. Pantalony, *Altered Sensations*, 102–4.

[46] The group found a change in tuning-fork frequency of 0.0109 per cent per degree. Joseph W. Horton and W. A. Marrison, 'Precision Determination of Frequency', *IRE Proceedings* 16 (1928): 137–54, on 139–40. Horton, Ricker, and Marrison, 'Frequency Measurement in Electrical Communication'.

[47] See fn. 26.

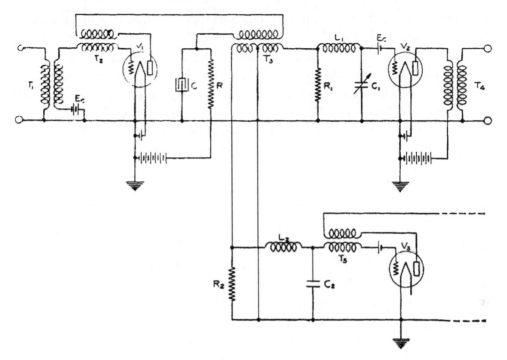

Figure 5.3 Horton, Marrison, and Ricker's 'harmonic producer' for the
tuning-fork standard of AT&T. $V_{1,2,3}$ are triode tubes showing (from left to
right) the grid, cathode, and anode. *Source*: Horton, Ricker, and Marrison,
'Frequency Measurement in Electrical Communication', 739.

system vibrating sinusoidally, while the multivibrator did not produce a sine
oscillation.[48] Moreover, the Western Electric group sought a mechanism to ensure that the target circuit vibrated exactly at a harmonic of the tuning-fork.
To this end, the members of the group devised a way to enforce the vibrations
of the target circuit using a cue from the source: a tuning-fork circuit.

In July 1921, Ricker put forward the basic principles of the harmonic producer. In Ricker's method, a triode circuit (V_1 in Figure 5.3), coupled to a
circuit that oscillated at the frequency of the tuning-fork through an inductive
coil T_1, was arranged to allow passage of electric current (or discharge) only

---

[48]The kind of oscillations produced by the multivibrator would be later mathematically analysed by Balthasar van der Pol. He found that, unlike the harmonic generator of AT&T, these
kind of 'relaxation oscillations' are useful for either multiplying or dividing an input frequency.
See Chapter 7.

when the voltage from T_1 was very close to its peak, that is only at a small fraction of the period of the input oscillation. Thus, it produced brief intervals of electric current at the frequency of the input oscillation. This brief current induced an electric current through the inductive coils T_3 in a second RLC circuit (that of L_1C_1 in Figure 5.3). Like a hammer on a piano string, the brief current induced voltage which generated a continuous oscillation in the L_1C_1 circuit. This circuit oscillated in one of its own natural frequencies, as determined by its coils, condenser, and resistor, in the same way as the natural frequency of a string is determined by its width and length. Unlike a piano hammer, however, the V_1 circuit induced voltage in exact periods, forcing the more flexible oscillations of the RLC circuits into a harmonic of the input frequency. Slight deviations from this frequency would be opposed by the induced voltage through T_3, which would restore the output to a harmonic of the income frequency. The internal frequency of the L_1C_1 circuit is fixed enough not to fall into another integer multiplication of the original frequency. The output of this generator can go through a third similar harmonic generator. The multivibrator, the alternative harmonic producer in use, did not possess a similar compensating mechanism. As will be shown in Chapter 6, this preference for a compensating mechanism would also characterize Marrison's frequency divider of the quartz clock. Through the variable capacitor the researchers changed the internal frequency of the circuit and with that the harmony. The third circuit, L_2C_2, has fixed variables to produce the tenth harmonic. In the scheme for reaching higher frequencies, the tenth harmonic becomes an income frequency for a second similar harmonic generator.[49]

The accuracy of the frequency to come out of the circuit hinged on that of the basic standard, i.e. the tuning-fork. Restoring to the common practice of examining frequency standards, Horton suggested comparing the vibrations of the tuning-fork with that of a steady clock, remarking that the new system should have 'the general characteristics of a good clock'. Unlike Dye, Horton was not satisfied with connecting the tuning-fork only to a tape chronograph; instead, he suggested controlling a clock mechanism by the tuning-fork.[50] In other words, he aimed at a clock. By June 1921 Bell's tuning-fork drove a

---

[49] Horton, Ricker, and Marrison, 'Frequency Measurement in Electrical Communication', 739–41; Ricker, NB 1097–2, esp. 41–2 (22 July 1922); Joseph W. Horton, 'Source of Waves of Constant Frequency', patent, US1560056, filed 1 May 1923 (issued 3.11.1925).

[50] Horton, Ricker, and Marrison, 'Frequency Measurement in Electrical Communication', 731.

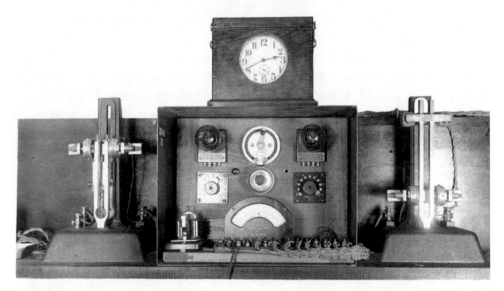

Figure 5.4 A clock connected to two tuning-forks in Western Electric's research department, from 14 November 1921. Note that the clock is connected (probably alternatively) to two forks of different dimensions. The left one is excited by magnets near its upper part and the right one by magnets near its lower part. *Source*: Courtesy of AT&T Archives and History Center.

clock mechanism of Horton's design, which included a synchronous motor and a commutator that reduced the input frequency (see Figure 5.4). Unlike the chronograph, the tuning-fork clock moved continuously. The same mechanism that transferred the tuning-fork vibration into a mechanical rotation for driving the clock made it possible to mark time signals on a chronograph for monitoring possible fluctuations over shorter intervals. The team compared this clock with the electric pendulum clock in their laboratory.

Like the system as a whole, the clock mechanism was based on well-known methods, in this case the phonic wheel employed for a similar end by Eccles, Jordan, and Dye. Still, it required technical research and especially development to accommodate it to the new precise system. The first motor was inefficient; by May 1922 the group developed a more complex synchronous motor, still on similar principles, which fitted better the low voltage produced by the tuning-fork circuit. Technical problems continued to hamper the system. In November the researchers had not yet compared the tuning-fork to a clock. By the next summer, they improved on the synchronous motor. 'Shortly thereafter' they built a new mechanism 'in which the 100-cycle motor was geared directly to the clock mechanism instead of operating through a stepping device' for reducing the frequency.[51]

Arguably, the coupling of the basic frequency standard to a clock was the most important novelty of the system. The idea and the means to accomplish it had precedents. Still, both the linkage of radio frequency to continuous time measurement and the construction of a clock on an electronically maintained tuning-fork (which suggests a steadier operation than Guillet's earlier microphone mechanism) were important original steps for the group. Dye followed a similar track but was satisfied with comparing the NPL's frequency standard to a timekeeper over limited intervals. Horton's group, however, conceived AT&T's frequency as a continuous reference for tuning and measuring frequencies of electric oscillations at the Bell System. The NPL, on the other hand, regarded its central standard as a means for calibrating other devices used as secondary standards; its continuous operation was not deemed important for its main purpose. Since Bell's frequency standard operated continuously, its researchers sought a continuous method for inspecting its performance.[52]

The comparison between the tuning-fork clock and exact time signals strengthened Horton and Marrison's confidence in the accuracy of their standard. In early 1923, using a chronograph, they found an average difference between the tuning-fork and the laboratory clock of 'about 6 parts in 1,000,000', well within AT&T's goal of one part in one hundred thousand.[53] Still, the group continued to improve the accuracy of its standard, reaching an accuracy of 3 parts in 1,000,000. Since this accuracy was higher than that of their laboratory electric pendulum clock, the researchers established it by direct comparison with the radio signals from the naval astronomical clock. The accuracy of the tuning-fork clock was only one order of magnitude lower than that of the BoS standard clock. Still in 1927, Horton's group suggested

[51]Marrison, NB 1112–2, pp. 38–41; Horton, Ricker, and Marrison, 'Frequency Measurement in Electrical Communication', 734–6. A report of R. V. L. Hartley to Harold Arnold, 28.11.1922 (loc 79 10 01 14, in Arnold's papers); Warren A. Marrison, 'The Evolution of the Quartz Crystal Clock', *Bell System Technical Journal* 27 (1948): 510–88, esp. 528 (quotation). The new motor might be the one mentioned in Marrison's notebook from October 1924, and examined in April the next year (NB 1444, pp. 42, 149–53).

[52]The tuning-fork of AT&T standard 'ran continuously from April, 1923, to May, 1927, except for four intervals totalling about three days'. Horton and Marrison, 'Precision Determination of Frequency', 139.

[53]Horton, Ricker, and Marrison, 'Frequency Measurement in Electrical Communication', 736.

improvements in the mechanism that would increase the accuracy of about twenty-fold; as mentioned, such accuracy was attained later by the NPL with its chronometer. The increase interest of the Bell group in horology can be seen in its use of the clock fork for exact time measurements. In January 1925, Warren Marrison at the laboratory recorded the timing of the total solar eclipse at different locations in the northeast USA. Stations at each of these locations sent a signal through telephone line to the laboratory in New York, where it was marked against the time signals from the tuning-fork.[54]

Notwithstanding their pioneering work on the electronic tuning-fork clock, frequency standards were the main aim of Horton's group. The clock was only a means to determine the consistency and accuracy of the frequency standard, and as such it was only part of the group's effort. The researchers also improved other parts of the system, such as the tuning-fork itself, its electronic circuit, and the thermostat used to keep it at a constant temperature. Although satisfied with the performance of the tuning-fork, Marrison also continued to explore alternative electromechanical vibrators as sources of constant frequency: steel bars and rods. Such an examination of alternatives is common in technical research. With its rich resources, AT&T often invested in the search for alternatives to methods that worked well. Even if they did not lead to useful methods for the corporation, patenting them could obstruct competing firms.[55] Yet, these attempts were not pursued as vigorously as the further study of the tuning-fork system and the development of frequency standards based on piezoelectric resonators, the subject of Chapter 6.[56] With their high accuracy, quartz clocks became the most popular device for exact timekeeping,

---

[54] Presumably the tuning-fork could mark division of seconds more exactly than a mechanism based on a standard second pendulum clock. Horton and Marrison, 'Precision Determination of Frequency', 141; Horton, Ricker, and Marrison, 'Frequency Measurements in Electrical Communication', 736; Marrison, 'The Evolution of the Quartz Crystal Clock', 528–30, second quote, 529; Marrison, NB 1444, pp. 97–105 (24.1.1925). Bureau of Standards, *Standards Yearbook* (U.S. Government Printing Office, 1927), 44.

[55] The high number of 'suggestions' recorded in the laboratory's notebooks and the diversity of their subjects, even within the same notebook, testify to the practice of examining alternative means also to methods in use. See, for example, notebooks of Ricker and Marrison.

[56] Davidson and Marrison, NB 1112–1, esp. p. 38 (Marrison, 29.5.1922—a report about earlier work); Ricker NB 1097–2, esp. pp. 21 (26.7—rod) 23, 30 (26.8., 2.12.1921—bar); Marrison, NB 1444, *passim* and pp. 55 (12.11.1924—bar), 73–8 (8.12.1924—torsional oscillations). On quartz see also the list of suggestions in NB 2161 and 2162. AT&T's archive has not kept Horton's notebook, so the discussion here relies more on Marrison's research. His notebooks, however, mention also the group's work in general without assigning a particular author.

replacing the pendulum clocks, and apparently obstructing the development
of tuning-fork clocks. Tuning-forks controllers became important again with
the minimization of electronics after the advent of the transistor. An electronic
tuning-fork watch developed in the late 1950s paved the way for the quartz
watch, which has since became ubiquitous, just as the electronic tuning-fork
clock paved the way for the quartz clock in the 1920s.[57]

5.5 Conclusions

The triode-maintained tuning-fork clock, the first electronic clock, resulted
from an accumulation of small steps. Innovations consisted of quite minor additions and modifications of previous methods, which in retrospect often seem
straightforward or even trivial. Once the triode became a central powerful
device of wireless technology, connecting the previously electromagnetically
maintained tuning-fork to the electronic valve was quite straightforward.
Although Abraham and Bloch did not connect the compound device to their
multivibrator, to Dye such a connection seemed quite trivial; since the idea
of comparing tuning-fork frequency to a clock was well known, it seems natural to connect its electronically maintained version to a chronograph. The
means to do so were known and only required some modifications: the phonic
wheel motor, the chronometer, and the mechanism for marking dots on tape.
Driving a clock dial, and not only a dot-marking mechanism from the phonic
motor, did not require much imagination, especially as the phonic wheel had
already been used for timers. Arguably the multivibrator was the only innovation not based on a previous device; yet it followed a known idea of
producing harmonics. This gradual process does not offer great 'Eureka' mo-
ments. It is precisely due to that absence that the process represents many
inventions, including important ones.[58] And, indeed, it was a process of invention, not merely a process of refinement and improvement, as the end result—a
highly accurate electronic clock and a system of radio frequency standards—is
clearly different from the original electromagnetic tuning-fork. Moreover, this
electronic device opened the way for overthrowing centuries-old pendulum
horology by the quartz clock.

---

[57] Carlene Stephens and Maggie Dennis, 'Engineering Time: Inventing the Electronic Wristwatch', *British Journal for the History of Science* 33 (2000): 477–97.

[58] The steam railway and the automobile are famous examples for such gradual inventions.

Carrying out these small steps was far from trivial. Each of them required specific cause to contrive the innovation, knowledge of and preferably experience with the related technics, and some ingenuity. Otherwise many more researchers would have suggested each innovation. To take an example: that two resourceful researchers like Abraham and Bloch did not connect their own multivibrator and their own triode-maintained tuning-fork suggests that the step required some ingenuity. That Guillet did not incorporate triodes in his new electromagnetic tuning-fork clock suggests the crucial role of the inventors' experience with electronic technics, in addition to their knowledge of the tuning-fork methods, of which also Guillet was an expert. The researchers discussed in this chapter designed their methods for specific aims, like measuring the amplification power of triodes, synchronization in secret signalling, measuring velocities of projectiles, and the frequencies of different radio waves and multiplex telephony channels. The crucial role of their specific aims is well illustrated by the difference between the otherwise equivalent usage of the tuning-fork frequency standard at the Bell System and at the NPL. Since only Bell required continuous reference to measure frequencies in its system, Horton's group constructed the first electronic clock, while Dye was satisfied with a chronometer.

Most, but not all, of these specific technological goals originated in efforts to improve methods of wireless communication for military, commercial, and government interests. Under the pressure of the Great War, academic scientists and engineers applied their expertise in exact measurements and in electronics for military aims, resulting in triode tuning-fork methods. Nevertheless, Dye, Horton, and their colleagues developed the electronic tuning-fork frequency standards for the novel needs of telecommunication as a mass technology. Precise knowledge of frequency was necessary for an efficient coordination and integration of large telecommunication systems, providing an immediate reason to construct accurate frequency standards. Exact frequency standards answered the commercial interest of AT&T and the social interest of the government in regulating and fostering radio communication. While standardization had a social function, its roots were in scientific practice. Scientists and scientific instrument makers had turned the tuning-fork into a high precision instrument for laboratory measurements of acoustics and time. It was scientists and academic engineers who transformed the tuning-fork from a research instrument into a central device of wide-scale telecommunication, which required precision previously limited to the scientific laboratory. While Horton, Dye, and their colleagues developed basic tuning-fork standards for the technosocial goals of their institutes, like good metrologists they sought to improve the precision of their standards, also beyond their practical goals. That the major use of the tuning-fork clock qua clock was to help investigate natural phenomenon (sun eclipse) suggests that the precise instrument still had a scientific value, even when it served the technical needs of a giant corporation.

6

THE INVENTION OF BELL'S QUARTZ FREQUENCY STANDARD AND CLOCK

The novel crystal frequency control technics suggested a better basis for determining higher frequencies than the electronically maintained tuning-forks. Their advantage was especially clear to members of AT&T's research branch due to their early acquaintance with piezoelectric oscillators and their usage. Crystal frequency control promised two advantages over the tuning-fork standards: higher resonance frequency and higher stability. Crystal resonators seemed especially useful as a secondary portable standard. Since crystal resonators naturally vibrate at radio frequencies, they make superfluous the complex prone-to-error harmonic producer that would otherwise be needed with the tuning-fork standards. In addition, they suggested a way to reach frequencies beyond the reliable range of tuning-fork systems (incorporating the harmonic producer), wavelengths which became particularly important for AT&T. Initially, however, two main obstacles hindered their use as a primary frequency standard: there were no means either for reaching the lower part of the useful electromagnetic spectrum from the piezo-resonator relatively high frequency or for directly comparing the resonator's vibration with that of a standard clock. Both obstacles stemmed from the contemporary inability to produce a lower stable frequency that is an exact integer division of a higher input frequency. While generating a known exact multiplication of an input frequency was a known art, no method was suggested for the reverse process. Controlling oscillations at lower frequencies and comparing a piezoelectric standard oscillator with a clock were deemed important enough for AT&T to assign the task of contriving a frequency divider to Joseph Horton's group and its sub-team headed by Warren Marrison. This chapter examines the process of invention of such a frequency divider at Bell Labs.[1]

---

[1] For simplicity, in this chapter I often refer to the Western Electric research department by the name of Bell Labs, which it gained in 1925, i.e. in the middle of the story told here. The

Sonar to Quartz Clock. Shaul Katzir, Oxford University Press. © Oxford University Press (2023).
DOI: 10.1093/oso/9780198878735.003.0007

This chapter analyses the process of invention and innovation within the modern research laboratory on the rich material, scientific and technological, and human resources provided by the corporation. It was part of the research and development enterprise to improve and implement the new frequency control method, as part of the fifth phase in the piezoelectric research. The efforts were directed at contriving and improving technics, rather than at scientific research, not even of an applied kind. Researchers at AT&T and other companies explored ways to improve frequency control and implement the technics in broadcasting senders, radio-telephony sender–receivers, and multiplex telephony, in addition to measuring standards.[2] Some of these studies preceded the attempt to employ the method for primary standards. Apparently, in contriving frequency standards, the researchers dedicated more time to the triode-based electronic circuits in which the piezoelectric oscillator was embedded than to the oscillator itself. The methods for dividing frequencies were mostly based on electronic technics. Marrison and the other researchers in his group worked as engineers and inventors rather than scientists, as Horton characterized the group's work.[3] Inventions like the frequency divider arose from the combination and modification of known methods, devices, and ideas, rather than from new findings.

The strategy of varying and combining extant devices was nothing new. It was common also among traditional 'independent' inventors of the nineteenth century, which the industrial research and development laboratories replaced as the main source for technical innovation.[4] Yet, the larger and richer industrial laboratories with their science-educated personnel provided better means for its implementation.[5] The present chapter examines the process of

organizational changes did not affect the groups discussed here. Researchers continued with the same work under the same supervisor at the same place.

[2] See Chapter 4, fn. 57.

[3] See p. 164 above.

[4] Thomas P. Hughes, *American Genesis: A Century of Invention and Technological Enthusiasm, 1870–1970* (New York: Viking, 1989), 138–9.

[5] Hugh G.J. Aitken, *The Continuous Wave: Technology and American Radio, 1900–1932* (Princeton, NJ: Princeton University Press, 1985), 188–9, 228; Ian Wills, 'Instrumentalizing Failure: Edison's Invention of the Carbon Microphone', *Annals of Science* 64 (2007): 383–409. Edison practiced variations and combinations at a large scale in his Menlo Park development laboratory in the early 1880s. It was not, however, a research laboratory; on the last point see David A. Hounshell, 'The Evolution of Industrial Research in the United States', in Richard S. Rosenbloom and William J. Spencer (eds), *Engines of Innovation: U.S. Industrial Research at the End of an Era* (Boston, MA: Harvard Business School Press, 1996), 13–85, on 19.

invention within industrial research laboratories by exploring the cognitive and experimental steps that enabled Marrison's group at AT&T to solve the important technical problem assigned to them. It focuses on the way modern, scientifically educated inventors working within a large industrial laboratory collected, combined, and adjusted the resources at their disposal to produce novelty. In Chapter 7, I turn to other groups that invented frequency dividers for similar and also somewhat different aims.

6.1 Piezo-resonators at AT&T

Thanks to their relationships with Cady (see section 4.3) researchers at Western Electric laboratory had gained information about his research on piezoelectric resonators already from 1919. Horton was among those attending a few of the meetings with Cady, at least since December 1920. A month later, i.e. still before inventing a method for frequency control, Cady sent the first in a series of resonators to the corporation. It probably reached Horton's frequency standard group and provided further opportunity for experimental study.[6] The familiarity of Arnold, Horton, and other researchers at AT&T with piezo-resonators facilitated their early consideration of piezoelectric crystals for the corporation's basic frequency standard.[7] Arnold thought that the corporation could employ piezoelectric resonators 'for wave length standards and for maintaining the frequency of electric circuits, particularly radio circuits'. Indeed AT&T would develop methods based on piezo-resonators for frequency standards and control (for radio and telephony), two technological aims that originated in its business interest. Research gained momentum at the corporation research branch from late 1923.[8]

In the early 1920s, AT&T sought some means for maintaining radio waves sent by separate transmitters in the same narrow frequency range. This technical need followed the corporation's strategy to provide unified service

[6]Chapter 4, fn. 37. In addition see entries in Cady, *Diaries*, kept at the Rhode Island Historical Society.

[7]Although a few researchers at the company, prominently Alexander Mclean Nicolson, gained experience with piezoelectric vibrators following their application for submarine detection (as discussed in Ch. 4), except for providing crystals to Horton's group these researchers were not involved in the company's research on frequency standards. E.g. Marrison, NB 1444, pp. 25, 30 (28.8.1924, 4.9.1924) on Rochelle salt crystals received by Nicolson.

[8]Arnold to O. [Otto] B. Blackwell, 26.12.1923, AT&T Archives, loc: 79 10 01 05. The interest in piezoelectric technology is evident in the extant notebooks of Marrison and the list of his suggestions that he provided at the beginning of NB 2161, from 4.5.1926.

to extended areas. For this, it developed a radio station, known as WEAF, to send the same programme to the whole nation, or at least to very large areas, through distant transmitters connected by the corporation's telephone lines. Yet, when two radio waves of close but unequal wavelengths meet, they interfere and produce thereby a beat whose frequency equals the difference between the frequencies of the two waves. This kind of interference, easily perceived as the audible beat of the resulting frequency, was technically known as heterodyne. A solution was to keep the transmitters within 50 Hz of each other, so that the heterodyne beat was supposedly lower than the audible range and was anyway filtered out by the loudspeakers. This solution, however, was difficult to implement. Westinghouse, which had tackled the problem with a regional station in 1921, synchronized its two transmitters by a wire connection, circumventing the need to determine the absolute frequency of their broadcast.[9] This solution, however, did not fit a large network of transmitters, as planned by AT&T. Instead, the corporation implemented the novel crystal frequency technology whereby each transmitter is constrained to the same narrow band by a piezoelectric resonator on site. The design of these resonators was assigned to Horton's group, with its expertise in maintaining electronic oscillations at constant and accurate frequencies. The group found that the precision sought required a device that would keep the crystals under constant temperature in order to prevent deviations in their frequencies. Marrison prepared such a crystal and its holder for WEAF in 1924.[10] Piezoelectric resonators and their employment both for frequency control (for WEAF) and for standards (for radio frequencies) were his main research subject from spring 1924.

In the realm of standards, the ability to reach high frequencies beyond the reliable range of the tuning-fork system became ever more important for AT&T. Within five years, between 1923 and 1928, the corporation raised the upper limit required of its frequency standards by two orders of magnitude to several hundred million (10^8) hertz, following the rise in frequencies of useful electromagnetic waves, as mentioned in Chapter 4. Above the range for broadcasting, short waves (up to about 30 MHz) were found to be useful, and to possess important benefits for long-distance communication. In the second half of the 1920s, researchers began considering the yet higher range

[9]Chen-Pang Yeang, 'Characterizing Radio Channels: The Science and Technology of Propagation and Interference, 1900–1935' (PhD, Massachusetts Institute of Technology, 2004), 327–56; Hugh Richard Slotten, *Radio and Television Regulation: Broadcast Technology in the United States, 1920–1960* (Baltimore: Johns Hopkins University Press, 2000), 24–6.

[10]Marrison to Dr. William O. Baker, 12 July 1977, AT&T Archives, loc: 86 08 03 02.

of very high and ultra-high frequency (VHF and UHF, that is above 30 MHz). AT&T had a special interest in this range as it was considered for radio–telephone links. Such links could complement its wire network by providing a connection to remote and mobile customers (like ships) and supplementary channels for intercity connections. Television, a heavy development investment for AT&T from 1925, was another potential (and eventually actual) user of higher frequencies. Thus, the expansion into higher frequencies opened ways for extending the markets for electric communication, which the corporation dominated.[11] The original tuning-fork standard had no longer covered the whole electromagnetic spectrum under use by the corporation, since the reliability of the harmonic generator decreased with the number of multiplications needed for high frequency. To reach the new frequency range, Horton's group added a cathode ray oscillograph, which allowed higher multiplications of the input frequency.[12] Still, piezoelectric resonators suggested a simpler and thus more useful method for determining and controlling very high and ultra-high frequencies.

Although establishing frequency standards was a distinct goal from controlling frequency, the two endeavours shared central methods, since they both relied on crystal frequency control oscillators. Even in the realm of measurements, crystal frequency control offered a more useful method than a comparison with a piezoelectric standard (as Cady initially suggested). The piezoelectric-maintained electronic oscillations, or shortly piezo-oscillators, were simpler to manipulate than the piezo-resonators, as the latter were embedded in circuits for frequency measurement. In particular, they allowed one to combine oscillations at several frequencies, each determined by a piezoelectric oscillator, to produce circuits oscillating at any desired (multiplication, addition, and subtraction) function of the input frequencies, as with the modulation of the tuning-fork standard. Thus, no later than May 1922, Horton's group started considering the replacement of the latter with quartz.[13] Marrison's notebooks suggest that much of the group's research pertained to

---

[11] Warren A. Marrison, 'Some Facts about Frequency Measurement', *Bell Laboratories Record* 6 (1928): 385; Lloyd Espenschied, 'The Origin and Development of Radiotelephony', *IRE Proceedings* 25 (1937): 1101–23; Russel W. Burns, 'The Contributions of the Bell Telephone Laboratories to the Early Development of Television', *History of Technology* 13 (1991): 181–213; Russel W. Burns, *Television: An International History of the Formative Years* (London: Institution of Electrical Engineers, 1998), 220–41.

[12] Joseph W. Horton and W. A. Marrison, 'Precision Determination of Frequency', *IRE Proceedings* 16, (1928): 137–54, 142.

[13] Marrison, NB 1112–2, p. 38 (29.5.1922).

both standards and control. This was true for the relatively general study of the piezoelectric resonators, like the measurements of thermal coefficient of quartz vibrations, i.e. the rate by which its resonance frequency changes with the temperature. Yet this was also the case for more practical methods, like the development of means to keep these resonators at a stable temperature. Piezo-electric oscillators enabled both measurement and control of oscillations at a broad range of wavelengths; apparently, the group used it for both goals. Thus, the researches toward frequency standards and frequency control converged in the work of Horton's group, leading to a wider field of expertise than each goal alone.[14] This convergence also allowed Marrison to pursue research without committing himself to replacing the corporation's basic standard. The combination of interests in standards and control was probably unique to AT&T. Other commercial companies, such as GE, RCA, and Westinghouse, studied, developed, and employed crystal frequency control for radio transmitters that they produced. Yet, they did not have a clear interest in measurements and standards. AT&T was the sole corporation that devoted much research to absolute measurement.[15] National research institutes that developed standards such as the BoS, the NPL and the PTR, on the other hand, did not develop methods for frequency control.

6.2 Comparing Piezo-oscillators with a Clock

By 1924, accuracy in measuring and controlling frequency had become crucial for the practical needs of AT&T. To ensure high accuracy Marrison's group sought the means to monitor the pace of quartz resonators whether used as a secondary or primary standard. Following the tradition of exact absolute measurements, the practice of monitoring the corporation's tuning-fork standard

---

[14]Cf. Marrison, NB 1444, e.g. on thermal coefficients, pp. 19 (12.8.1924), 81–4 (18.12.1924). Unfortunately, AT&T Archives do not hold contemporary notebooks of other members of the group.

[15]These companies employed the technique to stabilize the high frequency of the otherwise unstable short wave transmitters, used for long-distance transmission. Christopher Shawn McGahey, 'Harnessing Nature's Timekeeper: A History of the Piezoelectric Quartz Crystal Technological Community (1880—1959)' (Ph. D. thesis, Atlanta, Georgia Institute of Technology, 2009), pp. 111–12, and *passim*. In 1928, Bell Labs were mentioned along with the Navy Department and the BoS as the only USA organizations making absolute measurements of piezo-resonators. I haven't found any hint of interest in the task from a commercial company outside the USA. J. H Dellinger, 'The Status of Frequency Standardization', *IRE Proceedings* 16 (1928): 582.

by a clock, and their consequent view that a frequency standard should 'have the general characteristics of a good clock',[16] Marrison suggested that the resonators be compared to an external clock. Although the idea became almost self-evident, its realization was not so simple. Between November 1924 and January 1927, he devised and recorded in his laboratory notebooks nine different methods for reaching a known frequency lower than that of a quartz resonator. With his group members, he constructed and experimented with most if not all of these methods.

Marrison's first recorded method was based on a feedback mechanism. This was an electro-mechanical method 'for synchronizing a rotary machine with a current at radio frequency'. Although the description was general for possible patent application, he clearly had a piezoelectric oscillator in mind (Figure 6.1). A synchronous motor (M) was 'driving a generator G which supplies current at 10 times the motor input frequency f_1 [of about 1000Hz]'. The frequency of the current was multiplied another ten times by a harmonic producer (HP). An electronic modulator ('Mod'—like those used in the general scheme of the corporation's tuning-fork standard) then subtracted the resultant frequency ($100f_1$) from the frequency F of an oscillator (O), which could be a piezoelectric resonator. The resulting current was fed

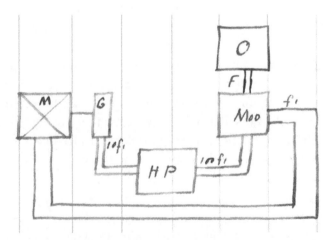

Figure 6.1 Marrison's suggestion for attaining an oscillation at a frequency 1/101 of the original from 15 November 1924. *Source*: NB 1444, p. 55. Figures from Marrison's NBs are courtesy of AT&T Archives and History Center.

[16] Joseph W. Horton, Norman H. Ricker, and Warren A. Marrison, 'Frequency Measurement in Electrical Communication', *AIEE Transactions* 42 (1923): 731.

back to the synchronous motor, forcing its frequency $(F-100f_1)$ on the motor. Through his earlier work on the tuning-fork frequency standard, Marrison had gained knowledge of the main devices involved in the new apparatus: synchronous motors, high-frequency generators, and harmonic producers. This background enabled him to modify the structure of the system to reduce (instead of increase) an input frequency. As with the tuning-fork clock, the devices motor 'could be geared to a clock or any suitable recording device to facilitate checking frequencies'. Thus, the new method coupled a timekeeper to the period of a vibrating quartz crystal, so in principle it could be used as a quartz clock. Yet, Marrison's device was not meant to mark time but to check the stability of a crystal vibration, an aim important enough to warrant the construction of such an apparatus.[17]

In July 1925, Marrison put on paper a modified 'suggestion', as devices that might have served as a basis for a patent were called at Bell Labs. In his modified device, he employed two motors instead pf one, probably to increase the reliability of the system. Conceived of as a way to validate the accuracy of secondary high-frequency standards, the apparatus did not seem to offer an alternative to the corporation's primary standard—it could not reach even a significant range of electromagnetic waves used by AT&T. The apparatus actually provided only two frequencies (the original and its 101st part). Although the lower frequency (f_1) could have been used as a basis for further manipulations, its indirect connection to the oscillator made it less reliable for that role.[18]

While wireless communication exploited ever higher frequencies, exact knowledge of frequencies below the common radio spectrum became important for multiplex telephony. The lower frequencies used for multiplex carriers, those in the range of 1,000–30,000 Hz, lay below the practical range of quartz resonators.[19] Finding ways to harness quartz crystals to determine frequencies in this range became a major occupation of Marison in 1925. It could answer

---

[17]Marrison, NB 1444, p. 55 (15.11.1924). In his history of the quartz clock Marrison implies that his November 1924 suggestion was a first attempt 'for a combined time and frequency standard', which did not become a basis for a clock only because a simpler method was proposed. Warren A. Marrison, 'The Evolution of the Quartz Crystal Clock', *Bell System Technical Journal* 27 (1948): 510–88, 537–9. The contemporary evidence, however, points to the use of the clock as 'a convenient method of checking the frequency over long intervals' (NB 1444, p. 177), rather than an attempt to establish a time standard.

[18]NB 1444, p. 177 (20.7.1925).

[19]On the use of this range for multiplex at the Bell System see E. H. Colpitts and O. B. Blackwell, 'Carrier Current Telephony and Telegraphy', *AIEE Transactions* 40 (1921): 205–300;

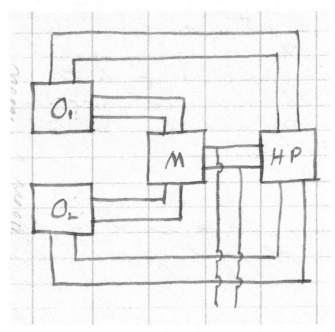

Figure 6.2 Marrison's two-oscillator suggestion from March 1925. In this circuit, Marrison replaced the motor of his earlier device with a second piezoelectric oscillator (O_2) that resonated at a frequency close to that of the first (O_1), and used the modulator (M) to subtract the frequency of one oscillator from that of the other. The resulting frequency, which is the difference between the two original ones and equals about one-tenth of them, was coupled to the piezoelectric oscillator (O_1) through a harmonic producer (HP). That oscillation could be further used for any end, such as measuring, control, and comparison with a clock. *Source*: NB 1444, p. 129.

two of AT&T's chief aims in this field: checking the accuracy of the frequency standard (by comparison to a clock) and in allowing them to control and measure oscillations at lower frequencies used by the Bell System.[20] In March 1925, he devised a way to reach frequencies at that lower range in a variation of the clock synchronizing circuit, which he had suggested four months earlier. (See Figure 6.2 and the explanation there.)[21]

T. E. Shea and C. E. Lane, 'Telephone Transmission Networks Types and Problems of Design', *AIEE Transactions* 48 (1929): 1031–44.

[20] See Marrison's notebooks for efforts to extend the range of quartz frequencies, and for attempts to use the same oscillator for a few frequencies. For an explicit mention of control at these frequencies see Hecht in Marrison, NB 2162, p. 169 (15.11.1926).

[21] NB 1444, p. 129 (19.3.1925).

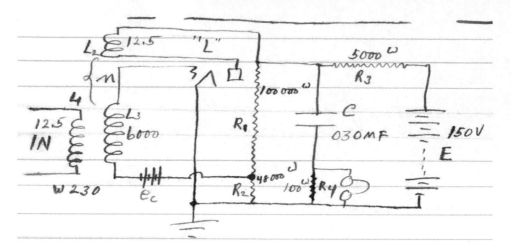

Figure 6.3 Marrison's electronic suggestion for reaching low frequencies, from June 1925. *Source*: NB 1444, p. 168.

Three months later, in June 1925, Marrison proposed a frequency divider on different working principles, reducing the frequency electronically rather electromechanically. His new idea was to control the oscillations of a target circuit, so as to ensure that its frequency would be an integer division of the original in a manner similar to how the tuning-fork controlled the oscillations of the RLC–triode circuits at an integer multiplication of its frequency. In analogy with the electronic harmonic producer used to control the frequency of a target circuit through periodic physical influence from the source circuit, Marrison invented am electronic 'subharmonic producer' that could generate oscillations at an exactly known division of an incoming frequency. Marrison was looking for a way to enforce an electric circuit to oscillate at a 'subharmonic' (i.e. integer division frequency) of a quartz resonator. The subharmonic producer was probably the first frequency divider that he actually constructed and experimented with.

Marrison's earliest 'subharmonic producer' promised to fulfil AT&T's needs but didn't fully meet them since it was unstable. In this 'producer', a triode circuit (the 'target' circuit) was coupled to a piezoelectric oscillator (the 'source')[22] in a way that made the former oscillate at a subharmonic of

[22] With a thought on patent claims Marrison referred to the source in a general way. Yet from the context it is clear that the radio frequency source in his mind (and the one actually used in its tests) was a piezo-oscillator.

the latter. Through an inductive coil (L_3 In Figure 6.3), the source-induced oscillating voltage in the circuit branch connects to the triode's grid, which controls the discharge in the triode tube (L), and thereby controls the current in the target circuit (on the right side). In the initial setting, no current flows through the tube, since the triode's grid is connected to a negative voltage (e_c), which blocks any discharge from the cathode. While there is no current through the triode, however, the battery E charges the condenser C. The condenser then applies (through the resonator R_1, which with R_2 forms a potentiometer) an opposing positive voltage on the grid. At some point the sum of the condenser voltage and the maximum of the alternating voltage induced by the source through L_3 is just sufficient to overcome the negative voltage on the grid and allow a discharge through the tube (since the rate of change of the voltage on L_3 is much higher than that on the capacitor C, the threshold is reached while the source voltage is maximal). The peaks of the high-frequency source voltage play a similar role to that of the lower-frequency voltage peaks in the tuning-fork harmonic producer. Once current flows, it reinforces itself by a feedback mechanism: the coil L_2, which is connected to the anode (and therefore becomes magnetically active when current flows in the triode), is coupled to the grid through coil L_3, and thus induces additional positive voltage on the grid. Due to this feedback mechanism, with the proper choice of parameters, current continues to flow 'until practically the whole of the charge on C has been dissipated'. At this point the system has run a full cycle and returns to its initial setting. The period of the cycle depends on the properties of the capacitor and coil in the RLC circuits. The capacitor needs to reach the breakthrough voltage depending on the relation between R_2 and R_1 in the potentiometer.

Before describing the proposal in his notebook, Marrison experimented with the device on the various components that required replacement, modification, and tuning. He found that 'by [manually] adjusting the various elements of the circuit the rate of change of C can be adjusted so that it will be discharged once for one, two, three etc. cycles of the high frequency circuit'. Yet, he managed to reach only the fourth 'subharmonic'. This satisfied some of AT&T's requirements, such as extending the range of frequencies controlled by piezoelectric crystals. It did not, however, satisfy other criteria, for it lacked, for example, the ability to drive a clock mechanism. In subsequent attempts to improve the mechanism, Marrison failed to reach higher subharmonics. Two years later, when he filed a patent for the device, it was still unstable in high subharmonics, and the inventor suggested 'using several circuits in tandem' in order to reach lower frequencies.[23]

Although he learned a great deal from his first harmonic producer of July 1925, Marrison clearly needed to modify his device. By 1926, he had conceived of a few other subharmonic producers. These more efficient circuits enabled him to compare the period of the piezoelectric oscillator with a clock, as well as to control lower frequencies and measure them. By July 1926, he devised a few methods to 'obtain from a high frequency (crystal) oscillator a sub-harmonic for the purpose of operating a synchronous clock'.[24] The device that he described on 23 July resembled the subharmonic producer that he had suggested in June 1925. It shared some components. In this arrangement, however, he separated the triode circuit coupled to the source from the circuit that oscillated at its subharmonic. This released him from the reliance on the grid's voltage, as was the case in his June 1925 suggestion, which did not allow high sub-harmonics.[25]

In order to realize this idea, he added to the circuit a neon tube, a well-known if less popular electronic device. It was known that once the insulation of a neon glow lamp breaks down, the current ionizes the gas in the tube and continues until it reaches nearly complete discharge. Thus, a neon tube accumulates charge until the voltage on its terminals reaches a known level and then quickly discharges it. The input oscillator (O on the left side of Figure 6.4) controls the target circuit by coils coupled with the grid branch of a triode (D), which is connected to a negative potential (induced by the battery e_c). In this arrangement, however, a 'peak of the input wave' overcomes the negative voltage, which is otherwise 'large enough to prevent plate [anode] current from following [from the cathode]'. 'Adjustments are made such that a definite quantity of space current flows [in the triode] for each wave' and consequently charges a capacitor (C). As with the previous method, C is charged until the voltage reaches a particular maximum value. Now, however,

[23] Marrison, NB 1444, pp. 168–9 (25.6.1925); Marrison, 'Subharmonic Frequency Producer', US patent 1,733,614, filed 20 August 1927. Unlike the source frequency the resulting alternating current is not sinusoidal (i.e. symmetric in time and in positive and negative current). Yet this was not an obstacle for using the output oscillation as an input for further manipulations.

[24] Marrison, NB 2161, p. 28 (23.7.1926). Marrison wrote that '[s]everal methods are possible for' operating a synchronous motor and presented a 'typical method' for that end. Unfortunately, the AT&T archives do not possess records of the team's crystal research from October 1925 to April 1926.

[25] NB 2161, p. 28 (23 to 31 Jul 1926).

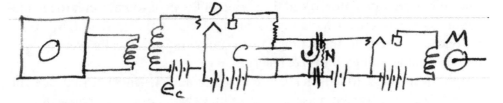

Figure 6.4 Marrison's 'typical method' for subharmonic producer using a neon tube (N), from 23 July 1926. *Source*: NB 2161, p. 28.

Marrison replaced the potentiometer, which set the breakdown voltage, with a neon tube. The neon tube generated current in another triode branch (on the right side of Figure 6.4), thereby dispensing with the coil-feedback arrangement. The discharge current 'may be amplified sufficiently to run the low frequency motor M'. Since the capacitor in this arrangement accumulates charge over successive very short intervals, Marrison concluded that breakdown voltage would be reached at one of these intervals, i.e. at an integer multiple of the input cycles, which would become the period of a triode circuit that incorporates a neon tube.

Theoretically, the circuit that Marrison made could have driven a synchronous motor, and thus a quartz clock. In practice, however, this and similar circuits were probably neither stable nor reliable enough for the task, and Marrison did not connect them to such a mechanism. Apparently, he found out that the breakdown voltage of the neon tube alone was insufficient to ensure that discharge would always begin at the same number of cycles, and not one cycle earlier or later. By July 31st he had suggested a remedy: connecting the neon tube to a coil and capacitor circuit that was 'tuned to the desired subharmonic'. (See Figure 6.5.) The tuning did not need to be highly precise,

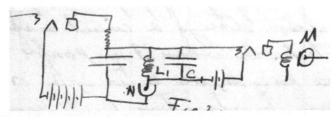

Figure 6.5 Marrison's second neon tube circuit from 31 July; only the right side part of Figure 6.4 is shown, while the left side of the circuit remained the same. Note that the neon tube is connected to an additional coil (L_1) and condenser (C). *Source*: NB 2161, p. 28.

since the source oscillator forced the period of the neon-tube circuit to an integer division of its own. By adding the tuning circuit to the neon tube, Marrison produced a system flexible enough to be tuned to a specifically precise value, but also stable enough for AT&T's needs.

Within two days, Marrison realized that he could dispense with a triode and a separate quartz oscillator as a source. In his newest design, Marrison directly exploited the physical properties of the quartz resonator, namely the sharp change in its capacity near resonance. He connected a quartz resonator in parallel to a neon-tube circuit that contained a resistor, capacitor, and battery (see Figure 6.6). Such a neon-tube circuit could oscillate at its own (unstable) resonance frequency. Marrison reasoned that 'in virtue of its variation of capacity with its [resonance] motion [the crystal] will control the frequency' of the circuit 'within narrow limits'.[26] 'The frequency of the neon tube discharge [would] be controlled when it is 1/n of that of the crystal when n is any integer.'[27] As an additional virtue, the circuit could 'be regarded as either a harmonic producer, a sub harmonic producer or both. The frequency

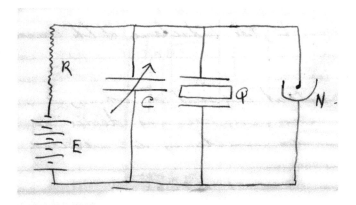

Figure 6.6 Marrison's neon-tube–crystal subharmonic producer of 2 August 1926. C is a variable capacitor, Q a quartz crystal, and N a neon tube. *Source:* NB 2161, p. 3.

[26] Recall (Ch. 4) that Cady had observed a discontinuity with capacity near resonance having very low value just above it. The lower capacity would fasten discharge in the neon tube. Marrison did not employ the recent suggestion of Van Dyke of modelling the resonator by an equivalent electric system, which had been published only in abstract form in 1925. On the model see Ch. 9.

[27] NB 2161, pp. 28 (signed 31.7.1926) and 32 (2.8.1926).

of the crystal is stepped down while the frequency of the neon tube oscillation is stepped up.' In its use as a subharmonic producer it could drive a clock mechanism. The dual role of the circuit points at the close connection between the research and development of subharmonic and harmonic producers. The circuit functioned well in tests, although its 'degree of control' decreased as the subharmonic frequency decreased.[28] Thus, in quite evolutionary manner the addition of a neon tube diverted Marrison from his strategy of constructing a separate subharmonic producer that divides the frequency of an extant oscillation. Here the manipulation is done on the circuit that embeds the crystal.

Three months later, Marrison modified the triode-based subharmonic producer with a capacitor and potentiometer which he had first proposed in June 1925 (Figure 6.3). Instead of coupling a quartz oscillator to the grid through coils, as in the original design, he connected a quartz resonator directly to the grid, dispensing with the use of a 'source' oscillator. The similar step of inserting the resonator within the circuit that he took with the neon tube probably inspired him to embed the resonator in the grid branch of this earlier circuit. More concretely he was aiming to increase the voltage induced by the quartz resonator, since low voltage made the grid more sensitive to fluctuations in the voltage due to other components of that branch of the circuit. These fluctuations were a probable source for the discharge that had occurred in the middle of the capacitor's cycle before it had reached its maximal voltage. This premature discharge had led to instability with large subharmonics in the original design and was therefore undesirable. In the modified design, the working principle remained the same. (See Figure 6.7 for more details.)

Tests made by George Hecht from Marrison's group showed that the method made it possible to reach the twenty-fifth subharmonic. Even though 'the stability decrease[d] rapidly as we go to higher orders (lower frequencies)', this was a clear advantage over the original design. As with other methods, one needed to tune the apparatus manually to reach a new subharmonic, but it kept each frequency without further human intervention. To improve stability Marrison suggested driving the quartz crystal with an independent tube circuit. His reasoning: 'By this means the crystal is kept vibrating at larger amplitudes than otherwise would be obtained and it has a greater controlling

---

[28] NB 2161, p. 34 (5.8.1926). Since the resonator possesses significance capacity, in checking the device Marrison dispensed with the independent capacitor. To make it possible to reach different values from the same circuit he used a variable resistor.

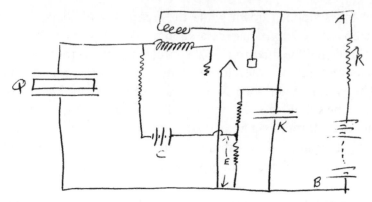

Figure 6.7 Marrison's subharmonic producer of November 1926. Here K is a capacitor and B a battery, while E represents the EMF available by the potentiometer. The oscillating voltage from the crystal (Q) suffices to overcome the negative voltage on the grid (from a battery) only as the condenser (K) charges up, which happens 'after the 1st, 2nd, 3rd (etc as desired) cycle of the crystal'. Once discharge begins, feedback coils keep sufficiently high voltage on the grid to ensure an abrupt discharge of the condenser, as in the design from June 1925 (Figure 6.3). *Source*: NB 2161, p. 62.

effect on the low frequency output.'[29] Here, Marrison went back to dividing the circuit into subcircuits, a general tendency in his design, although in this case he did not return to a separate piezoelectric oscillator until two months later, when he used a new method, which AT&T would adopt for its frequency standard.

The First Quartz Frequency Standard and Clock

By 27 January 1927, Marrison had invented a new method, dubbed 'A Sub Multiple-Frequency Control System'. The new system was deemed reliable enough for use by AT&T in its frequency standard, as it furnished the means to reach frequencies useful for driving a clock to monitor the accuracy and stability of the resonator's frequency; Marrison's new frequency divider enabled the construction of AT&T's central frequency standard on a quartz resonator. The timekeeper was not only an important part of AT&T's first crystal standard, but was also the first operating quartz clock. Although technically it did not lead to later devices, Marrison's achievement is nevertheless significant.

[29]NB 2161, pp. 62 (10.11.1926), 63 (17.11.1926); NB 2162, pp. 169–71 (15.11.1926).

Enjoying insights and experience gained from two years of research, Marrison's new method combined a few earlier ideas in modified form with one important novelty. As with most of his earlier methods, a separated quartz oscillator controlled the voltage on a grid of a 'control tube' by coils without being integrated into the circuit. The current from the tube controlled a circuit tuned to oscillate by its own capacity and inductance at the desired (sub)frequency (labelled 'Submultiple Oscillator' in Figure 6.8), an idea that Marrison had first used in his neon tube-controlled circuit of 31 July 1926. In order to use its frequency (e.g. to run a clock) Marrison coupled the submultiple oscillator to an output circuit. The important novelty that he incorporated was a self-regulating mechanism that adjusted the frequency of the submultiple oscillator to the desired one by changing its inductance through a special coil (labelled 'Control Inductance' in Figure 6.8). This mechanism relied on feedback from the output frequency, which included a harmonic producer that multiplied the oscillation back to the higher input frequency. The role of the harmonic producer was to compare the submultiple oscillation with the input frequency. This resembled the working principle of Marrison's first proposal to synchronize a motor with a high-frequency oscillator, from November 1924. The new design, however, was totally electromagnetic (except for the crystal, of course).[30]

In the feedback mechanism, Marrison coupled the output of the harmonic producer to the grid branch of the control tube through an inductive coil in series with the coil coupled to the piezoelectric oscillator ('Master Oscillator' in Figure 6.8). Its design was based on a theoretical understanding of the combination of the alternating voltage from two sources. The combination of their induced voltage determined the existence and strength of current through

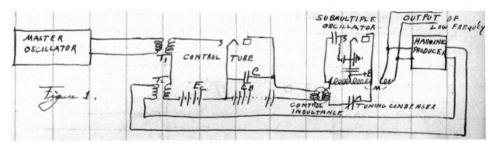

Figure 6.8 Marrison's 'Sub-Multiple-Frequency Control System' of January 1927. *Source*: NB 2161, p. 73.

---

[30] NB 2161, pp. 73–4 (27.1.1927).

the tube. While the piezo-oscillator induced a steady voltage wave, possible changes in the frequency of the submultiple oscillator altered the wave induced by the harmonic producer and thereby changed the phase relations between the two waves. Constructive or destructive interference would thereby change the voltage and consequently the discharge current through the tube. When the submultiple frequency was a little too high the phase difference between the two waves would increase and the positive voltage on the gird would decrease, leading to lower current in the anode (see Figure 6.9 for more details).

Here entered the peculiarities of the 'control inductance', a ring of magnetic core having two windings, one connected to the plate (anode) and one to the submultiple oscillator. The material at the core had a high and variable permeability, which meant that the magnetic field strength (**H**, called also magnetic intensity) of the magnet was not a linear function of the external magnetic field (**B**). A change in the permeability in one part of the control inductance ring sufficed to change the permeability and thereby the inductance of the whole. Thus, when the direct current in the control tube winding decreased, it would increase the inductance of the whole inductance ring, including the part connected to the submultiple oscillator. Its higher inductance would lower the frequency back to its desired value. A decrease in the output frequency would cause a reverse process: raising the voltage on the grid, resulting in a lower inductance in the submultiple oscillator and an increase in its frequency to the desired value.[31] Thus, the arrangement functioned as a feedback mechanism that held the output frequency as a constant division of the input frequency.

The methods and practices of Bell Labs provided Marrison with the resources required to devise this self-regulating circuit. A few researchers at that laboratory had already suggested exploiting the strong response of a core of variable permeability on the magnetic and electric behaviour of coils and circuits. Although different from its use in Marrison's clock, such usage was suggestive in his search for circuits with a self-regulating mechanism. In addition, in late 1922, Ralph V. L. Hartley, Horton's direct supervisor, suggested a harmonic generator based on coupling two circuits by a common

---

[31] In the initial suggestion, the voltage from the piezo-oscillator alone was insufficient to overcome the negative potential of the battery E_c, as was the case in Marrison's earlier suggestions. In the operating version, which the group developed in the coming weeks, the circuit was tuned to allow current at the peeks of the oscillator's positive voltage, and the voltage from the harmonic oscillator was used only to vary the value of the average current through the tube; Warren A. Marrison, 'Frequency Control System', US patent 1,788,533, filed 28 March 1927.

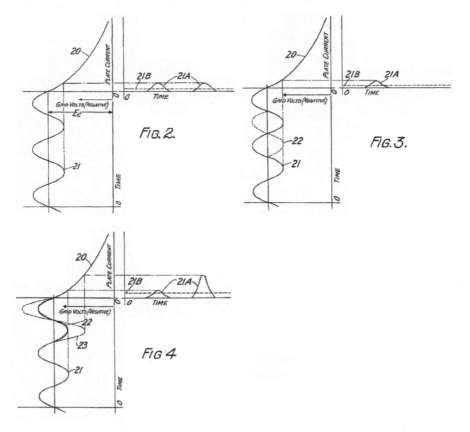

Figure 6.9   The effect of the harmonic producer on the control tube current—a graphical presentation: the left figure ('Fig 2') shows the grid voltage (left side) and the resulting current on the triode ('plate current' on the right side) under the influence of the voltage induced by the master (piezoelectric) oscillator alone. In this case, the maximal voltage on the grid branch of the control tube would suffice to produce a small stable periodic current in the triode, as shown in the right side of the figure. The two other figures show the effect of connecting the harmonic producer to the circuit when the frequency of the submultiple oscillator is either faster ('Fig 3') or slower than the subharmonic ('Fig 4'). When the frequency of the submultiple oscillator increases, the phase of the electric wave produced by the harmonic producer advances in relation to that of the master oscillator. As shown for an extreme case in 'Fig 3', in that case the two waves have destructive interference, which nullifies the current at the plate. Since the harmonic producer sends a wave only once every few cycles of the piezoelectric oscillator, the current would be zero only at these cycles and would keep its normal value at the other cycles as in 21A in the figure (corresponding to 21 of the grid voltage on the left side). If the output frequency were to decrease, it would retard so as to approach the master oscillator phase ('Fig 4'). Their constructive interference would lead to a higher voltage on the grid (23), resulting in a stronger current in the control tube, as shown on the right side of the figure. When synchronized, the two voltage waves would keep an in-between and constant phase difference that maintained a constant current in the tube. *Source*: Marrison, 'Frequency-Control System'.

magnetic core with windings in each circuit. The high inner magnetic permeability of the core blocked an electric voltage, except for short intervals when
the magnetic flux changed. Hartley relied on a specific material developed by
Western Electric, dubbed 'permalloy', due to its very high permeability relative to that of other ferromagnets. Starting in 1913, Gustaf Elmen from the
research branch developed the alloy to provide high permeability for the needs
of high loads in the Bell System's long telephone wires. Once the alloy was
available, researchers at AT&T, such as Hartley, sought ways to implement it
in other technical contexts. Their regular meetings and other interactions provided knowledge about such resources and methods. The size of AT&T and its
comprehensive care for the full system of telecommunication were crucial for
developing the alloy for use in one realm (cables understood by electromagnetic theory) and later for its implementation in others (electronic systems).
Marrison employed permalloy in July 1925, using it as a balance tilted by a
current initiated by speech in a 'suppressed carrier system'. Apparently, Marrison's own ideas of using the control inductance in the subharmonic producer
inspired at least one additional researcher at Bell Labs, William A. Knoop,
who used it to couple a few circuits in a synchronization system, patented in
1928.[32]

Marrison's new subharmonic producer performed very well (Figure 6.10).
By October 1927, Marrison and Horton had 'found that the frequency of a
given current can be controlled by a current of higher frequency with much
greater stability than by a current of lower frequency'.[33] The inventor had
already filed a patent in March.[34] Interestingly, AT&T decided to patent Marrison's earlier suggestions for subharmonic producers only after achieving
the newer more stable method. It might have been that only after frequency
division had seemed practical did AT&T try to defend it against potential

---

[32] R. V. L. Hartley, NB 1060–1101–3 (30.12.1922); quotation from Marrison NB 1444, p.
175 (10.7.1925); S. Millman, ed., *A History of Engineering and Science in the Bell System:
Physical Sciences (1925–1980)* (New York: Bell Telephone Laboratories, 1983), 801–8; William
A. Knoop, 'Synchronizing system', US patent 1,747,248, filed 10 February 1928, patented 18
February 1930. Apparently uses of variations in magnetic permeability were rare outside AT&T;
for an exception, see Quincy A. Brackert, 'Receiving system', US patent 1,567,566, filed 12 April
1921. Brackert worked at Westinghouse.

[33] They implied that this conclusion had been reached before the invention of the circuits
described in their paper, i.e. before March 1927, but I haven't found an earlier indication for
such a strong claim; Horton and Marrison, 'Precision Determination of Frequency', 143.

[34] Marrison, 'Frequency Control System'. This patent included a few improvements on the
suggestion from January.

Figure 6.10 Bell's standard frequency oscillator without shield. *Source*: Warren A. Marrison, 'A High Precision Standard of Frequency', *IRE Proceedings* 17 (1929): 1101–22, on 1111.

competition, or that it had not wanted to disclose that it was working on the problem to prevent others from entering the field. Anyway, its strategy did not block successful competition (as discussed in Ch. 7). With the application in the patent office, the Marrison group continued to examine the different components of the crystal-based frequency standard system and made further improvements. They found that the new subharmonic producer allowed them to produce frequencies at a ratio of fifty to one, twice as much as attained earlier and probably with steadier results.

Although by using two dividers in series it would have been feasible to reach a frequency of 100 Hz to drive a synchronous motor, the group preferred dividing the frequency into two steps: one electronic and one mechanical. Horton and Marrison explained that 'it has been found advantageous to secure a current having a frequency one-tenth that of the standard and to use this current to operate a 5000-cycle synchronous motor'. They designed the motor with 100 teeth so it 'revolve[d] at 50 revolutions per second'. The simple thing was to connect this vibration to a clock mechanism, but since they had 'already available clocks designed to operate on a 100-cycle current [from the

tuning-fork standard], it was thought to be simpler to use these as they were.'
Therefore, instead of operating a clock mechanism directly, they connected
the 50-cycle mechanical motor to an electric generator producing a 100-Hz
current that drove a 100-cycle clock.[35] A vacuum tube in the latter circuit
also supplied electric oscillations at that low frequency. Similar conversions
between electric and mechanical oscillations had already characterized their
tuning-fork standard. In later designs Marrison dispensed with the generator
for driving a clock mechanism. He continued, still, to use the mechanical re-
duction of frequency to produce exact electrical low-frequency oscillations at
100 and 10 Hz, which he further employed to reach integer multiplications
of these frequencies in order to reach the range of frequencies used by the
corporation.[36]

Improving the Precision of the Clock

In the summer of 1927, Marrison's group installed at AT&T the first frequency
and time standard based on a quartz resonator. A quartz clock was an integral
part of this system, ensuring its accuracy. The group could have stopped its
development, since the system had attained the accuracy required for AT&T's
telecommunication system, yet the success of the standard encouraged the
researchers at Bell Labs to carry out elaborate research aimed at the clock's
further improvement. They put much effort into increasing the precision and
stability of their system. One reason for their effort was their expectation that
the continuous increase in the useful radio range would lead to a demand for
higher accuracy in frequency measurement and control. Still another motivation seems to follow the field of precision measurement and that of horology in
particular to which the laboratory entered with its new standards. A success in
this well-established field promised prestige for the researchers and the corporation alike, which was important for the public image of the laboratory and

---

[35] Horton and Marrison, 'Precision Determination of Frequency', 148–9. In this paper, presented on 13.10.1927, they write that 'this standard has been in operation . . . [for a] short time'.
The motor was probably introduced in March–April; see NB 2162, pp. 188, 189 (31.3.1927,
8.4.1927, entries by J. L. Whittaker, a researcher at Marrison's group).

[36] Marrison, 'A High Precision Standard of Frequency', 1112 (like a few other publications, it
appeared also at *Bell System Technical Journal*, 8 (1929): 493–514); Warren A. Marrison, 'The
Crystal Clock', *Proceedings of the National Academy of Sciences of the United States of America*
16 (1930): 496–507.

AT&T in general and for the researchers' career.[37] Moreover, it seems that the quest for higher precision gained its own momentum—problem solving for the sake of the challenge, especially in such a competitive field as horology, where precision was the name of the game, and this pushed the group to pursue higher accuracy.

Although not a direct interest of AT&T's, metrology and methods of keeping time in particular occupied a growing portion of Marrison's work. The field was not new to him; he had participated in time measurements for their own sake at least since he had taken part in the measurements of the 1925 eclipse (Ch. 5).[38] By April 1929, the quartz clock had reached an error rate 'in the order of 0.01 second a day [about 1 in 10^7]'. This was twice as accurate as the pendulum clock at the Bureau of Standards. The most accurate contemporary clock based on the new Shortt free pendulum mechanism, the culmination of 250 years of pendulum horology, had an error between one-hundredth and one-thousandth a second a day.[39] So although the quartz clock could not yet replace the best pendulum clock in 1930, Marrison suggested using it to display sidereal and mean solar time for astronomical needs from the same time-keeping element, which one could not do with a pendulum clock.[40]

To approach such an accuracy and to make their system more useful, Marrison and the researchers at Bell Lab had to improve the individual components of his quartz system. They examined, reconsidered, and suggested possibly better designs of the crystals and their cuts, the mounting and gaps of the crystal oscillator and its circuit, the subharmonic producer, and the synchronous motor, as well as improved means of producing a wide range of frequencies from the resonator, and a better method for comparing the quartz oscillator to an astronomical clock. Marrison deemed the crystal itself to be '[b]y far the

---

[37] On the usage of scientific and technical achievement for public relations see David Philip Miller, 'The Political Economy of Discovery Stories: The Case of Dr Irving Langmuir and General Electric', *Annals of Science* 68 (2011): 27–60. In addition to publication through professional journals and meeting, AT&T reported about some of its successes in the field in the press (see Marrison's file in the AT&T Archive). Later AT&T mentioned its precise quartz clock in advertisements and also in a short film probably from *c.*1950 at https://www.wpafilmlibrary.com/videos/102297 (accessed 6 July 2014).

[38] Marrison, 'The Evolution of the Quartz Crystal Clock', 528–30; Marrison, NB 1444, pp. 97–105, (24.1.1925).

[39] Steven J. Dick, *Sky and Ocean Joined: The U. S. Naval Observatory 1830–2000* (Cambridge, UK: Cambridge University Press, 2002), Ch. 11, esp. 461–2.

[40] Marrison, 'A High Precision Standard of Frequency', 1112; Marrison, 'The Crystal Clock'.

most important element in a crystal-controlled oscillator' and dedicated much effort to improving its stability.

From an early stage the researcher recognized the resonator frequency's sensitivity to variations in temperature, and thus kept the crystal resonator in a specially controlled container. That was also the practice of the laboratory with its tuning-fork standard. Identifying the effects of temperature variation as the greatest single source of error, Marrison worked in two directions. First, he suggested improvement in the container's thermal regulators, as he did with the tuning-fork system. Second, he attempted to reduce the sensitivity of the quartz resonator to variations in temperature. In May 1927, he contrived a successful stable quartz cut. He reasoned that since 'plates of quartz cut in the plane of the optic and electric axes usually have positive temperature coefficients and . . . plates cut in the plane of the optic axis but perpendicular to an electric axis have negative coefficients', a cut between these directions, or a vibration that involves both, would have a practically zero coefficient. In order to get the desired effect, Marrison applied knowledge of physics concerning the elasticity of quartz and its dependence on temperature. Still, despite the scientific research of piezoelectric vibrations in the early 1920s (discussed in Ch. 9), its findings were insufficient to provide precise guidance for solving the technical problem. To implement his idea, Marrison, thus, carried out further experimental research on the elasticity of quartz (the value of the temperature coefficient at angles to known cuts is not simply a known function of the values in those cuts) and also a practical search for appropriate angles and shapes of the crystal cuts carved at the laboratory. Experimentation showed the exact angle of the cut. It also revealed, without a theoretical reasoning, that rings 'have a temperature coefficient lower than disks of the same diameter and thickness' (Figure 6.11). With these rings the researchers attained at room temperature a coefficient of 'less than one part in a million per degree C'. As a by-product, 'the ring shape permit[ted] of an improved method of mounting in which there is very little friction to the holder'.[41]

[41]Marrison, 'A High Precision Standard of Frequency', quotations on 1103, 1104, 1106; Fredrick R. Lack, 'Observations on Modes of Vibration and Temperature Coefficients of Quartz Crystal Plates', *IRE Proceedings* 17 (1929): 1123–41. Marrison, NB 2161, for example pp. 92–3 (10.5.1927): 'Temperature Coefficient of Quartz Control' (first appearance of the oblique cut idea), 99, 126, 129, 130, 147 (24.5, 25.8, 3.10, 4.10, 16.12.1927), also in Marrison's next 'Suggestion Book', NB 2658, which he opened on 14.6.1928.

Figure 6.11 Georges Hecht, Marrison's assistant, finishing the first set of 'zero-temperature-coefficient' quartz rings. *Source*: Courtesy of AT&T Archives.

Marrison's ability to use the results of previous and contemporary researchers points to the professionalization and specialization of science and technology in the twentieth century. His method of applying compensating effects, for example, resembles Wilhelm E. Weber's development a century earlier of reed pipes that kept constant pitch under changing airflow. Weber had also studied two opposing effects, in his case the effect of the airflow on the metal reed and on the air column within the pipe, and found that they can neutralize each other. Whereas Marrison built on the work of other researchers, Weber had had to study much of the corresponding information about the acoustics of reeds and air columns himself.[42]

Bell Labs offered rich resources for designing and constructing the new cut of the quartz crystal that Marrison needed. The facilities at the laboratory made it possible to cut and polish the crystals in house. Marrison was able

[42] Myles W. Jackson, *Harmonious Triads: Physicists, Musicians, and Instrument Makers in Nineteenth-Century Germany* (Cambridge MA: MIT Press, 2008), 114–18. Weber's technological endeavour was part of more general research on the phenomenon under question—acoustics. Marrison, on the other hand, concentrated on the use of piezoelectricity (and earlier elasticity of forks) for precision measurements.

to deploy his research team to extend the empirical knowledge on which his ideas lay, and to examine their practical application. Moreover, he could use the results of another group within Bell Labs. Following Raymond Heising's observation of a discontinuous change of frequencies with the dimensions of crystal cuts, his assistant Frederick R. Lack examined the effect of specific directions of cuts and temperature on the frequencies of crystal resonators. Heising suggested a special cut, different from Marrison's, that exploited the observed phenomenon to reduce the crystal's thermal coefficient to practically nil. Already in July 1927, Heising had filed a patent suggesting the use of two crystal cuts, one with a positive and one with a negative thermal coefficient 'to reduce undesired variations in the frequency of the controlled oscillator due to temperature changes'. Heising began by examining the effect of two perpendicular pairs of electrodes in the same quartz plate, and only later found a direction that combines coefficients of opposite signs in one crystal cut. He, however, had been working, not on a better standard, but on components for radio transmitters. Thermally stable crystal cuts were valuable both for such circuits, which lacked thermal control, and for the corporation's frequency and time standard. Working at the same company, Marrison and Heising easily exchanged knowledge and experience about the resonators, which advanced the research of both groups.[43]

Although the rich resources and specific needs of AT&T made Marrison's development more likely, other researchers devised resonator rings too, although not for temperature stability. A few months before Marrison, Erich Giebe and Adolf Scheibe at the German PTR devised a resonator ring. They did not aim at high thermal stability, but at obtaining a wide range of frequencies from piezo-resonators, including relatively low ones (they reached down to 800 Hz). They also sought to observe their resonance in special electric discharge containers illuminated by the strong vibration that they developed.

[43] Marrison recalled his collaboration with Heising concerning 'Crystals and Quartz Clock', Marrison to Baker, 12.7.1977. In 1929, he and Lack referred to each other's papers, which appeared in the same issue (see fn. 41). Raymond A. Heising, 'Crystal-controlled Oscillator', US patent 1,840,580, filed 25 July 1927. According to their collaborator on quartz crystals, Warren P. Mason, Marrison preceded Heising in suggesting a special cut from the same crystal (Warren P. Mason, 'Low Temperature Coefficient Quartz Crystals', *Bell System Technical Journal* 19 (1940): 74–93, on 75). 'Raymond A. Heising—Member Board of Direction, 1929', *IRE Proceedings*, 17 (1929), 6; 'Raymond A. Heising: Board of Directors-1947', *IRE* Proceedings, 35 (1947), 1179; Lloyd Epsenschied, 'R. A. Heising, Former President of IRE, Dies at 76', *Specturm, IEEE* 2 (1965): 222.

In 1927, the German pair used torsional modes of vibrations, previously un-explored, and utilized quartz rings cut in the plane of their electric axes, which vividly showed their modes of vibration (see Ch. 9).[44] Unlike Giebe and Scheibe, David Dye at the NPL sought more stable piezoelectric resonators. Yet he was probably more concerned about the resonators' mechanical stability since he had designed a mechanism for compensating for variation of frequency with temperature. Like Marrison, Dye found a solution in a quartz crystal ring. As discussed in Chapter 7, like Marrison, in the mid-1920s Dye developed a frequency standard based on quartz to replace the tuning-fork standard of his laboratory. It is not clear, however, when he began using a quartz ring for his standard. Moreover, Dye cut his crystal perpendicular to the electric axes, i.e. perpendicular to Marrison's cut.[45]

The variation of the resonance frequency with temperature reduced the accuracy of all quartz frequency-measuring devices. Apparently, however, Marrison was the only one to engage in manipulating the crystal to make it less sensitive to temperature variations. Researchers in other organizations (GE, Westinghouse Company, Navy Department, and BoS) concentrated only on the means of temperature control.[46] One can point out a few factors that led Marrison to seek also the means to reduce the temperature sensitivity of the resonator. In early 1928, the BoS sought to attain an accuracy of 1 in 100,000 (more than one order of magnitude higher than was necessary to meet the current Federal Radio Commission's requirement for

[44] By early 1927 Giebe and Scheibe concluded that 'the influence of temperature is utmost small [äußerst gering]', ('Die Tätigkeit der Physikalisch-Technischen Reichsanstalt im Jahre 1926', 275); E. Giebe, 'Leuchtende piezoelektrische Resonatoren als Hochfrequenznormale', *Zeitschrift für technische Physik* 7 (1926): 235. E. Giebe and A. Scheibe, 'Piezoelektrische Kristalle als Frequenznormale', *Elektrische Nachrichten-Technik* 5 (1928): 65–82 (received in November 1927).

[45] L. Essen, 'The Dye Quartz Ring Oscillator as a Standard of Frequency and Time', *Proceedings of the Royal Society of London A: Mathematical, Physical and Engineering Sciences* 155 (1936): 498–519; Walter G. Cady, *Piezoelectricity: An Introduction to the Theory and Applications of Electromechanical Phenomena in Crystals* (New York: McGraw-Hill, 1946), 77; Ray Essen, 'Greenwich Time: From Pendulum to Quartz 1', *Horological Journal* 154 (2012): 198–201.

[46] Dellinger, 'Frequency Standardization', 581. In the paper, sent in March 1928, Dellinger stated that '[t]emperature control is necessary for reaching an accuracy of 1 to 100,000'. The four organizations took part in a research project coordinated by the BoS to develop a radio frequency standard. McGahey, 'Harnessing Nature's Timekeeper', 157–60.

broadcasting), while AT&T aimed for accuracy of 1 in 10,000,000.[47] The increasing demand for precision left extant temperature control devices insufficient, and that called for additional solutions. That the corporation sought simple portable standards and the means for controlling circuits in addition to a primary standard provided another reason for seeking more stable crystals. The inclusion of a quite bulky temperature control unit made a portable device cumbersome, and even impractical. Marrison, thus, suggested that for such ends crystals cut in the new manner could be employed 'without any form of temperature control'.[48] Whether portable standards and frequency control devices stimulated Marrison, or whether they were only an afterthought, this is another example of the benefits that Bell Labs gained from pursuing distinct but connected research aims of standards and control.

Ironically, following his scrutiny of the frequency standard system, Marrison replaced the component of his circuit that had initially enabled the construction of an operating quartz clock: the subharmonic producer based on controlled inductance. Initially content with its performance, by April 1928, he judged that it has 'not been as good as desired' and suggested improvements in the control circuit, without altering its basic principles. Although Marrison expressed satisfaction with the improved design, sometime during the following year he replaced it with a circuit that worked on different principles. Originally, Marrison's strategy in his electromagnetic dividers was to construct a circuit with a rather stable (but adjustable) internal frequency, and to limit its oscillations to a subharmonic of a controlling circuit, in similar to the harmonic producer controlled by the tuning-fork circuit.[49] In his new design, he abandoned the idea of a stable submultiple circuit and settled for an inherently unstable oscillator, i.e. a circuit able to vibrate at many frequencies. Such a circuit would oscillate at a frequency 'which is a small multiple or submultiple of' the control frequency.[50]

In replacing the subharmonic generator, Marrison adopted an approach that originated and was developed outside the Bell System. In July 1927, James

---

[47] Dellinger, 'Frequency Standardization'; Marrison, 'Some Facts about Frequency Measurement', on p. 385.

[48] Marrison, 'High Precision Standard', 1106.

[49] The circuit in which the resonator was embedded in the main circuit with a neon tube from 2.8.1926 did not follow this strategy.

[50] Quotations from NB 2161, p. 184 (23.4.1928); Marrison, 'A High Precision Standard of Frequency', 1112.

Clapp from the MIT Shortwave Research Laboratory described the use of a similar inherently unstable circuit to reach exact lower frequencies from a quartz oscillator. Clapp employed Abraham and Bloch's well-known harmonic producer, the multivibrator, to generate not only harmonics but also subharmonics. In the same year, Balthasar van der Pol, from the research laboratory of the Dutch company Philips, pointed out that the tendency to oscillate at a particular subharmonic is a general property of circuits with a non-linear relationship between current and voltage, of which the multivibrator is only one example. With this solution and theoretical understanding, Marrison could replace his rather complex, if ingenious, submultiple circuit with a simpler circuit. Despite the replacement of a key component in the system by a method that originated outside the corporation, Bell Labs maintained its lead in the quartz time standard for a few years.[51] With the exception of the quartz resonator itself, the system as a whole was more important than the specific details of its signal components. Thus, the quartz clock required more than the invention of one component (like the subharmonic generator); it also required the development of a few ingredients and their assemblage. Notwithstanding Marrison's prime role, this system originated from a network of physicists, engineers, and inventors, both inside and outside the Bell System. Chapter 7 will discuss the parallel inventions of frequency dividers by researchers at MIT, Philips, the Italian Navy, the NPL, and the Tokyo electro-technical institute.

6.3 On the Process of Invention at AT&T Research Laboratories

Inventing the quartz clock system, Marrison followed a twisted road. Table 6.1 summarizes the devices that he suggested for a time and frequency standard, their sources, and the relationships among them. Quite a few of the devices that he suggested for dividing the frequency of piezoelectric resonators did not evolve from their immediate predecessor. Instead, he adopted ideas from the tuning-fork standard in use by AT&T at the time, and from other systems. Other devices that he contrived evolved from previous ones. Yet sometimes the processes of modification and elaboration led to entirely new designs. For example, the circuits that Marrison devised during July and August of 1926

---

[51] Marrison, 'The Crystal Clock'. Still, in the early 1930s the national laboratories of Britain and Germany, which had a stronger interest in accuracy, exceeded the precision of AT&T's clock.

began with the utilization of a separate piezo-oscillator and ended with em-
bedding a piezo-resonator within a main neon-tube circuit without a separate
'subharmonic' producer. At other times, Marrison returned to an older design
in a modified form. For example, in November 1926, he considered a different way of connecting the piezo-resonator to the triode, in a circuit otherwise
equivalent to the one he had suggested in June 1925. Marrison's technics for
reducing frequencies can be grouped into two approaches: mechanical and
electronic. He began with the former and turned to the latter in March 1925.
Still, a month later he returned to a mechanical method. Although not ultimately used, his earlier methods provided ideas that he incorporated in later
circuits. For example, the 'inductance control' method, which he used in the
1927 quartz clock, shared a logical structure with his first suggestion from
November 1924, although one was mechanical and the other electronic. In
both cases Marrison remultiplied the lower output frequency and compared
it with that of the piezoelectric oscillator. His ideas are worth recording in detail because they reveal how industrial inventors worked, using the resources
of their laboratories to proceed along many different tracks in pursuit of a
given goal.

Oftentimes novelty emerges from implementing known ideas in a new
context. Many of Marrison's innovations followed from variations on previous technics and adaptations of methods and procedures from one realm
(e.g. tuning-forks) to another (e.g. quartz resonators), or from one device
(e.g. piezo-resonator connected to a neon tube in August 1926) to another (e.g.
piezo-resonator connected to a triode in November). Replacing the tuning-
fork with the quartz resonator required considerable changes to AT&T's
frequency standard, but Marrison's group continued to use parts from the
preceding standard system (e.g. the clock mechanism), and to combine ideas
and technics from other parts of that system. For example, in November 1924,
Marrison employed the technique of driving a synchronous motor by electric oscillations, a technique previously used with the tuning-fork standard.
Replacing this mechanical method with an electronic one, a few months later
he devised a subharmonic producer analogous to the harmonic producer used
with the older standard. As he had done with its predecessor, he controlled
the frequency of a target circuit, tuned to oscillate at the desired frequency,
by a periodic impact from the source circuit. In these and other cases, novelty
resulted from combining known and usually common components and modifying their specific properties. Though known, the components that he used
were state-of-the-art devices available at the corporate laboratory, from the
common triodes, through the less ordinary neon tubes, to the relatively new

Table 6.1 Marrison's suggestions for reaching a known lower frequency from that of a quartz resonator, November 1924–January 1927

Date	Kind	Operating principles	No. of frequencies obtained[a]	Highest division of frequency	Possible uses	Preceding method	Background knowledge[b]
Nov. 24	Synchronous motor	Mechanical + modulator	1	101	Clock[c]		Clock mechanism, harmonic producer, and modulator of tuning-fork standard
Mar. 25	Two piezo-oscillators	Modulator	1	About 10	Clock	Nov. 24	
June 25	Subharmonic producer	Controlled tube-capacitor circuit	3	4	Attaining lower frequencies		Harmonic producer (as in the tuning-fork standard)
July 25	Two synchronous motors	Mechanical + modulator	1	About 20	Clock	Nov. 24	
23 July 26	Subharmonic producer	Controlled tube-capacitor circuit with neon tube	Unknown	Unknown	Clock and lower frequencies	June 25	General use of neon tube in electronics
31 July 26	Subharmonic producer	Controlled tube-capacitor with neon tube and its own LC circuit	Unknown	Unknown	Clock and lower frequencies	23 July 26	
2 Aug. 26	Subharmonic producer	Piezo-resonator–neon tube	Unknown	Unknown	Clock and lower frequencies	31 July 26	Experience with piezo-electric resonators
Nov. 26	Subharmonic producer	Piezo-resonator connected to controlled tube	24	25	Clock and lower frequencies	June 25	
Jan. 27	Submultiple oscillator	Magnetic control inductance, feedback	49	50	Clock and lower frequencies	June 25; Nov. 24; 31 Jul 26	Use of variable inductance at Bell System

[a] Above the input frequency/ies. [b] Beyond the knowledge that was used in the preceding methods. [c] That is, the method made it possible to drive a clock mechanism.

piezoelectric resonators and oscillators and the uncommon coil with a high permeability core. Their combinations resulted in novel devices.

Inventors of mechanical and electromechanical technics carried out much of their creative thinking with visual drawing on paper, or in their 'mind eye'.[52] Like them Marrison seemed to combine and modify components of his system in his mind, but there are no signs that concrete drawings or visualization of their working principles played a central role in his thinking, clearly not on paper. Central components of his circuits and their working mechanism did not lend themselves easily to visualization. His systems allowed only for schematic illustrations, far from mimetic drawing (e.g. coils that faced each other in reality were far apart in the figure, e.g. Figure 6.3). No drawing showed how inductance is changed in the 'control inductance' or how a condenser or a neon tube accumulates charge. Thus, they could not help in the creative step of their inclusion in a new design. Schematic visualizations might have helped Marrison and other inventors in thinking about their systems, and he continued to use figures. Yet precise engineering drawing (in mind or on paper) of the devices and their intrarelations lost much of their importance in the novel electronic technics.

The new way to cut a quartz crystal that Marrison developed, the high point (but not the end) of the research on stabilizing the crystal resonator, exemplifies his debt to theoretical knowledge. Although he contributed to the knowledge of piezoelectricity and elasticity by empirically finding the exact direction of a zero-thermal coefficient in a quartz crystal, fostering science was only an atypical by-product of his group's research. It more often relied on scientific knowledge for its technical developments. Members of Bell Labs learnt about new developments and approaches sometimes before they were published, through direct exchange with external researchers, like Cady, and through inner seminars. While outside developments were discussed within Bell Labs, the circulation of ideas in use at AT&T was also important, like those concerning combining crystal cuts in different ways to reduce frequency variation with temperature, and the use of permalloy in coils to control electric circuits. The laboratory's large staff allowed its researchers to experimentally explore a sizeable number of their ideas. Like his colleagues, Marrison also had rich material resources at his disposal. These included, among other

---

[52] Eugene Ferguson emphasized the importance of visual thinking in technical design, Eugene S. Ferguson, *Engineering and the Mind's Eye* (Cambridge, MA: MIT Press, 1992); Eugene S. Ferguson, 'The Mind's Eye: Nonverbal Thought in Technology', *Science* 197, no. 4306 (1977): 827–36.

things, state-of-the-art electronic tubes,[53] the means to cut and polish crystals at the laboratory, and the permalloy for his coils. The general strategy of combining and varying extent technics and components was well known and used by inventors. Yet the modern corporation's facilities made it particularly potent. The rich intellectual and material resources allowed the modern corporate inventor to employ various devices and methods, to modify them to specific needs, to combine them, and to examine the performances both of the individual components and of the composite devices. While Marrison, Horton, and other researchers at Bell Labs showed ingenuity, these resources were crucial for their inventions and their success in constructing the first quartz clock; its genesis illuminates how inventors in industrial research labs created novelty. Chapter 7 looks at the process of inventing equivalent technics for frequency division at other laboratories, and at other paths to these technics.

---

[53]Cady, for example had to borrow tubes from Western Electric, e.g. H. D. Arnold to B. W. Kendall, 13.2.1918, Arnold to Cady, 5.6.1919, AT&T Archive, Loc. 79 10 01 03.

7

DEVELOPMENT OF QUARTZ CLOCK TECHNICS
OUTSIDE AT&T

AT&T was just one of the institutions that developed exact and stable high-frequency dividers needed for the construction of a quartz clock. As mentioned, other national, academic, government, and commercial organizations sought some means to divide the frequency of the piezo-resonator due to various goals connected to their interests in standards, control, and manipulation of radio frequencies. This chapter examines the simultaneous invention of these different technics for dividing high frequencies, and their use for frequency standards and clocks. It follows research at the NPL, whose tuning-fork standard was discussed in Chapter 5, the Italian Navy's *Istituto Elettrotecnico e Radiotelegrafico,* Clapp's research at MIT's short-wave laboratory and its continuation at the facilities of the small General Radio Company, the research laboratory of Philips, and Japan's radio research committee, showing that they exploited different (but related) material, technical, and intellectual resources in reaching practical dividers.

That a few related but different goals and technical-scientific resources led to practically equivalent methods testifies to a vibrant field of electronic research. This field included physicists and engineers undertaking a wide spectrum of research from the pursuit of general laws and understanding the phenomena to the design of a particular instrument, within academic, national, and commercial laboratory settings in a few countries. To capture the historical process that led to the quartz clock, the present chapter suggests a comparative examination of parallel and related developments at the mentioned research sites and their interactions, transcending national and institutional boundaries. It further examines the different kinds of research and their contributions to the development of the quartz clock technics.

Differences in research projects and institutional settings notwithstanding, the physicists and engineers who developed quartz frequency and time standards all worked with relatively new kinds of research institutions aimed at

Sonar to Quartz Clock. Shaul Katzir, Oxford University Press. © Oxford University Press (2023).
DOI: 10.1093/oso/9780198878735.003.0008

tightening the connection between scientific and technological knowledge, on the one hand, and practical methods and devices, on the other, discussed in this book's Introduction. In one way or another, these institutes were established to facilitate the transfer of knowledge from science (especially physics) to technology and to foster the growth of technological knowledge primarily by carrying out research of more general scope than that usually carried out by independent inventors and production laboratories. Still, the present analysis of research carried out at these variant sites reveals significant differences between the groups. These included differences in the importance attributed to searching for general rules, in the role of scientific knowledge and theoretical and mathematical derivations, and in the role of empirical research, as well as differences in the kind of solutions that would satisfy researchers. As it involved researchers of different kinds of institutions, the story discussed in this chapter and Chapter 6 provides an apt case for illustrating these differences and understanding the various strategies vis-à-vis the pursuit of knowledge in organizations primarily devoted to fostering technology.

7.1 Quartz Standards at State Laboratories

While the integration of one communication system motivated AT&T's research on quartz frequency standards, the control of national and international networks of different senders, primarily in broadcasting, motivated the research of other organizations, such as national laboratories. National research institutions in North America and Europe sought reliable and practical systems of frequency measurement to enable supervision of the various users of electromagnetic waves. One clear goal was to provide the technical means to ensure that the wavelength of any particular sender would not drift from its allocated (absolute) value. How much a station might drift from its frequency was a question of technical possibilities and the requirement of the technological system in which it was embedded. Higher accuracy would reduce interference and allow division of the spectrum (for broadcasting, commercial, and military uses alike) into more channels, and thus increase the traffic through the ether—a clear aim of regulating agencies. Heterodyne interference, as we have already seen, pointed at a specific goal, a variation lower than 50 Hz, which would allow adjacent stations to use the same frequency in sending either identical or different signals. By 1928, this became a pronounced aim. Dellinger, from the BoS, explained that to meet this goal one needed to reach an accuracy of 1 in 10^5, one order of magnitude higher than the extant standard.[1]

Regulations and internal needs drove also other laboratories to develop more accurate means for frequency measurement and control. Although well-kept tuning-fork standards reached high accuracy, both national and industrial laboratories decided on quartz standards as the more convenient method for attaining accuracy on site, by the transmitters. Their reasons were similar to those that led AT&T to developing quartz standards. Unlike the tuning-fork, which required a quite complex multiplication process, the piezoelectric resonator immediately produced oscillations at the radio range. This simplified the measuring instruments based on the quartz resonator, and allowed them to be made portable and cheaper, facilitating, thereby, their distribution and use. The internal high frequency of the quartz resonator also facilitated its use for higher frequencies that became increasingly useful during the 1920s. In addition studies indicated that the internal stability of the quartz resonator might be higher than that of the tuning-fork. In their research and development of methods for regulating frequency, a few of these laboratories sought some means to either reach a larger range of frequencies from a particular piezo-oscillator or check its frequency by a clock standard. For these ends they developed some means to divide frequency, which could and eventually would control clocks. Still, their paths, as well as their motives for these inventions varied considerably. I will begin with national research institutes.

Vallauri and Vecchiacchi

Two governmental laboratories, the British NPL and the Italian military's Istituto Elettrotecnico e Radiotelegrafico della Marina—Livorno, developed methods for reaching lower frequencies in order to compare the quartz standard to a clock. The Italian group reached its method following its work on international comparison of frequency standards. Its road to a subharmonic producer, which allowed for the construction of a quartz clock, was quite different from that of Marrison's group. The Italians did not seek such a method from the beginning of their research but developed it only following an unexpected observation in their electronic circuits. Following Cady's personal

---

[1] J. H. Dellinger, 'The Status of Frequency Standardization', *IRE Proceedings* 16 (1928): 579–90; Hugh R. Slotten, *Radio and Television Regulation: Broadcast Technology in the United States, 1920–1960* (Baltimore: Johns Hopkins University Press, 2000), on 43–51.

journey of 1923 (section 4.2), the American BoS organized a more thorough international comparison of its own frequency standard and those of four European laboratories to ensure and improve the accuracy of these standards and ensure that they were sufficient to prevent cross-border radio-wave interference. Between 1925 and 1927 the participating laboratories compared their standards by measuring the resonance frequency of the same quartz oscillators.[2] In this case the BoS did not circulate a material object in order to distribute a particular value, as was done with other standards, such as electric resistance and voltage, *after* the establishment of the magnitude. It rather circulated an object to compare different methods of measurement and the values of parallel standards. As with the case of the electricity units, the goal was to reach consensus. Yet, this was not done to promote a particular value of the standard or a method of any particular laboratory, but to help attain an accepted value.[3]

In Germany and Britain, these standards were the responsibility of national research institutions (PTR and NPL). Military research institutes, which took general responsibility for standards also beyond the armed forces, represented Italy, France, and Canada, which did not have national physical institutions. The comparative project encouraged participating laboratories to measure frequencies of quartz resonators in absolute units. Initially, all participating laboratories made their measurements 'with the aid of an intermediate device (tuning-fork, alternator, or oscillator) which produces an audio frequency, this being measured in terms of the clock'.[4] The Livorno Institute used a tuning-fork meter as an intermediate. Giancarlo Vallauri, the founder and director of this small radiotelegraphy institute, had developed a tuning-fork meter in 1923. Vallauri, a Navy technical officer and professor for electric technology in nearby Pisa and from 1926 in Turin, was an expert on wireless

[2]Dellinger, 'Frequency Standardization'. In an additional round the BoS sent oscillators to Canada and Japan. Christopher Shawn McGahey, 'Harnessing Nature's Timekeeper: A History of the Piezoelectric Quartz Crystal Technological Community (1880—1959)' (Ph.D. thesis, Georgia Institute of Technology, 2009), 159–60.

[3]Simon Schaffer, 'Accurate Measurement Is an English Science', in M. Norton Wise (ed.), *The Values of Precision* (Princeton, NJ: Princeton University Press, 1995), 161; Joseph O'Connell, 'Metrology: The Creation of Universality by the Circulation of Particulars', *Social Studies of Science* 23 (1993): 129–73. See also the introduction to Chapter 5.

[4]Dellinger, 582; Walter G. Cady, 'An International Comparison of Radio Wavelength Standards by Means of Piezo-Electric Resonators', *IRE Proceedings* 12 (1924): 309.

communication and terrestrial magnetism and an important figure in the Italian technical radio community.[5]

In preparation for measuring the BoS piezoelectric oscillator of September 1926, Vallauri designed a method for improving his laboratory tuning-fork meter with his former physics student from Pisa and current assistant, Francesco Vecchiacchi.[6] Initially, they did not aim for a new quartz standard. They rather aimed at reaching exact values at higher precision. Incorporating a harmonic producer, the original meter reached a wide spectrum of frequencies including the range of the quartz resonators. Without a modulator, however, the relatively coarse discrete spectrum of this meter did not have the precision to display the measured frequency in the required accuracy.[7] In order to determine a frequency with a value between two known harmonics, Vallauri and Vecchiacchi proposed an 'interpolation method', modifying a scheme previously employed by Charles B. Jolliffe and Grace Hazen at the BoS. The latter utilized the heterodyne interference between the known and unknown waves, which produced an audio wave equal in frequency to the difference between them. They measured the resulting beat's frequency by a sonic wave meter, and added or subtracted its value from the value of the closest harmonic in their meter. Employing the heterodyne effect in another common manner, the Italians dispensed with the audible measurement. They compared the piezo-oscillator with a regular wave meter—an RLC tube circuit with a variable capacitor, precalibrated by the tuning-fork ('oscillatore locale' on the right side of Figure 7.1). The tuning-fork was compared to a standard clock. To compare the wave meter with the piezo-oscillator they adjusted the capacity of the wave meter to reach an exact synchronization with the piezo-oscillator detected by the vanishing of the audio beats. At this stage, they observed the

---

[5] Edoardo Savino, 'Vallauri, Gian Carlo', in *La nazione operante: profili e figure*, 2nd edn (Milan: Vicolo Pattari, 1934), 341–2.

[6] Vallauri mentioned the use of the method in measuring the piezo-oscillator sent by the BoS, carried out in September 1926 and March 1927 (the dates are mentioned by Dellinger). Another, direct, method was used in the August 1927 measurement (see below). An earlier use of the direct method, which Vallauri does not mention, is implausible, especially as he mentioned a similar and slightly earlier method by the BoS and does not try to claim originality. Giancarlo Vallauri, 'Confronti fra misure di frequenza per mezzo di piezorisuonatori', *L'Elettrotecnica* 14 (1927): 445–52, on 449. Francesco Carassa, 'Francesco Vecchiacchi', *Rendiconti del Seminario matematico e fisico di Milano* 27 (1957): 19–21. I am grateful to Massimiliano Badino and Giuseppe Castagnetti for their help with the Italian sources.

[7] To avoid this problem Horton and his group had included a modulator in their tuning-fork standard system back in 1922; see Chapter 5.

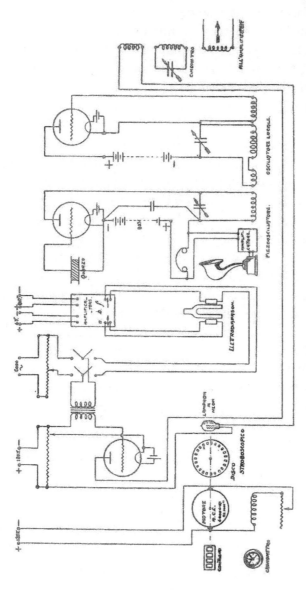

Figure 7.1 Vallauri and Vecchiacchi's method for comparing the frequency of a quartz oscillator ('piezo-oscilatore') to that of a chronometer (on the bottom left side of the figure) through a tuning-fork ('eletrodiapson' in the middle of the figure), and an 'oscillatore locale' used as a wave meter. Note the loudspeaker connected to the piezoelectric oscillator for detecting heterodyne interference.

change in the wave meter's capacity in comparison with its state at the closer precalibrated marked frequency. Using the basic theory of RLC circuits, they calculated the change in the internal frequency of the wave meter that resulted from the alternation of capacity and added that to the frequency marked on the wave meter.[8]

The actual measurements led to an unexpected finding. The two circuits reached synchronization not only once when their frequencies were equal, but also a few more times for each particular setting. Vallauri and Vecchiacchi inferred that synchronization is also achieved when the two circuits reach harmonics and also subharmonics of their original frequencies. Moreover, synchronization was obtained when the relation between the two frequencies was not only an integer but also a fraction (probably when both related to the same lower frequency as a common dominator). Vallauri and Vecchiacchi realized that due to this property they could utilize harmonics and subharmonics of the quartz oscillator to calibrate wave meters within the whole radio spectrum. Developing such a calibration method, they discovered that under proper conditions the piezo-resonator *enforces* either its own frequency or its harmonic or subharmonic on the wave-meter circuit. This finding must have surprised them more than their earlier findings. To be more specific, they found that the circuit containing the quartz crystal controls the oscillating frequency of the coupled circuit ('oscillatore locale' in Figure 7.1) when it has enough energy, and when the natural frequency of the controlled circuit is close enough to one of its harmonics or subharmonics. In other words, they constructed a novel harmonic and subharmonic producer.[9]

Marrison put much effort into devising such a practical subharmonic generator. Yet, it is unclear if, and to what extent, the Italian team knew about other attempts to obtain a wider spectrum from the piezo-oscillator. There is no indication that the aim had been on their research agenda before. Their success seems to be a by-product of their work for the international comparison, and their discovery of the controlling effect of the circuits seemed to have been an unintended consequence rather than the result of a quest for such a method. Unlike Marrison, they did not offer any explanation (e.g. in terms of process) for the effect of the piezo-oscillator that they observed, a further indication that such a method was not their central interest. Still, they were

---

[8]Vallauri, 'Confronti fra misure di frequenza per mezzo di piezorisuonatori'; Charles B. Jolliffe and Grace Hazen, 'Establishment of Radio Standards of Frequency by the Use of a Harmonic Amplifier', *Bureau of Standards Scientific Paper* 530 (1926): 179–89.

[9]Vallauri, 'Confronti fra misure di frequenza per mezzo di piezorisuonatori', 449.

familiar enough with contemporary telecommunication technics to appreciate
the potential utility of harnessing the piezo-oscillator as a new radio-frequency
standard and to develop the effect into a working method for calibrating wave
meters.

Vecchiacchi carried out the method a step further by applying it to compare directly the quartz oscillations to a clock. Between March and August
1927 he conceived of a way to apply the above-mentioned method of division
for that end. The new method allowed for direct measurement without inter-
polating from the observed value to the calculated frequency of the crystal.
Vecchiacchi's method included four similar stages of division, each by five to
eight times, reducing the frequency to the order of 100 Hz. At each stage a
source of known frequency coupled to an RLC tube circuit, similar to the
wave meter used in the previous method, enforced the frequency of the coupled circuit to one of its precise harmonics. The outcome at each stage served
as the source frequency for the next one (see Figure 7.2). Vecchiacchi did not
rely only on the enforcing power of the source circuit. To make the internal
frequency of each circuit close enough to the subharmonic value, he manually
adjusted the capacity in the subharmonic system until he did not hear any interference in telephone headsets, one of which was connected to each circuit.
The last circuit in the series drove a small synchronous motor, itself connected
to a chronograph. Comparing the marks of the motor and a pendulum clock
on the chronograph, the observer measured the period of the former and by
a simple numeric function calculated the frequency of the quartz resonator.
For each measurement Vecchiacchi ran the system for only 135 seconds. This
sufficed to determine the frequency of the piezo-oscillator. The method, thus,

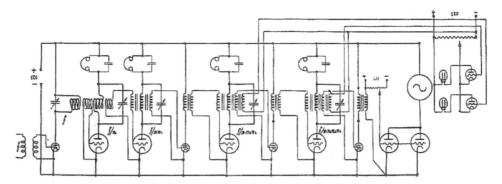

Figure 7.2 Vecchiacchi's method for a direct comparison of a quartz oscillation
to a clock; f at the bottom left-hand side is the source frequency.

allowed for direct determination of the piezo-oscillator, which could serve as a basic standard for further frequency measurements.

The Italian team, however, did not attempt to develop a continuous source either of standard frequencies or of a clock mechanism. Their system seemed too fragile and sensitive for use along long intervals, especially as it relied on manual tuning of components that were not highly stable, e.g. variable capacitors. Vecchiacchi devised it as a way of determining frequencies of particular circuits at particular places and times, and so it remained. His method did not promise the same kind of stability and robustness that Marrison sought in developing similar subharmonic producers. The Italians did not seek such stability, as they did not try to establish a primary frequency standard on the quartz resonator. For the needs of frequency standards they relied on the tuning-fork. Still, as Vallauri asserted, the Livorno Institute provided a solution to the problem of frequency division, which could be practical for radio frequencies.[10] This, however, was only one among a few solutions suggested at the time.

Dye

Time measurement occupied David Dye since he had begun developing the NPL's tuning-fork frequency standard early in the decade (Chapter 5). By 1928 to the latest, he worked on a method for directly comparing a piezo-oscillator to a standard clock. Until his early death in 1932, Dye refined and improved the various parts of the NPL's quartz standards, devising, among other things, a more stable crystal cut, and constructing one of the most accurate quartz clocks to date. The details of his initial system, however, are unknown.[11] Further, it is unclear when exactly he launched his research into

[10] Giancarlo Vallauri, 'Confronti fra misure di frequenza per mezzo di piezorisuonatori', *L'Elettrotecnica* 14 (1927): 682–4. This paper from 25/9/1927 is a supplement to the one from July under the same title; further information is available in Francesco Vecchiacchi, 'Applicazione all'oscillografo catodico della demoltiplicazione statica di frequenza', *L'Elettrotecnica* 15 (1928): 805–14, on 807.

[11] The details of the timekeeping system were published only after it was further refined by his former assistant Louis Essen. By 1936, '[t]he clock . . . enable[d] short intervals of time to be measured with a precision better than ±0.001 second and intervals of the order of several months with a precision of ±0.002 second.' This was similar to the accuracy of the most precise pendulum timekeepers. L. Essen, 'The Dye Quartz Ring Oscillator as a Standard of Frequency and Time', *Proceedings of the Royal Society of London A: Mathematical, Physical and Engineering Sciences* 155 (1936): 498–519, on 518; E. [Edward] V. A. [Appleton], 'David William Dye. 1887–1932', *Obituary Notices of Fellows of the Royal Society (1932–1954)* 1, (1932): 75–8.

using quartz as a primary frequency standard and comparing it with a clock. Returning from a journey in Europe in January 1928, John Dellinger from the American BoS reported that Dye had devised a method enabling such a comparison. He added that Dye employed a multivibrator to that end. His method was likely similar to that of Clapp's (described in the subsection 'Clapp'). By that date despite his 'complete satisfaction' with the tuning-fork standard, Dye regarded the quartz resonator as a better basis for the laboratory primary standard. Consequently, he concentrated his research on the piezoelectric frequency and time standard, delegating the tuning-fork work to his assistant, Lewis Essen.[12] Still, it is unclear whether he had begun working on connecting the quartz resonator to a clock before or only after learning about Marrison's success or about Clapp's method, even if the former is somewhat more plausible.[13]

Dye's adoption of the quartz standard followed extensive scientific and practical research on crystal oscillations that showed their stability. While Vallauri's team probably had not engaged in the study of piezoelectricity before 1926, Dye had already begun studying piezoelectric crystals following Cady's 1923 introduction of the resonators to the NPL. The piezoelectric resonators, Dye explained, 'possess the particularly valuable properties of simplicity and portability, and at the same time enable selected frequencies to be reproduced with the precision of a primary standard.' Continuously improving the tuning-fork standard, Dye chose the piezo-resonators as preferable secondary standards. Like the researchers at Western Electric, Dye saw the advantage of quartz standards for high frequencies, which became the central range of practical wireless service for civil and military communication and for broadcasting. To answer the needs of the British users of these wireless services and

---

[12] *Annual Report of the National Physical Laboratory for the Year 1927* (Teddington: National Physical Laboratory, 1928), 91 (the same publisher for consequent NPL reports mentioned in the footnotes that follow). On the division of labor from 1928, Louis Essen, 'Timekeeping', in Ray Essen (ed.), *Birth of Atomic Time: Includes the Memoirs of Louis Essen* (Peterborough: FastPrint Publishing, 2015), 27–36.

[13] Dye acted as the chairman of the commission responsible for standards at the October 1927 meeting in which Horton and Marrison presented their clock. Yet, Dellinger mentioned his visit to the British laboratory in summer 1927, when he probably had learnt about Dye's method. It is still possible that Dye had learnt about the advancement of the Bell Laboratory, or about the work of other researchers, like Clapp and van der Pol and van der Mark (see the subsection 'Van der Pol') through other channels and that they had encouraged the research at NPL. *NPL 1927 Report*, 3–4; Dellinger, 'Frequency Standardization', 582 and 592; this is a published version from March 1928 of a talk given in January 1928.

their regulating bodies, Dye's small group prepared a series of resonators for radio frequency from 40 to 1500 kHz, which encompassed the practical range at the time.[14]

As he put piezoelectric resonators to practical use, Dye launched more thorough and general research on their physical properties. This programme led him, among others, to a theoretical description of the behaviour of a piezoelectric resonator in terms of an equivalent circuit of resistance, inductance, and capacity, variables familiar to physicists and electrical engineers (discussed in Chapter 9). His assistant, Paul Vigoureux compiled a comprehensive summary on piezoelectric oscillators, a study which led him to a few new results in addition to a 1931 monograph. The group at Bell Labs did not follow similar research, which went beyond their immediate practical aims. In Horton's terms, Dye acted also as a scientist, while Horton and Marrison only as engineers. Their example suggests that rank-and-file physicists in the corporation, the largest private employer of scientists, did not practice as scientists. This raises doubts about Shapin's claim that since most scientists found employment in industry, industry became the new home for science.[15] As a researcher at a national laboratory, Dye had to respond to the direct interests of the regulatory bodies, the military, and commercial companies, which presented their demands and wishes from the NPL through the British Radio Research Board. Yet he needed also to respond to the general interests of the public in advancing knowledge open to all. In this vein, publications and the extension of knowledge were regarded as aims of the NPL staff.[16] This dual role allowed him some freedom to follow his professional interests. Thus, Dye's research at a national laboratory seems to fill an intermediate position between the practical emphasis that characterizes the industrial (and military) laboratories and the tendency toward general knowledge pursued by academic research institutes.[17]

---

[14] *NPL Reports* 1925 (quotation on p. 11) and 1924 (p. 79).

[15] Steven Shapin, *The Scientific Life: A Moral History of a Late Modern Vocation* (Chicago: University of Chicago Press, 2008), 93–110.

[16] *NPL Reports* for 1923 to 1928, quotation from 1925 report (p. 11). On Dye's research in response to Radio Research Board request, see reports for 1926 (pp. 11–12) and 1928 (p. 13). D. W. Dye, 'The Piezo-Electric Quartz Resonator and Its Equivalent Electrical Circuit', *Proceedings of the Physical Society of London* 38 (1926): 399–458.

[17] Notwithstanding, a small number of researchers at Bell Labs, as at a few other industrial research laboratories, enjoyed a similar freedom to follow questions that were further than the immediate technological quest. Clinton Davisson and Lester Germer, for example, carried out research on electron diffraction during the same years. Arturo Russo, 'Fundamental Research

7.2 Academic and Commercial Developments

While researchers at Bell Labs, the NPL, and the Italian naval laboratory developed methods for frequency division in improving central frequency standards, other laboratories obtained such methods in searching for ways to manipulate frequencies. Clapp at MIT sought ways to reach a wide range of radio frequencies from one reliable piezoelectric oscillator, for the needs of small users of wireless communication. Koga at the Tokyo electro-technology institute sough a frequency divider for his scheme of secret signalling, probably for the benefit of the military. Van der Pol at Philips looked further to the properties of triode circuits in general, which his corporation developed at the time, but enjoyed also experimenting with measuring short intervals for laboratory use. All designed methods that could be used to drive a quartz clock. The laboratories in which they worked did not seek to establish standards but were still interested in controlling frequencies for wireless transmission, and in Clapp's case also in their measurements.

Clapp

In his search for a means to measure a wide range of frequencies from one piezoelectric oscillator, James K. Clapp devised a subharmonic producer and consequently employed it for a new quartz clock. His research points to another distinct technological question that led to quartz clock technology. Clapp's aim differed from the aims of Marrison, Dye, and Vecchiacchi, who were attempting to find ways to determine the accuracy and stability of piezoelectric oscillators. Clapp's general motivation and the context of his research also differed. He studied and taught in MIT's electrical engineering department, where he received a Master of Science degree in 1926. In July of the same year he began working at the short-wave research laboratory established by the department and Colonel Edward Green at the latter's estate. Under the direction of Edward Bowles, a professor at the institute, the laboratory enjoyed independence in its choice of research subjects related, as the name suggests, to two-way wireless communication and broadcasting. The laboratory staff addressed questions of interest to small broadcasters and radio amateurs. It suggested methods that they could use, and published them in *QST*, a popular amateur radio journal. In presenting his method for reaching a wide spectrum from one oscillator, Clapp also addressed such an audience,

at Bell Laboratories: The Discovery of Electron Diffraction', *Historical Studies in the Physical Sciences* 12 (1981): 117–60.

providing concrete instructions for constructing the apparatus and for using the method in one's own laboratory. Yet, he deemed the method held enough interest to publish it in a scientific journal, which served as a venue for presenting scientific instruments and laboratory methods.[18]

Like the large corporations and the national laboratories, small and independent users of wireless communication needed methods for accurately measuring radio frequencies, in order to conform to the regulations and to improve their use of their allocated spectrum. With the equipment of a small laboratory (unlike the better-equipped laboratories of Bell and NPL), the decreasing reliability of tuning-fork calibration with high harmonics became a considerable problem in the broadcasting range. It became a more acute problem with short waves, which was left to amateurs. Including the short-wave range, the quartz oscillator was 'roughly in the middle of the harmonic spectrum', making it preferable for reaching the useful spectrum, and replacing the tuning-fork with calibrating simple wave meters.[19] Yet, such a use of the piezo-oscillator posed the technical problem of attaining many frequencies from that of one quartz oscillator, affordable to amateurs. For researchers at AT&T the problem amounted to finding a means for using the piezo-resonator to control electric oscillations at the sought-after frequencies (for both standards and control). For Clapp, it was enough to find a manual method for calibrating the output oscillators to the sought-after frequencies, also without an automatic control. Still, the similarity to Marrison's practical aim led to a similar solution, i.e. devising a method for dividing the incoming frequency by a known factor—a subharmonic producer. This was not the only approach to solving the problem; others, for example researchers at the PTR, induced a few resonance modes from the same crystal.[20] Clapps' subharmonic producer, however, differed from Marrison's and seems to be independent of the latter's. The MIT researcher submitted a paper describing his method for publication on 15 February 1927, i.e. a few weeks after Marrison reported his own solution

[18] Alex Soojung-Kim Pang, 'Edward Bowles and Radio Engineering at MIT, 1920–1940', *Historical Studies in the Physical and Biological Sciences* 20 (1990): 313–37, on 317–19; 'Clapp, James K[liton]', American Men of Science, 5th edn (1933); James K. Clapp, ' "Universal" Frequency Standardization from a Single Frequency Standard', *Journal of the Optical Society of America* 15 (1927): 25–47. On the journal see Terry Shinn, *Research-Technology and Cultural Change: Instrumentation, Genericity, Transversality* (Oxford: Bardwell Press, 2008),47.

[19] Clapp, ' "Universal" Frequency Standard', 30.

[20] E. Giebe and A. Scheibe, 'Piezoelektrische Kristalle als Frequenznormale', *Elektrische Nachrichten-Technik* 5 (1928): 65–82, and the discussion in Chapter 9.

in his notebooks and before the AT&T researcher filed a patent on his method. Clapp's paper appeared in July.

Clapp's approach may seem naïve, but it was nevertheless successful. He assumed that the same mechanism that allowed one circuit to control another at its harmonic should also allow controlling the second circuit at a subharmonic. In other words, he tried a frequency multiplier also for dividing an input frequency. He employed the standard device—Abraham and Bloch's multivibrator, discussed in Chapter 5. From earlier research about the device, he concluded that 'the most important property of the multivibrator is that it may be so readily controlled by the frequency standard. The very instability which formerly restricted the use of the device is now highly desirable, since the multivibrator may easily be forced to operate on a fixed frequency by the introduction of a small harmonic voltage from a fixed frequency source' (29). Although it had not been suggested before, the extension of that observation to subharmonics seemed so obvious to Clapp that he did not even explicate it, let alone provide a justification. Clapp did not differentiate between the ability of the device to be controlled by a higher or a lower frequency. Apparently obliviousness to possible problems with such an extension played to his advantage, as he was able to construct a working method.

Clapp suggested a practical method of calibrating a wave meter with one quartz oscillator. Following a common practice with the tuning-fork, he coupled a multivibrator to a piezo-oscillator by coils (L_n and L_1 in Figure 7.3). He assumed that the piezo-oscillator would constrain the multivibrator to oscillate at a harmonic or subharmonic of its own frequency. By varying the capacities in the multivibrator, its frequency would jump from one subharmonic to the other. Using the device for measurements required manual calibration with the help of a heterodyne circuit. Like Vallauri and Vecchiacchi, Clapp tuned the heterodyne circuit to synchronization by listening to the heterodyne beats disappear. In Clapp's method, one first calibrated the heterodyne circuit by the crystal oscillator, marking the fundamental (resonance) frequency of the crystal, and its harmonics (not subharmonics). The heterodyne circuit also enabled one to measure the proportion between the frequencies of the crystal and that of the multivibrator (i.e. the order of subharmonic). The n subharmonic in the multivibrator would synchronize with the heterodyne circuit n times while the original oscillation would synchronize only once. By counting the number of times the subharmonic synchronizes between two successive harmonics of the original (which are marked on the

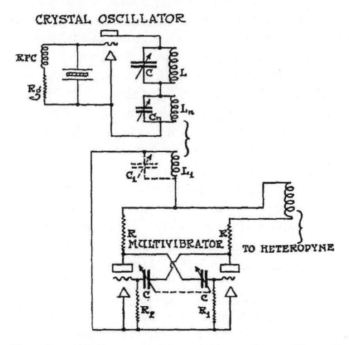

Figure 7.3 Clapp's multivibrator–subharmonic producer: The coils L_1 and L_n are coupled; all the capacitors (c) are variable capacitors. *Source*: Clapp, '"Universal" Frequency Standardization', 35.

heterodyne circuit) one can, thus, determine the order of the subharmonic.[21] Then one can use the known value of the subharmonic frequency of the multivibrator as the basis for further measurements. For example, again with the help of heterodyne, one can mark the zero beat points (i.e. synchronization) between a wave meter and the multivibrator, calibrating thereby the former in intervals equal to the subharmonic frequency. In this manner of using an intermediate electric circuit and the heterodyne method to calibrate wave meters by a piezoelectric oscillator, Clapp followed the Pierce 1923 method of using his oscillator to calibrate frequency meters (Section 4.4). Unlike Clapp,

[21] For example, the 4th subharmonic of the 100-kHz oscillator would synchronize with the heterodyne as the original oscillator would at 100 kHz, but would return to that state also at 125, 150, and 175 kHz and again with the quartz oscillator at 200 kHz.

however, Pierce did not try to control the frequency of the intermediate circuit and to produce either harmonic or subharmonic oscillations.[22]

Although it was not a short procedure, the calibration of a wave meter by Clapp's method could be carried out with relatively simple means at a small laboratory that lacked the facilities of Bell Labs or the NPL. Clapp explicated how this could be done, specifying the ingredients he used (like the commercially available vacuum tubes), and ways to attain the needed reproducible conditions. By suggesting an inexpensive method that could be carried out by many researchers, Clapp showed a 'democratic' approach. His attitude toward temperature control is a good example of this democratic approach. Marrison developed elaborate mechanisms for ensuring that the crystal remained at a steady temperature. At the time Clapp sent his paper, Dellinger recruited four government and industrial laboratories to perfect temperature control.[23] Clapp, however, was satisfied with a simple means:

> A glass container, with long arms containing the leads to the crystal, may be immersed in an ice bath, (or oil bath with thermostat control) so that reproducible conditions may be easily realized. Comparison of the standard crystal against an auxilliary [sic] crystal will yield the temperature-frequency curve of the standard; once this has been determined, it is not necessary to make use of the bath each time the standard is used.[24]

Of course, the accuracy attained by Clapp's method could not compete with that reached by Marrison or Dye. His aim, was quite different from theirs—seeking an inexpensive and practical method for measuring a wide range of frequencies at small laboratories. With his method, practitioners at such laboratories could attain much higher accuracy than with their RLC wave meters, even if it was still much lower than the state of the art at the large, well-equipped laboratories.

Although Clapp reached a 40th subharmonic with the multivibrator, he admitted that above the 20th, the method was less reliable. Thus, he suggested the use of two multivibrators in cascade to reach low frequencies, where the second oscillates at a subharmonic of the first. With the quartz crystal used by Clapp, of 138.5 kHz, the two producing the 20th subharmonic each can

[22] Clapp, ' "Universal" Frequency Standardization'. George W. Pierce, 'Piezoelectric Crystal Resonators and Crystal Oscillators Applied to the Precision Calibration of Wavemeters', *Proceedings of the American Academy of Arts and Sciences* 59 (1923): 81–106.

[23] See Chapter 6, fn. 46.

[24] Clapp, ' "Universal" Frequency Standardization', 32.

result in a frequency of 346 Hz. Vannevar Bush, a professor of electrical engineering at MIT, famous today for his later work on the differential analyser and his role in science policy, suggested using the method to drive a small synchronous motor. 'The motor may drive a suitable clock mechanism, allowing the frequency of the standard to be compared directly with standard time.' It seems, however, that Clapp had not connected his system to a clock before sending his paper for publication. He promised a future paper on the subject, but that paper would appear only in 1929, after he moved to the General Radio Company.[25]

Since 1925 General Radio (GR), which focused on measurement instruments for radio frequencies, had produced and marketed piezoelectric oscillators to radio stations and small laboratories, and quartz plates to radio amateurs. As mentioned in section 4.3 Cady had presented his piezo-resonators to Melville Eastham, the founder and manager of the company in 1921. Seven years later, the company expanded and reorganized its research and development activities, hiring five new engineers, more than doubling the number of its engineering staff. Clearly, piezoelectric oscillators was at the centre of its interest, as it brought not only Clapp but also J. Warren Horton from AT&T to fill the practically new position of chief engineer. Regarding their experience, it is more than likely that GR hired them with the specific aim of producing a frequency standard calibrated by a clock. With Lewis M. Hull, Clapp continued the research that he had carried out at the MIT short-wave laboratory on driving a clock by a quartz oscillator. Hull was on leave from Radio Frequency Laboratories Inc., Boonton, a company for which he had developed related wireless technics after completing his PhD in physics in 1922.[26]

By November 1928, Clapp and Hull installed a system that they designed for producing a wide range of frequencies from a central quartz oscillator standard, itself compared by a clock mechanism to the time signals of the US Naval Observatory. This was an internal laboratory system which resembles the tuning-fork standard that Horton had installed for the Bell System in

[25] Clapp, ' "Universal" Frequency Standard', 41–6.

[26] Arthur E. Thiessen, *A History of the General Radio Company, 1915–1965* (Concord, MA: General Radio Company, 1965), 29–31. McGahey, 'Harnessing Nature's Timekeeper', 184–8; 'Hull, Lewis Madison', in Who's Who in Aviation, 1942–43 (1942); 'Lewis M. Hull', in *IEEE Global History Network* (http://www.ieeeghn.org/wiki/index.php/Lewis_M._Hull, last accessed 18.10.2012). Between 1922 and 1928 Hull applied for more than 15 patents relating to wireless technics; see the Espacenet patent database.

the early 1920s. It served to calibrate piezo-oscillators as secondary frequency standards, which GR marketed for radio stations. As Clapp suggested in 1927, two multivibrators in cascade reduced the frequency of a master piezo-oscillator. For the new more accurate aim, Clapp and Hull kept the quartz oscillator in a constant-temperature unit, with a thermal regulator. The multivibrators 'submultiplied' the frequency from the 50 kHz of the quartz they chose, to 1000 Hz, used to drive a synchronous motor of a clock mechanism. Apparently, GR did not sell the clock system as such. A year later it began selling a quartz-based clock designed for measuring short intervals.[27]

In the summer of 1928 when Hull and Clapp first compared their piezo-oscillation to the Navy time signals, the idea of such a comparison or a quartz clock was not new. The clock of Bell Labs with its various elements was well known; Hull and Clapp did not fail to mention it. Moreover, Horton, one of the central contributors to that clock, supervised the research of Hull and Clapp. Yet, their subharmonic producer, a crucial component in such a clock, was Clapp's original idea and design, conceived before the announcements of alternative methods. The accuracy of the system, however, did not hinge on the subharmonic producers but on the quartz crystal and the stability of its physical conditions. Within ten days of initial testing Hull and Clapp found a daily deviation of no more than 8 parts per million. This was still considerably higher than the average error of 1 part in ten million, which Marrison announced in the same year (Ch. 6). Given the simpler means at the GR laboratory (e.g. simple thermostat, no cut to reduce temperature fluctuations) and its more modest aims, the higher accuracy at Bell Labs is not surprising. Nevertheless, in comparison with earlier available frequency standards, GR's system attained an impressive accuracy, clearly above the needs of calibrating standards for broadcasting stations.[28]

While initially Clapp presented the ability of the multivibrator to produce stable and exact subharmonics as a straightforward result of its instability,

---

[27] The secondary standards were 'piezo oscillators including quartz plates mounted in holders with adjustable air gaps. The plates are adjusted to within one-tenth of one per cent by grinding, and the final setting is made by adjusting the air gap with a screw provided for this purpose which is then locked.' Lewis M. Hull and James K. Clapp, 'A Convenient Method for Referring Secondary Frequency Standards to a Standard Time Interval', *IRE Proceedings* 17 (1929): 252–71, on 268. Unlike Clapp's earlier paper they published this in an engineering journal. James K. Clapp, 'A New Frequency Standard', *General Radio Experimenter* 3, no. 11 (April 1929), on 1–2, 4 (available at https://www.ietlabs.com/pdf/GR_Experimenters/1929/GenRad_Experimenter_April_1929.pdf).

[28] Hull and Clapp, 'Referring Secondary Frequency Standards'.

in 1929 he and Hull devoted most of their paper to justifying the method, mainly by describing its experimental validation. A few factors might have led to the new interest in justification: the higher accuracy of the clock relative to Clapp's earlier calibration scheme; the involvement of Hull; the need to justify the method in comparison with other methods, such as Marrison's; doubts that had been raised about the validity of the method; and the longer time they had to examine the method. Shortly before Clapp published his first paper, van der Pol had advanced a theoretical discussion of subharmonic producers based on what he called 'relaxation oscillations' (more about this in the subsection 'Van der Pol'). At first glance van der Pol provided an explanation and thus a justification for using the multivibrator for a 'step-down' frequency converter, which Clapp had developed without such a theoretical basis. Although regarding the system as a kind of relaxation oscillator, Clapp and Hull observed that van der Pol's analysis is valid only for small voltage oscillations, while their method required high voltage 'far beyond the region of the characteristic' function of the multivibrator's vacuum tubes. They concluded that 'no adequate mathematical analysis of the action of the multivibrator has been given'.[29] Instead of such an analysis they carried out experimental research to study the conditions under which the multivibrator could serve as a reliable subharmonic producer.

A central aim of their research was to explain the particulars of the design that improve the stability and the degree of the subharmonic, like their finding that an asymmetric multivibrator (i.e. where the triodes differ from each other) is preferable to a symmetric one, and that the additional valve used as a 'control voltage' further stabilized the outcome. They, thus, examined the effect of changing variables of the system like the voltage, capacities, and resistances in the two branches of the multivibrator on the stability of the output frequency. This was a quantitative study, where they recorded the values of the various variables, and examined whether it could provide workable conditions. It did not, however, lead to a mathematical law as they did not quantify the effect on the multivibrator. Their goal was not to examine the system as such, but to point out the range in which it could be reliable for their specific aim. For example, they stated that with a 'control voltage' between 1.8 and 2.8 V 'the fundamental frequency of the multivibrator assumes discreet [sic] values, these being submultiples of the control frequency'. When the control voltage is lower, they found that for some capacities of the multivibrator it took discrete values, but for capacities of in-between values, it oscillated in

---

[29] Ibid, 256.

an uncontrolled manner. Higher voltage 'drew' the multivibrator towards the source frequency, instead of its subharmonic. 'The net result of these effects is the appearance of a fairly definite *optimum value* of control voltage for any given harmonic order.' Note that this value depends on the harmonic order. Furthermore, they did not suggest any general rule for determining what values would be useful for a particular end. Still they succeeded in establishing 'that this state of oscillation of the multivibrator . . . is in fact an oscillatory state wherein the appropriate multivibrator harmonic bears a constant and permanent phase relation to the injected oscillation.'[30]

Clapp and Hull, like Marrison, were satisfied with such confirmation of the assumption used in their method. Even as they exhibited interest in the behaviour of the system for its own sake, this remained secondary to practical concerns. Observing that a general understanding of the multivibrator would require lengthy research and many resources, Clapp and Hull gave up such an aim for a more limited one of finding the operational principles of the particular system, which they used. Their endeavour is an example of engineering research, as it focused on a particular device—the multivibrator—in its specific use as a subharmonic vibrator. This engineering research was used to provide instruction for design (making the multivibrator asymmetric) and for operation (pointing at the proper voltage). While historians have recognized these roles of engineering research, its role of justifying the use of a particular method, central in this example, has not received similar recognition. Perhaps justification gained prominence in this case due to its particular importance in the context of measuring technologies.[31]

Koga

While Clapp, Vecchiacchi, and Marrison designed subharmonic producers to facilitate measuring techniques, Issaku Koga contrived a 'frequency transformer' to manipulate with electromagnetic waves. Such manipulation was also one of Marrison's aims, as he attempted to obtain oscillations of relatively low frequency for multiplex telephony. Koga, however, was not interested in multiplex, but in secret wireless signalling, for which he independently

---

[30] Ibid., respectively 258, 259, 265 (emphasis in the original).

[31] Following Michael Polanyi, Walter Vincenti regards the search for operating principles as characterizing engineering in relation to scientific research, even if it is not exclusive to the former. Likewise, research aimed at justifying a particular operation is often done also in science. Walter G. Vincenti, *What Engineers Know and How They Know It: Analytical Studies from Aeronautical History* (Baltimore: John Hopkins University Press, 1990), esp. 112–36, 209.

invented yet another frequency divider. His suggestion arrived at the journal *IRE Proceedings* in June 1927, three months after Marrison filed a patent for his subharmonic producer and a month before Clapp's paper appeared. Koga studied electrical engineering at Tokyo Imperial University. He worked at the city's municipal electro-technical institute (officially translated as Tokyo City Electrical Research Institute) with his dissertation supervisor Tsunetaro Kujirai on projects related to Kujirai's role on the Japan's Radio Research Committee. Among others, the committee was looking for secret signalling methods, presumably at the request of the military, and Koga suggested a solution.[32]

In 1926 Koga devised a scheme for secret signalling; however, it required a means for obtaining an electrical wave whose frequency is exactly half the frequency of an incoming wave.[33] According to his later recollections, he came up with a method for dividing frequencies serendipitously. He found out that when he coupled a RLC triode circuit whose natural frequency approached half the frequency of an input circuit, the coupled circuit would abruptly change its frequency to exactly half the original frequency. Extending his initial observation, he concluded that the input oscillator enforces or attracts the output oscillator to one of its subharmonics. Koga's finding was, thus, very similar to that of Vallauri and Vecchiacchi from the same year, and he explained it in similar terms. Like Vecchiacchi, he used the effect to construct a frequency divider. Unaware of the Italians' simultaneous work, Koga supported his finding by a 1922 observation of a similar synchronization between triode circuits of nearly equal frequencies, by the British physicist Edward V. Appleton. Appleton concluded that 'if two oscillating triode systems are nearly in resonance, they vibrate with one frequency', but did not develop that to harmonics of each other. Koga further found that his 'frequency transformer' could yield also simple fractions like 2/3 or 3/2 of the

[32] Isaac Koga, 'A New Frequency Transformer or Frequency Changer', *IRE Proceedings* 15 (1927); 'Isaac Koga 1899–1982', *URSI, Information Bulletin*, No. 222 (September 1982): 88–91; Isaac Koga, 'Frequency Demultiplication and the Origin of Frequency Shift Keying System', *Journal of the Institute of Electronics and Communication Engineers of Japan* 56 (1973): 1335–40(in Japanese). I am grateful to Shigehisa Hirose for providing me with the text and an English summary along with additional helpful information about Koga, and to Kenji Ito for further information.

[33] As discussed in Chapter 5, secret communication motivated also Eccles and Jordan to develop technics related to tuning-fork standards. Due to the developments of radio telecommunication in the decade that separated the two efforts, Koga did not use electromechanical methods like the phonic wheel and sought-after electronic technics.

original frequency.[34] These values could also be useful for secret signalling. Yet in print, Koga did not refer to possible military usage, but rather to their application for low-frequency radio communication.

In the basic arrangement, Koga used a common and relatively simple triode circuit, whose resonance frequency could be changed manually by a change-able capacitor, as the controlled (output) circuit. Like the Italians, Koga relied on manual calibration of the controlled oscillator to reach synchronization with a subharmonic of the income frequency. Once the 'natural frequency' of the controlled circuit approaches (by adjusting a variable condenser) an integer division of the controlling frequency it took that frequency. In the example given, the controlled oscillator takes a division of the incoming fre-quency when it is tuned to 1.5 to 0.15% from its value (higher division required closer tuning). A piezoelectric oscillator of the Pierce–Miller kind (on which see section 4.4; here left side of Figure 7.4) serving to control the frequency was coupled to the plate (anode) of the triode (V_2) of the controlled oscil-lator, which, therefore, oscillated at a division of the piezoelectric oscillator. It could control up to the tenth division of the 70-kHz resonator. Koga sug-gested the use of this 'frequency transformer' for controlling low frequencies by a quartz resonator. He further commented that '[i]f we use this process in

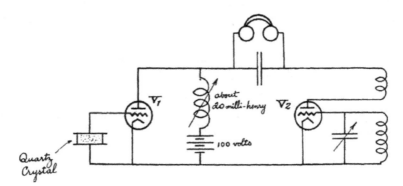

Figure 7.4 Koga's 'frequency transformer' for reaching low frequencies from those of a quartz oscillator (on the left side). The changeable capacitor on the right side (connected to the triode—V_2) can manually determine the power of the division. *Source*: 'New Frequency Transformer', 671.

---

[34]Koga, 'Frequency Demultiplication'; Koga, 'A New Frequency Transformer'. Edward V. Appleton, 'Automatic Synchronization of Triode Oscillators', *Proceedings of the Cambridge Philosphical Society* 21 (1922): 231–48, on 232.

a series of steps, we will be able to get still lower frequencies'.[35] Such lower frequencies could have been used for driving a clock mechanism, but there is no indication that Koga considered that possibility at the time. Since he was not concerned with high accuracy, or standards, comparing his circuits to a clock made little sense.

Yet, the experience probably stimulated his later study of piezoelectric resonators and their stability, discussed in Chapter 9. Among others, in the early 1930s he devised quartz cuts that kept stable vibrations. In the middle of the decade he constructed a quartz clock using one of his cuts. He employed, however, a multivibrator (as suggested by Clapp) rather than his own transformer as a frequency divider, suggesting that his method was less reliable. Yet it seems that the method was practical if not preferable. In 1952, the new exact Japanese clocks reduced the frequency in two stages, first using Koga's method and then a multivibrator.[36]

Van der Pol

In June 1927, when Koga's article reached the *IRE Proceedings*'s editor, Balthasar van der Pol and Jan van der Mark from the Dutch electric company Philips patented another subharmonic producer. They based their method on a neon glow lamp tube as an unstable oscillator, which oscillated only at subharmonics of an input frequency. The glow lamp, thus, played a role similar to that of the multivibrator in Clapp's simultaneous invention. Their road to the invention, as well as their motivation and specific aims, however, differed considerably from those of Marrison, Clapp, and Koga. Their method originated in a research programme on triodes, rather than in a search for some means to measure and control frequencies. This research was theoretical rather than experimental in character and was connected to a general interest of Philips in electronic technics. Yet its conclusions and the practical methods derived from research that relied on empirical findings, among others from work on exact measurements of time.

In their method, the two Dutch researchers connected a neon tube (Ne in Figure 7.5) in series to a constant voltage source of high frequency ($E_0 \sin \omega t$) (could be a piezo-resonator), and connected the two components in parallel to a variable condenser (C); these two branches were connected in series to a

[35] Koga, 'A New Frequency Transformer', 672.

[36] *Progress in Radio in Japan* (Japanese National Committee for URSI: Tokyo, 1963): 8–9.

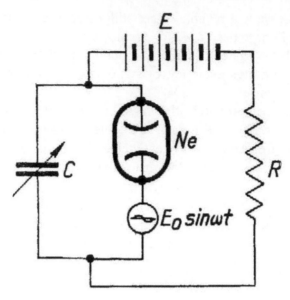

Figure 7.5 Van der Pol and van der Mark's circuit for 'frequency demultiplication'. *Source*: 'Frequency Demultiplication', 363.

resistance (R) and a source of direct voltage (E). As Marrison had done a year earlier (in his July–August 1926 suggestions discussed in Ch. 6, pp. 186–9), they exploited the capability of a neon tube to accumulate charge and then to discharge it when the voltage reaches a known level. With proper adjustment of the variable capacitor (which alters the voltage that could be accumulated in the neon tube), they determined the number of cycles of the input frequency needed to discharge the neon tube. Each discharge generated a current in the embedded circuit (of R and C in the figure). Thus, the period of discharge determined the frequency of the electric oscillation in the circuit. '[I]t [wa]s found that the system is only capable of oscillating with *discrete frequencies, these being determined by whole submultiples of the applied frequency*' (364). Changing the variable condenser caused jumps in the frequency from one level to another, without passing through in-between values (as shown in Figure 7.6). In this manner they could reach as high as the 40th subharmonic, and in some cases, the 200th.[37]

Their system resembles Marrison's attempt at a divider based on a circuit connecting a piezoelectric resonator to a neon tube from August 1926 (Ch. 6,

---

[37]Balthasar van der Pol and J. van der Mark, 'Frequency Demultiplication', *Nature* 120 (1927): 363–4; Jan van der Mark and Balthasar van der Pol, 'Improvements in or Relating to

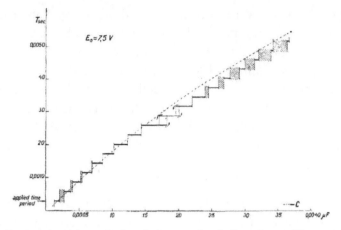

Figure 7.6 Jumps in the periods of the resulted electric oscillator of van der Pol and van der Mark's neon tube circuits as a function of its capacitor's capacity. The grey areas represent ranges of unstable frequency. *Source*: 'Frequency Demultiplication', 363.

Figure 6.6). However, unlike Marrison they did not rely on any property of the input alternating voltage above its stable frequency, whereas Marrison relied on the steep change in the capacity of the embedded piezo-resonator to ensure the jumps from one subharmonic to the next. Thus, despite the similarities in the methods, van der Pol and van der Mark did not use the input system to change the properties of the circuit that included the neon tube. Instead, they relied on the internal ability of the neon-tube branch of the circuit to synchronize its oscillations to a particular subharmonic of the input frequency, a property which Marrison had not recognize. Their divergent understandings of the system reflect the different research projects from which they arrived at their neon-tube subharmonic producers.

The electronic triode and its usage formed the main subject of study of van der Pol. It had already been a central theme of his research before September 1922, when Philips hired him to lead the study on the subject, a central interest of the company. The former incandescent lamp-based company was expanding its research laboratory as part of its strategy to diversify its product to knowledge-based fields like radio and X-ray equipment. He had received a PhD in physics two years earlier, after studying at Utrecht

Electric Frequency-Transforming Devices', GB 296829, filed 14 June 1927 (complete specification 14 March 1928); the inventor's name is not mentioned, but it appears in the American patent US 1,927,425 (filed 29 May 1928); also in the German, French, and Belgian patents.

and supplementing his education in England where he trained at the cele-
brated Cavendish Laboratory of J. J. Thomson in Cambridge. He also gained
experience in technologically related subjects while working with the physicist and radio expert John A. Fleming in London. Before joining Philips's
research department, van der Pol continued his research as an assistant at the
laboratory of the venerable H. A. Lorentz in the Haarlem Teylers Museum,
publishing in scientific and engineering journals. His assistant, van der Mark,
had just joined Philips in February 1927, after studying physics and working
as an assistant to the famous experimentalist Pieter Zeeman at Amsterdam
University.[38]

The two physicists from Philips devised their submultiplication method
relying on van der Pol's identification of a special kind of periodic phenomenon, which he coined 'relaxation–oscillation'. In 1926, he showed that
this kind of oscillation appears in systems of positive feedback, i.e. systems
that augment their own amplitudes of oscillations within some physical limits
(systems with 'negative resistance'). His interest in such systems originated
from attempts to understand theoretically the complex behaviour of self-
coupled triode circuits. That was the aim of a research programme that he
carried out with his former colleague at the Cavendish Laboratory, Appleton, who shared his interest in physics questions related to radio (Koga,
as we saw, relied on Appleton's finding to support his own finding that
led to his 'frequency transformer'). Appleton and van der Pol described
the dynamics of a triode circuit by a nonlinear differential equation. Often
physicists avoided such nonlinear equations, which cannot be solved analytically, and reverted to a simpler case or assumption, which allowed for a
linear approximation. Appleton and van der Pol, however, could not evade
the complexity of the equation regarding the problem at hand—the mathematical understanding of a particular device. Confronting an equation of

---

[38]F. Kees Boersma, 'Structural Ways to Embed a Research Laboratory into the Company: A Comparison between Philips and General Electric 1900–1940', *History and Technology* 19 (2003): 109–26; Marc J. de Vries, *80 Years of Research at the Philips Natuurkundig
Laboratorium (1914–1994): The Role of the Nat.Lab. at Philips* (Amsterdam: Pallas Publications, 2005) (with contributions by Kees Boersma), esp. 37–40; M. L. Cartwright, 'Balthazar Van Der Pol', *Journal of the London Mathematical Society* 35(1960): 367–76; 'Pol,
Balthasar van der', in J. C. Poggendorffs biographisch-literarisches Handwörterbuch für
Mathematik, Astronomie, Physik, Chemie und verwandte Wissenschaftsgebiete, Vol 6, pp.
2040–1; 'Jan van der Mark', https://web.archive.org/web/20190606093845/http://vandermark.
ch/vdmark-moran/195.html, accessed 25 May 2011.

a kind that had been little examined before, van der Pol produced fruitful results.

In his mathematical analysis, van der Pol generalized his and Appleton's equation for the triode to a dimensionless equation valid for any system of a similar dynamics, writing it as

$$\ddot{v} + \varepsilon\,(1 - v^2)\,\dot{v} + v = 0,$$

where v is a linear expression of the amplitude, and ε is a positive parameter of the system. This became a famous expression in nonlinear dynamics, known as 'van der Pol's oscillator'.[39] Although original in this form, the equation had a few predecessors written for specific electrical systems with a positive feed-back loop.[40] The equation cannot be solved analytically. Therefore, van der Pol developed an approximate solution for a small value of ε. For the other cases he illustrated the results by graphical numerical solutions for three specific values of ε: 0.1, 1, and 10. The case where the self-coupling of the system is stronger, i.e. where ε is high (e.g. 10), was the most interesting of these as it led to a hitherto unknown kind of solution. This case yielded 'a purely periodic solution of non-sinusoidal nature, the time period of which is expressed by the time of relaxation of the system [i.e. the time it takes to reach a steady oscillation]. For this reason the term *relaxation-oscillation* is suggested for this phenomenon.'[41] Van der Pol considered the theoretical prediction of this novel kind of periodic solution to be an important insight into the dynamics of non-linear systems. This result, he thought, might be applicable to quite a few fields and various problems. In his early publications on relaxation–oscillation he indicated a few electric and electronic devices that present this behaviour, including the neon glow lamp that he later used and the valve with two grids (tetrode), a generator connected to dynamo (unmentioned by van der Pol the

---

[39] Balthasar van der Pol, 'Over Relaxatietrillingen', *Physica* 6 (June 1926): 154–7; Balthasar van der Pol, 'Über "Relaxationsschwingungen"', *Jahrbuch der drahtlosen Telegraphie und Telephonie, Zeitschrift für Hochfrequenztechnik* 28 (1926): 178–84; Balthasar van der Pol, 'On "Relaxation-Oscillations" ', *Philosophical Magazine* 2 (1926): 978–92; Giorgio Israel, 'Technological Innovation and New Mathematics: van der Pol and the Birth of Nonlinear Dynamics', in *Technological Concepts and Mathematical Models in the Evolution of Modern Engineering Systems* (Birkhäuser, Basel, 2004), 52–77.

[40] Jean-Marc Ginoux and Christophe Letellier, 'Van der Pol and the History of Relaxation Oscillations: Toward the Emergence of a Concept', *Chaos: An Interdisciplinary Journal of Nonlinear Science* 22, no. 2 (April 2012): 023120.

[41] Van der Pol, 'On "Relaxation-Oscillations" ', 987, emphasis in the original.

phenomenon had been recognized already in the 1880s),[42] and also a totally mechanical example of a steam engine and suggested it. In February 1927, he added pipe organs. A year later, he and van der Mark suggested an analysis of the heartbeat as a system of relaxation–oscillation, a possibility that he had conjectured two years earlier. At that stage, they thought it is applicable not only to diverse natural systems but also to the 'reoccurrence of epidemics and of economical [sic] crises'.[43]

Despite the potential usefulness of the new notion of relaxation–oscillation, its relationship to concrete physical systems was not self-evident. Regular triode circuits, which intrigued van der Pol, do not show the phenomenon of relaxation–oscillation (their equation has a low value of ε). Cross-connected the triodes formed an example of relaxation–oscillation in Abraham and Bloch's multivibrator, which was the first example that van der Pol examined in some detail. Since the mathematical analysis of the equation was limited, he employed the example to support his new concept. One indication that the multivibrator fitted was its period as noted by its inventors, which agreed with the expectation of the relaxation time of an RLC 'relaxation oscillation'. With this insight in mind van der Pol analysed the multivibrator, showing that its oscillations can be expressed by his general equation (van der Pol oscillator). This equation allowed him to propose the arguably first account of the manner in which the multivibrator's 'oscillations are *maintained*'.[44]

The theory, however, did not account for the most useful property of the multivibrator—its ability to adjust its oscillations to incoming frequencies, since it did not consider the case of oscillations under impressed periodic voltage. In other words, it explained why the multivibrator keeps a stable frequency, but not why it oscillates at discrete values, which are integer multiplications of an input frequency. Van der Pol, therefore, turned to an experimental study to confirm the ability of the multivibrator to 'pick up' [*mit-nehmen*] the frequency of an input voltage. He reported the conclusions of this experiment in February 1927. He further supported his claim with the experience of Dye's tuning-fork frequency standard system, discussed in Section 5.3,

---

[42] Ginoux and Letellier, 'Van der Pol and the History of Relaxation Oscillations', 2.

[43] Quoted in Giorgio Israel, 'The Emergence of Biomathematics and the Case of Population Dynamics A Revival of Mechanical Reductionism and Darwinism', *Science in Context* 6 (1993): 469–509, on 476. Van der Pol, 'Über "Relaxationsschwingungen" '; Balthasar Van der Pol, 'Über "Relaxationsschwingungen" II', *Jahrbuch der drahtlosen Telegraphie* 29 (1927): 114–18.

[44] Van der Pol, 'Über "Relaxationsschwingungen"', 182–3; quote from van der Pol, 'On "Relaxation-Oscillations" ', 988 (emphasis in the original),

where the multivibrator multiplied the frequency by up to 20 times. In his brief discussion of the issue, however, van der Pol did not refer to the case in which the 'picked-up' frequency was a fraction of the input frequency.[45] He referred to this possibility only later after he had produced a frequency divider with van der Mark.

In the same publication, van der Pol pointed out a particular use of the neon tube, which he had mentioned previously, for multiplying the frequency of a tuning-fork circuit by William Eccles and Winifred Leyshon from July 1926. Leyshon and Eccles suggested replacing the multivibrator in Dye's tuning-fork-controlled method with a neon-tube circuit, which they judged to be 'a much simpler construction'. Leyshon had studied with Eccles at Finsbury College and for her PhD at London University, acquiring knowledge and skills about electronic oscillations and the tuning-fork, which she employed in her joint work with Eccles. Unlike her teacher (whose work I discuss in Section 5.2), however, her main professional interests were not in radio. Leyshon found employment at the physics laboratory of the London School of Medicine for Women. In this context she sought ways of producing flashes at constant periods, useful in physiological studies, for example, of muscles. In other words, she wanted to exploit the stable period of oscillation for measurements. Her interest was similar to an early interest of Dye's, who utilized the tuning-fork to measure projectiles' velocities during the war. Differences in aims notwithstanding, Eccles and Leyshon's method probably inspired van der Pol and van der Mark in constructing their June 1927 neon-tube subharmonic producer. Later, in early 1928, the British researchers also designed a neon-tube circuit to reduce the frequency of a quartz resonator, similar to that used by the Dutch. In the British version the controlled circuit oscillated at one-sixth of the crystal's frequency.[46]

[45] Van der Pol, 'Über "Relaxationsschwingungen" II', 118 ('mitgenommen').

[46] Winifred A. Leyshon, 'On the Control of the Frequency of Flashing of a Neon Tube by a Maintained Mechanical Vibrator', *Philosophical Magazine* 4 (1927): 305–24, quote on 307; W. H. Eccles and W. A. Leyshon, 'Some New Methods of Linking Mechanical and Electrical Vibrations', *Proceedings of the Physical Society (London)* 40 (1928): 229–33; W. A. Leyshon, 'Forced Oscillations in Self-maintained Oscillating Circuits', *Philosophical Magazine* 46 (1923): 686–98; W. H. Eccles and W. A. Leyshon, 'Some Thermionic Tube Circuits for Relaying and Measuring', *Journal of the Institution of Electrical Engineers* 59, (1921): 433–6; Grace Briscoe and Winifred Leyshon, 'Reciprocal Contraction of Antagonistic Muscles in Peripheral Preparations, Using Flashing Neon-Lamp Circuit for Excitation of Nerve', *Proceedings of the Royal Society of London. Series B*, 105 (1929): 259–79; on Dye *NPL 1919 Report*, 50.

Previously, van der Pol had not worked on issues related to frequency standards and control. In general he seldom occupied himself with designing specific devices or methods. His main task was to provide more general understanding of electronic devices, which would furnish other researchers at Philips with tools for improved design. In addition, his employer promised him an exceptional freedom in pursuing his own interests, similar to that enjoyed by Irving Langmuir at General Electric. Langmuir's research, which led to breakthroughs in the understanding of metal filaments and to a more efficient incandescent lamp, was regarded as a model for the power of combined scientific and technological research to achieve useful results that could be difficult to attain by either kind of research alone. The constraints on the work of the Dutch physicists were, thus, much looser than those on Horton and Marrison, whose research was directed toward the practical needs of the Bell System. They represented different kinds of industrial researchers. Unlike the research of Horton or Marrison, much of van der Pol's research was regarded as a contribution to physics, which he took care to publish in physics journals.[47]

Still the industrial environment in which he worked encouraged van der Pol to devise methods based on his findings (as GE did for Langmuir), even while assigning much of the practical research to an assistant.[48] Philips had a clear interest in applications related to subharmonic producers. It had developed crystal-control transmitters already in 1925. As a radio producer and user with international inspirations Philips could have benefited from a frequency divider in a few ways, e.g. in controlling different frequencies by one quartz crystal. Van der Pol and van der Mark were versed enough in current technics to appreciate the potential uses of a subharmonic producer and thus to work on its realization. Although not explicitly stated in van der Pol's earlier publications, his empirically based conclusion that a relaxation oscillator adjusts its period to that of an input oscillation suggested that it could tune itself also to *sub*harmonics of the basic frequency. That, at least, was how van der Pol and van der Mark explained the working principle of their neon-tube 'frequency demultiplication' method. They did not suggest, however, any further application of the method for comparing high frequencies to a clock. This would have drawn them into the realm of precise measurements, which was

---

[47]Boersma, 'Structural Ways to Embed a Research Laboratory into the Company', esp. van der Pol's letter from 24.4.1922, quoted on 125 (fn. 45); de Vries, *80 Years of Research at the Philips Natuurkundig Laboratorium*, 36–62.

[48]Leonard S. Reich, 'Irving Langmuir and the Pursuit of Science and Technology in the Corporate Environment', *Technology and Culture* 24 (1983): 199–221.

not their expertise. Moreover, Philips did not seem to have a specific reason to establish its own system of absolute standards. Unlike AT&T, it did not run a telecommunication network.

The Effect of van der Pol's Work

To the best of my knowledge, no quartz timekeeper has been constructed using van der Pol and van der Mark's neon-tube method. Still their work shaped the development of the quartz clock, even if it was due more to van der Pol's general results than to the particular method that they developed. The theoretical concept of relaxation–oscillations and the empirically based claim that such a system often synchronizes with the harmonic or subharmonic of an input frequency suggested the possible design of quite a few methods to produce subharmonics as needed for standards and clocks. In later works van der Pol explicitly claimed that Clapp's use of the multivibrator to divide frequencies is an example of a relaxation oscillator. Koga's method, however, was an example of sinusoidal oscillator, as was the case (although unmentioned by van der Pol) with Vallauri and Vecchiacchi's and with Marrison's subharmonic producers. Both kinds show 'automatic synchronization', where the system takes the frequency of an external impressed force or voltage. Yet their ability to synchronize to subharmonics of the impressed frequency was established only empirically. Moreover, relaxation oscillators presented the phenomenon over wider frequency ranges and could obtain larger division ratios, which suggests its advantage for most applications.[49] Marrison's success to reach high ratios can be attributed to the feedback mechanism that he added to enforce the sinusoidal oscillations within the desired frequency.

Thus, van der Pol's concept reinforced the inference from the simultaneous invention of a few subharmonic producers that other solutions are also possible. Therefore, any particular solution lost much of its importance, and patents lost much of their value, even more so since Clapp's method, support by van der Pol's concept, employed a well-known and widely used device, the multivibrator. In these circumstances, AT&T could not control the quartz clock technics and their market. This was probably one reason that the corporation did not enter the new market. It did not sell quartz clocks.[50]

[49] Balthasar van der Pol, 'The Nonlinear Theory of Electric Oscillations', *IRE Proceedings* 22 (1934): 1051–86, .esp. 1080–1.

[50] In 1930 AT&T declined to sell a quartz clock to scientists from California to use in case of earthquakes. Instead it directed them to General Radio (on its clock see below). AT&T archives

Nevertheless, Marrison's project was a successful enterprise for AT&T. It furnished a highly precise and accurate system of frequency standards for the company's internal needs of coordinating the many signals of the corporation's network, required by its business strategy. Thus, it fulfilled the initial aim of the frequency standard project. Internal integration of the corporation's many communication channels was more valuable to the telecommunication giant than the small market for the quartz clock. That Marrison continued to dedicate a few years to improving the quartz clock indicates the satisfaction that his research gave the corporation.[51]

General Radio entered the niche market left vacant by the telecommunication monopoly. As mentioned, like the other makers of the early quartz clock, GR first constructed a clock for its own use, building on the expertise of Clapp at MIT's short-wave laboratory and Horton at AT&T. In 1930, however, less than two years after constructing its own device for calibrating its commercial wave meters, the company began marketing its 'syncro-clocks'. An exact clock seemed a reasonable extension of the company's products of precision instruments and components for radio and electronic users. Indeed, it advertised the clock as particularly useful for measuring short temporal intervals in the laboratory. Large laboratories like those of the NPL, PTR, and Bell, which were not interested in selling clocks, however, carried out consequent research on improving the accuracy of the quartz timekeepers, exceeding the accuracy of the pendulum clock during the 1930s. Like GR, which continued improving its clocks, these laboratories developed also portable (but still quite massive) quartz clocks, which they also sold to external customers.[52]

7.3 Concluding Remarks to Chapters 5 to 7

Like the automobile, the creation of the quartz clock required the invention of a system. At the centre of this system stood the piezoelectric resonator, obviously its most important component. Unthinkable without Cady's discovery

---

loc. 80 002 02 08, especially letters of A. Day to Arnold (17.5.1930) and of H. P. Charlsworth to F. B. Jewett (9.12.1930).

[51] Marrison's success in providing state-of-the-art clocks contributed also to the reputation of the scientific and technological research at Bell Labs. As it was regarded as an asset of AT&T's, it provided the corporation an additional reason to fund this research.

[52] Warren A. Marrison, 'The Evolution of the Quartz Crystal Clock', *Bell System Technical Journal* 27 (1948): 510–88, on 545–60; on GR see McGahey, 'Harnessing Nature's Timekeeper', 212–18.

that quartz crystals display a very sharp and stable electric resonance, the clock was still impossible without valve electronics. As a system, the quartz clock evolved from the scientific and technological research of many individuals on a few topics in quite different civilian and military institutional settings, in various countries and languages. Understanding its origins, like those of many other complex and influential technics, thus, requires an international and interinstitutional history.

Since the integration of its communication system, a strategic business aim of AT&T's, depended on highly accurate and stable frequency standards, the corporation invested extensively in their research and development, resulting in the first quartz frequency standard and clock (Ch. 6). Still, the development of this system benefited also from additional interests of the corporation that led to specific technical needs not only for primary standards but also for frequency control and for secondary standards. The need for controlling transmitter frequencies to enable national radio broadcasting was a central motivation for studying and improving piezoelectric oscillators and for devising methods for their manipulation. Multiplex telephony, a main interest of the corporation, provided an additional reason to devise methods for attaining exact lower frequencies from a quartz oscillator (in addition to their use for monitoring a frequency standard). For an efficient exploitation of the novel high radio frequencies in Bell's use, the laboratory developed secondary piezoelectric standards, useful beyond the practical range of the tuning-fork standards. The meeting of the needs for both standards and control was unique to Bell Labs, stemming from the size and universal character of Bell's telecommunication services. National laboratories like the NPL shared Bell's interest of coordinating, if not integrating, large and diverse communication networks, for which they saw a similar need in a highly accurate quartz frequency standard. Dye and his colleagues, however, did not share AT&T's interest in controlling frequencies, which encouraged the researchers at the corporation to explore the use of piezo-resonators also when their value as a primary standard was unclear. Thus, the unique combination of research on standards and control placed Bell researchers in an advantageous position to create the first quartz clock.

AT&T developed the new frequency standard to answer technical needs that followed its business interests as shaped by its monopolistic position within the American market. Government bodies, like the NPL, answered similar technical goals for the regulating interest of the state. In both cases, as it is often with standards, sociotechnical interests led to the establishment of the new frequency standard. Moreover, in this case technology presented a need not only for having an accepted and transferable measurement standard, but also for its high precision. To increase efficiency, electronic telecommunication required accuracy previously used only in scientific research. In this aspect, electronic technics differed from the earlier electric technics. While earlier telegraphy motivated much of the research on the standards of electric resistance, physicists strove for higher accuracy to answer scientific questions concerning the relations between electromagnetism and light.[53] Technology, rather than science, however, motivated the quest for high accuracy in frequency measurements. The new quartz clock was a by-product of this quest. Ironically, it was devised as a monitoring method for frequency because the pendulum clock was regarded as the most accurate standard, but in a few years the quartz clock exceeded the accuracy of the former timekeeper.

While AT&T and the NPL developed quartz clocks for coordinating radio and telephony systems, subharmonic producers that enabled the comparison of the standard with a clock, and thus the construction of a new clock, emerged also from other concerns and immediate aims of electronic telecommunication. Although the other inventors of subharmonic producers did not build a full clock system, the multiple inventions of such technics point to a general conflation of needs and resources. As Table 7.1 suggests, individual groups exploited their experience with diverse methods and different means in devising these inventions. Moreover, they followed distinct immediate aims, set for different goals, and achieved variant methods, which still enabled a similar result. The Italian naval laboratory shared AT&T's and NPL's interest in frequency standards for the radio range, but it seems that only the external impulse of the American BoS drove it to employ piezoelectricity for this aim. The BoS itself saw in the quartz oscillator a way of comparing different standards more than a basis for an independent standard. Clapp sought a simple and cheap method for determining radio frequencies by small laboratories. Koga was looking for a method of secret signalling. Van der Pol sought a better understanding of triode circuits, which would facilitate manipulating electric oscillations for various uses. After its first applications for telecommunication,

---

[53] Schaffer, 'Accurate Measurement Is an English Science'; Bruce J. Hunt, 'The Ohm Is Where the Art Is: British Telegraph Engineers and the Development of Electrical Standards', *Osiris* 9 (1994): 48–63. Gooday points out a pressure for higher stability, which allows for higher accuracy, from technology in the case of secondary standards of industrial use. These, however, were much less accurate than the primary laboratory standard (in this case of electric resistance). Graeme Gooday, *The Morals of Measurement: Accuracy, Irony, and Trust in Late Victorian Electrical Practice* (Cambridge, UK: Cambridge University Press, 2004), 60–1.

Table 7.1 A comparison of the inventions of subharmonic producers

Institute: researcher	Immediate aim	General goal	Earlier experience/ idea's origin	Means
Bell: Marrison	Precise frequency standard of wide range	Coordinating the Bell System	Tuning-fork standard, parallel work on PE, also as secondary standard	Electronic control with a feedback mechanism (control coil)
Livorno: Vecchiacchi & Vallauri	Measure quartz frequency, high-frequency standard	Frequency radio standards	Tuning-fork meter/ BoS's standards comparison	Empirical discovery that piezo-oscillator controls a coupled oscillator
NPL: Dye	Precise frequency standards for radio range	Regulating electro-magnetic communication	Tuning-fork standard, secondary PE standards, PE research	Multivibrator
MIT Short-Wave Lab: Clapp	Obtaining many frequencies from one piezo-oscillator	Developing inexpensive radio methods	Radio wave manipulation	Multivibrator
Tokyo Electrotech Inst: Koga	Dividing a wave frequency in two	Secret signalling	General use of electronics	Empirical discovery that one circuit controls another
Philips: van der Pol & van der Mark	Unclear if they had a specific one	Applications of vacuum tubes in communication	Theoretical study of tubes, experience of others, Eccles & Leyshon's method	Neon tube
Physics lab, London Women Med. School: Leyshon	Producing flashes of constant period	Physiological experiments	Eccles's work on tuning-fork	Neon tube

the researchers of GR developed the technics for laboratory measurement of short time intervals. Thus, the creation of methods for dividing the original resonator frequency was not a case of solving a common problem, as one may think of in the case of a 'classical' simultaneous invention, but of solving a few distinct though connected problems. A close scrutiny of other cases might show that this is often the case with 'simultaneous' invention.[54]

Telecommunication provided the immediate context and the general electronic tools for the work on frequency measurement. With extensive research into the triode and its oscillations for usage in radio and telephony, telecommunication stimulated the growth of a dense and expanding technological field. Exploring the properties and uses of oscillators, the field provided methods for measuring and comparing frequencies, such as the heterodyne method, and also tools for manipulating these frequencies, such as the multivibrator. Moreover, the interconnections between different electronic technics also allowed for the application of techniques developed for other questions about frequency measurement (e.g. the triode mechanism and the neon tube). The different, but still connected, electronic methods for producing subharmonics provided a similar level of stability. Yet, when it came to the degree of accuracy, the resources and specific goals of the different groups led to diverging results. High accuracy was essential for establishing a primary frequency standard, as required for the Bell System, and the British network. It was therefore sought and reached only by Marrison and Dye and their collaborators. From an early stage they realized that accuracy would be determined by the stability and uniformity of their basic resonator, which was first the tuning-fork and then the quartz crystal, which could not be improved solely by electronic means, even if the latter sufficed for devising frequency dividers. The material

---

[54]This confluence of similar methods that originated from different needs deserves further historiographical attention. The development of the practical incandescent lamp by Edison and Swan (and a few other inventors) can be seen as a classic case of simultaneous invention following a common goal of constructing an electric light for in-house use (even if also in this case, as probably in most, Edison conceived of a lamp durable at a higher voltage than Swan's, leading to differences in their designs). See Robert Friedel and Paul Israel, *Edison's Electric Light: Biography of an Invention* (New Brunswick, NJ: Rutgers University Press, 1988) on 90–1, 115–17. An example closer in subject, time, and kind to the subharmonic provider is the invention of methods for using the triode as an amplifier and oscillator by Lowenstein, Armsrong, and Meissner; see Sungook Hong, *Wireless: From Marconi's Black Box to the Audion* (Cambridge MA: MIT Press, 2001), 155–7, 181–9. On simultaneous invention see Shaul Katzir, 'Scientific Practice for Technology: Hermann Aron's Development of the Storage Battery', *History of Science* 51 (2013): 481–500.

and cognitive resources of both institutes provided the researchers means for refining the accuracy of the resonators, but in different ways, reflecting the different character of research in the industrial and the national laboratories.

That Horton's group enjoyed an advantageous early familiarity with Cady's findings about the piezoelectric resonator originated not only in the contingencies that made the relationships of the research branch with Cady especially strong, but also in the corporation's efforts to maintain a large scientific-technological network at the disposal of its researchers. The laboratories' own large staff and its rich material resources allowed its researchers to enjoy a large shared experience and to explore experimentally a sizeable number of their ideas. These ideas, however, were directed towards specific improvements of practical methods, rather than exploring unknown areas. The case where Marrison attained novel information about the phenomena, his finding of the exact direction of a zero thermal coefficient, was a by-product of his development of a temperature stable resonator. Also in this case he based his work on extant scientific knowledge of piezoelectricity and elasticity. Dye at the NPL, on the other hand, not only designed piezoelectric resonators for frequency standards, but also engaged in the study of their properties and suggested a way to understand them. Establishing standards and expanding knowledge seemed to be complementary aims of his government laboratory. Enjoying a special status at Philips, van der Pol explored the behaviour of triodes and mathematical equations related to them. While the particular practical bearings of these researchers were unclear, their connections to the technics in use by the corporation were evident, and they indeed led van der Pol to an unexpected practical subharmonic producer. On the other hand, the members of Horton's group at Bell had little freedom to pursue more general, in this sense scientific, questions that did not relate directly to their immediate technological aims. They were, however, well versed in scientific and engineering literature and stood to benefit from developments in these fields. In particular, their prior graduate scientific training facilitated their study and use of piezoelectricity. Moreover, a background in scientific study and research helped Marrison as well as Dye to transfer not only knowledge and methods but also an interest in high precision from physics to technology. This background encouraged them to attain an accuracy formerly known only in sciences such as astronomy and geodesics.[55]

---

[55] For earlier examples of transferring the ideal of exactitude from physics to technology see Kathryn M. Olesko, 'Precision, Tolerance, and Consensus: Local Cultures in German and British Resistance Standards', in Jed Z. Buchwald (ed.), *Scientific Credibility and Technical*

The historical literature on measuring standards often concentrates on the determination of fundamental standards that define a physical magnitude, and the controversies about the proper methods for measuring them. These issues, however, were not central in the development of the tuning-fork and quartz frequency and time standards. The quartz clock emerged from a project aimed at measuring and disseminating a known magnitude, which had an accepted reference in the astronomical clock. Perhaps the ability to refer to a fundamental accepted standard helped in transcending local and national boundaries and in evading potential problems of reliability and the construction of trust, prominent in other histories on measuring standards. In this case, methods travelled quite easily between different places and local cultures. Tuning-fork technics developed independently by French and English physicists for military use were easily implemented by the National Physical Laboratory, and crossed the Atlantic to the American BoS and Bell's industrial laboratory and back to the Italian Navy. The piezoelectric resonator, as a standard, moved from a university laboratory to industrial and national laboratories in the USA, Europe, and Japan. Moreover, its acceptance as a useful frequency standard did not seem to raise any significant dissent, even if some laboratories preferred the tuning-fork standard. Methods travelled relatively easily due to the common experience with electronics, which provided also the means for working with the piezoelectric resonator; to the portability and the availability of quartz specimens that could be used as a piezoelectric resonator, which was resilient to transportation hazards and to changing physical conditions (e.g. Earth's magnetic and gravitational fields); and lastly to the circulation of objects, and mutual visits. While historians have pointed out that in other cases the main role of such visits was to settle disagreements, here circulation was important for prompting the use of the piezoelectric technics rather than for reaching a consensus.[56]

Standards in 19th and Early 20th Century Germany and Britain (Dordrecht: Kluwer Academic Publishers, 1996) 117–56, espe. the discussion of Philipp Brix on p. 122. Shaul Katzir, 'Hermann Aron's Electricity Meters: Physics and Invention in Late Nineteenth-Century Germany', *Historical Studies in the Natural Sciences* 39 (2009): 444–81.

[56] For this historiographical position see for example the editorial essays in M. Norton Wise (ed.), *The Values of Precision* (Princeton, NJ: Princeton University Press, 1997), esp. 8–12, 226–30.

8

DEVELOPMENT OF NEW APPLICATIONS BEYOND SONAR AND FREQUENCY CONTROL

The successful employment of piezoelectricity in sonar ushered in many further applications of the hitherto hardly used phenomenon for various ends. The discovery of the abrupt changes in the electric properties of crystal near resonance and the invention of crystal frequency control, widely used by researchers related to telecommunication, further contributed to the spread of its application. Piezoelectricity thus became a field also applicable to various ends beyond its main applications. This short chapter surveys the endeavour of contriving technics based on piezoelectricity beyond submarine detection and frequency control. This endeavour was one of the three main enterprises in the fifth phase of research on the phenomenon, in the scheme suggested here, i.e. that of a field recognized as technically useful. Chapters 5–7 discuss one of the other enterprises of the fifth phase, namely research and development of frequency control that led to the quartz clock. Chapter 9 examines the further study of the phenomenon within physics. Since a detailed history of the applications discussed here would divert the narrative from this book's main thread, they receive only a brief exposition. This exposition aims at portraying the main ways at which researchers attempted to utilize piezoelectricity and suggests what led them to these attempts, regarding the general knowledge of its phenomena and acquaintance with its uses.

Crystal frequency control suggested a highly useful method more accurate and easier to use than extant alternatives, and in some implementations (as in direct control of high frequencies) without a viable alternative. It had an evident advantage over alternatives. Yet, beginning in 1917, engineers, scientists, and inventors began applying piezoelectricity to other ends in commercial devices and measuring instruments, also when the advantage of methods based on the phenomena was less clear. Sonar and crystal frequency control made piezoelectricity more familiar to scientists and engineers (the latter had rarely known it previously) and offered methods for its use, and thus inspired

Sonar to Quartz Clock. Shaul Katzir, Oxford University Press. © Oxford University Press (2023).
DOI: 10.1093/oso/9780198878735.003.0009

the development of other applications. Furthermore, piezoelectric resonance suggested a new effect that could be technically utilized. Researchers thus considered the phenomenon when they looked for ways to solve technical difficulties or to improve on available technics, appreciating knowledge and techniques acquired in the earlier phases of the study of piezoelectricity. In market technics they employed dynamic effects, i.e. piezoelectric vibrations and resonance. Although researchers could not avail themselves of the unique properties of piezoelectric resonance before 1919, they could have suggested using vibrating crystals, as Rutherford and Langevin had done, also before its wartime study. Moreover, in post-1917 measuring instruments they employed also the static effect, which was well known but had been hardly applied previously.

Probably the economically most important technics was frequency filters, which exploited the unique properties of piezoelectric resonance. As mentioned above (Section 4.1), Cady's first method for manifesting the stable resonance frequency of quartz was a kind of filter between two circuits. Yet, Cady aimed at a frequency standard rather than at a filter, and presented his invention in this way. Moreover, for the practical use of a filter, his suggestion suffered from a severe disadvantage: it allowed only for a very narrow band of frequencies to pass through. Filters had been used in electric communication to eliminate interferences with higher frequencies (harmonics) and to separate between messages sent at different frequencies along the same wire in telephony, in a system of multiplex telephony first used by AT&T in 1918. This kind of employment required that filters allow for the passage of a relatively wide band of frequencies. With advancement in the knowledge of piezoelectric resonance and the exceeding needs for precision in telecommunication, engineers developed ways to use a piezoelectric resonator for a wider band of frequencies. In the same year, 1927, Lloyd Espenschied from AT&T and Clarence Hansell from RCA filed patents for somewhat different methods for reaching this goal.

Espenschied had been working on carrier telephony system since the 1910s. This system relied on filters consisting of capacitors and coils in a manner coined 'lattice arrangement' (also referred to as bridge circuits). It became evident, however, that due to inherent resistance of the coils such a filter could not provide a sharp 'cut-off' frequency that would allow the passage of virtually all the oscillations below (or above) a particular frequency and prevent all oscillations above (or below) the same frequency. The piezoelectric resonator, well known to Espenschied from his work at Western Electric's research branch, suggested to him a way to replace the coil and capacitor (or one of them), and to allow a clear cut-off due to the sharpness of the effect.[1] The same properties of the quartz crystal inspired Hansell, who worked on frequency control at RCA laboratory for its system of international communication based on wireless short waves. From his patent application, it seems that he was interested in the means for modulating high-frequency waves used in the system, e.g. 'to utilize only one of the two side bands', which suggested a more economical transmission of signals. At these frequencies 'the percentage difference between the two side bands is exceedingly small, and separation becomes difficult'. The piezo-resonator provided a means for a highly selective filter. Actually, both inventors realized that it was too selective for their purposes and suggested means to make the filter wider: by employing a few resonators of nearby resonance frequencies, and by adding capacitors or coils to the circuit.[2] Yet, for practical uses in carrier telephony the wave band allowed to pass through the filter was still too narrow. Consequently, researchers at Bell Labs continued developing new methods, building on coil–capacitor methods and on Espenschied's and Hansell's ideas. A circuit invented by W. P. Mason in 1929 became the basis of the first piezoelectricity-based filter used by AT&T.[3]

The simultaneous invention of the crystal filter by Espenschied and Hansell points at the importance of their shared resources, namely their expertise with piezoelectric resonance. It also indicates similarities in their aims, although these were not the same. Filtering had a few distinct applications within the evolving electronic telecommunication, due to the various technics of frequency manipulation, themselves becoming increasingly useful with the fast advancement in electronics and short wave. As seen in Chapter 7, regarding the case of frequency dividers required for the construction of the quartz

---

[1] In this replacement, Espenschied followed (knowingly or not) Cady's use of a piezo-resonator instead of capacitor and coil in the crystal wave-meter. For the history of the piezoelectric filters see McGahey, 'Harnessing Nature's Timekeeper', 2009, 192–209. Lloyd Espenschied, 'Electrical Wave Filter', US patent 1,795,204 (A), filed 3 January 1927, issued 3 March 1931; J. Tebo, 'Lloyd Espenschied—Radio Imagineer', *Communications Society* 10, no. 4 (September 1973): 3–6; J. E. Brittain, 'Electrical Engineering Hall of Fame: Lloyd Espenschied', *Proceedings of the IEEE* 95 (2007): 2259–62.

[2] Clarence W. Hansell, 'Filter', US patent 2,005,083 (A), filed 7 July 1927, and issued 18 June 1935, 1; Kenyon Kilbon, 'Pioneering in Electronics: A Short History of the Origins and Growth of RCA Laboratories, Radio Corporation of America, 1919 to 1964' (1964), esp. 21; R. G. Kinsman, 'A History of Crystal Filters', in Proceedings of the 1998 IEEE International Frequency Control Symposium (Cat. No.98CH36165), 1998, 563–70.

[3] Buckley, 'The Evolution of the Crystal Wave Filter'.

clock, such a process in which different particular goals in the same field led to simultaneous invention is not rare as one might think.

Sonar had stimulated attempts to further exploit piezoelectricity already before the invention of frequency control. As mentioned, already during the war Nicolson had developed techniques used in submarine detection for telephone microphones and speakers. Electroacoustic continued to attract researchers during the interwar period, suggesting innovative electromechanical transducers for acoustic waves, to fit various particular uses.[4] In this case the effect was employed to answer an established technical problem, rather than to solve a new one, as was the case with sonar, and at least partly with frequency control. In other words, inventors took up the effect to suggest alternative solutions to non-critical problems. Arguably, earlier in solving such non-critical problems they had not turned to a previously unapplied phenomenon, like piezoelectricity, since the risk and effort were too high. They took it up only after it had been successfully employed to address critical technical problems.

Still, in other areas of application inventors attempted to exploit previously unexploited properties of piezoelectric crystals. For example, beginning in 1925, working in France Edgar-Pierre Tawil relied on the changes in the refraction of polarized light in vibrating crystals to suggest optical relays to be used for television. In resonance, the crystal becomes transparent and could allow the passage of light to the screen. It could also be used to deflect or refract a beam of light according to its electric state.[5] Due to the novelty of his application, Tawil embarked on research about piezoelectricity and its influence on optical properties of quartz crystals, which is discussed in Chapter 9. Eight years later, Alexis Guerbilsky, an engineer and inventor working in Paris, proposed another kind of piezoelectric-based electro-optical relay (this one using plates of non-uniform thickness) to be used in sound recording and for microphones.[6] A year earlier the Berliner engineering Dr. Johannes Gruetzmacher, to take another example, invented a different method for a relay useful for wireless directional control, e.g. for

---

[4]Cady's textbook mentions about 25 different papers on the subject, without a claim for completeness. Cady, *Piezoelectricity*, 690–1.

[5]Tawil, *Comptes rendus 183* (1926): 1099; Pierre Edgar Tawil, 'Procédé et dispositifs pour moduler l'intensité d'une source lumineuse', FR601732, filed 7 August 1925, *L'imprtial* 19/7/27, p. 1

[6]*Comptes rendus 196* (1933): 1871–3. Following a colleague's suggestion, he explained how the same crystal plates could be used also as dynamometer, *Comptes rendus 197* (1933): 399–401.

airplanes (he utilized the effect of a crystal on an air gap).[7] While I could not find how Tawil became acquainted with piezoelectricity, it is clear that Guerbilsky and Gruetzmacher acquired considerable experience with piezo-resonators before attempting their application for relays. Gruetzmacher had filed a few patents on devices based on crystal frequency control. Guerbilsky worked together with Chilowsky on methods of television. Guerbilsky and Chilowsky did not adopt Tawil's piezoelectric relays, but employed piezoelec-tric resonators to respond to particular frequencies. Although Chilowsky did not collaborate with Langevin on piezoelectric transducers, he obviously fol-lowed the development of the method, which was based on a patent that he shared with Langevin. By 1937, Guerbilsky established a direct connection with Langevin, working in his laboratory on a theory of plates of nonuniform thickness like those that he had used in his invention.[8] Personal connections seem to have played an important role in the transmission of knowledge also in this case.

Late in the interwar period, William Spencer of the British Electrical and Musical Industries (EMI) utilized piezoelectricity in delay lines, employing piezoelectric transducers to convert electromagnetic waves into supersonic waves and back. Due to the lower velocity of the elastic waves, which were reflected by special means, the signal was delayed. Delay networks had al-ready been used in systems of high-frequency signalling, 'for example when the signals are passed through separate channels, the outputs of which are subsequently combined and the time of transmission through one channel is different from that in the other.' At the late 1930s such systems became important also for radar.[9] The use of piezoelectricity as a transducer of electro-magnetic into ultrasonic waves was quite common practice at the time. Beyond its use in sonar it was employed in the research on ultrasonics both within and outside the scientific laboratory, where the mechanical power of ultrasound

[7]Johannes Gruetzmacher, 'Steuerungseinrichtung, insbesondere Wechselstromrelais', DE613413 (C), filed 7 April 1932, and issued 16 April 1936.

[8]Constantin Chilowsky and Alexis Guerbilsky, 'Method and Apparatus for Television', FR644240 (A), issued 4 October 1928; Alexis Guerbilsky, 'Lames piézo-èlectriques d'épaisseur non uniforme', *Journal de Physique et le Radium* 8 (1937): 165–8.

[9]Percival William Spencer, 'Delay Device for Use in Transmission of Oscillations', US patent 2,263,902 (A), filed 8 February 1938 (in UK), and issued 25 November 1941, 1; Walter Kaiser, 'What Drives Innovation in Technology?', *History of Technology* 21 (1999): 107–23.

was employed.[10] I will return to the use of piezoelectricity for experiments on ultrasonic phenomena in Chapter 9.

The phenomenon found more diverse uses in measuring instruments and methods. A few reasons led to its more common employment in this field. Researchers (both scientists and engineers) were more willing to work with delicate and complex apparatuses that need constant care (and thus unacceptable in technics of wider use), which was often the case with technics based on piezoelectricity. Their need for methods that would allow them to measure physical properties that allude most other instruments due to their magnitude (large or small) or conditions under which they should be measured and their strive for highly sensitive and reliable measuring devices recommended the phenomenon to them. Lastly, that scientists were better acquainted with piezoelectricity (compared to engineers) made them more ready to examine its usage. The construction of measurement instruments began shortly after the discovery of the effect, as mentioned in Chapter 1. It was also the first kind of application that Cady suggested for the sharp electric piezoelectric resonance. With further encounters and publicity of the phenomena it found many more uses in scientific and technical laboratories and field work.

Piezoelectric crystals were employed to measure very high and instantaneous pressures of explosions and in many methods for recording pressure, acceleration, and vibrations, including in seismographs; they were employed to measure elastic moduli and electric capacity of various materials, very small distances, and magnetostriction, and were used to scan in micrometers, and for chronometers (also beyond the use of crystal frequency control for the quartz clock) and thermometers, a device aimed also at implementation in non-laboratory devices. In some cases, the piezoelectric crystal was embedded within more or less robust instruments whose user did not need to master the effect (e.g. seismographs). In other cases, it was part of a more fragile experimental setting (e.g. measurement of magnetostriction).[11] Measurements of pressure and acceleration usually relied on the static effect, but most other measuring technics employed the dynamic effect. Scientists and

---

[10] There were also suggestions for using the power of ultrasonic waves for other aims, for example to produce emulsions, sols, and powders from metals (by B. Claus in 1935).

[11] This list of uses was drawn mainly from publications mentioned in *Science Abstracts* from 1920 to 1939 in its section on physics (thus, this part refers mostly to laboratory uses), and from Cady, *Piezoelectricity*, 667–98. Cady provides a list of 62 papers published during the period on 'piezoelectric transducers for the measurement of pressures, accelerations and vibrations'.

engineers found in the effect a promising source for technical solutions to extant problems and for improvement on known technics.

The spread of applications of the phenomenon due to its earlier employment is nicely exemplified by its use to measure high-pressure explosions, especially inside guns already during the war. The idea probably occurred independently to J. J. Thomson in Britain, and to Gordon F. Hull and Oscar Wood in the USA. Apparently, the research and development on both sides of the Atlantic began only towards the end of the war. Although the technics relied on the static effect, and so in principal could have been conceived at the beginning of the war, the three scientists suggested it only after the piezo-electric sonar had become known among researchers mobilized to the war effort. Thomson, as mentioned, was a member of the BIR central committee. Hull and Wood worked on ballistics, as active researchers in Washington, DC, the headquarters of the NRC with connections to other scientists working in the area; it is unlikely that they were ignorant of the research on submarine detection.[12] This application led to research on the linearity of the piezoelectric effect in previously unexamined high pressures, an example of a scientific study stimulated by interest in improving technics. It is, thus, discussed in Chapter 9, within the analysis of the changes in the field's scientific study.

In sum, developers of diverse technics began considering piezoelectric phenomena for solutions to their problems due to the combination of two main causes: improved acquaintance with the effect and its richer repertoire of phenomena that could be technically useful. The latter included methods already in use in other technics (like piezoelectric vibrations) and new phenomena discovered after its early application, such as piezoelectric resonance.

[12] A. B. Wood, 'Admiralty Experimental Station, Shandon (Gareloch) Dunbartonshire', *Journal of the Royal Naval Scientific Service* 20 (July 1965): 42–78, esp. 51–2. J. C. Karcher, 'A Piezoelectric Method for the Instantaneous Measurement of High Pressures', *Scientific Papers of the Bureau of Standards* 445 (1922): 257–64.

9

THE TRANSFORMATION OF A RESEARCH FIELD

Crystal frequency control and quartz clocks, sonar, and other practical applications of piezoelectricity, discussed in the previous chapters, transformed the research in the field.[1] The volume of its empirical and theoretical studies increased a few fold from the pre-applied phase. Its study attracted a growing number of scientists, who mostly examined new or previously neglected topics. The consequent, extensive disciplinary research is one the three enterprises that characterized the fifth phase of research and development of piezoelectricity in the schematic sequence sketched in the Introduction to this book. This transformation makes piezoelectricity an advantageous case for examining the change in disciplinary study of a phenomenon due to its successful technical application. That the phenomenon attracted few researchers in its pre-applied phase (Ch. 1), that the volume of its interwar research still allows for a comprehensive survey, and that it attracted strong societal interests due to frequency control technics make it particularly apt for studying the way technical relevance shapes disciplinary research.

Technology affected research in physics through two main routes. The more willingly acknowledged and easily perceived way was through improved and new laboratory technics, i.e. instruments and methods, which transformed

---

[1] The present chapter reviews dozens of scientific studies by a corresponding number of scientists. In following the scientific literature, I have relied on the abstracts of papers published in *Science Abstracts* from 1920 to 1939. The mentions of works without additional references are based on these abstracts. I quote the full papers or other contemporary sources only when they contributed considerably to my knowledge of the original work. In studying the careers of the scientists involved I have relied on the rich database of the World Biographical Information System (WBIS) Online (http://db.saur.de/WBIS/) and *J. C. Poggendorffs biographisch-literarisches Handwörterbuch für Mathematik, Astronomie, Physik, Chemie und verwandte Wissenschaftsgebiete*, which is partly included in this database, and on the list of their publication in Thomson and Reuters' 'Web of Science' (http://apps.webofknowledge.com/), which is based on *Science Abstract*. As with the references to scientific work, I refer to other sources regarding the biographies only when they added to the information or claims that appear here.

Sonar to Quartz Clock. Shaul Katzir, Oxford University Press. © Oxford University Press (2023).
DOI: 10.1093/oso/9780198878735.003.0010

experimentation and led to new results in various fields. No less important was the subtler way by which the interests of users and developers of technics directed an increasing share of topical research that would plausibly help technology, understood here as the knowledge concerning technics. The interests of technics users, 'technological interests', led physicists to study questions relevant for improving technics useful for influential users, such as state agencies and corporations. This hidden process often evaded contemporaries and has not received due recognition in the historiography of interwar science.[2] The sharp transformation of piezoelectricity into an applied field during the First World War makes it especially useful for discerning this process in the interwar period, a central concern of this chapter. Although technological interests shaped research in physics to a further extent than in earlier periods, the phenomenon was far from new. Neither did it abate later. On the contrary, their influence has strengthened and seems today stronger than ever. It is thus important to understand the process by which aims of commercial, social, and military powers to improve technics have shaped research in science.

This chapter examines the research of piezoelectricity within the discipline of physics following its applications until the beginning of the Second World War. My main focus is on the effect of the new discoveries and applications on the kind of questions and manners by which physicists and other scientists studied the field. The chapter reviews the main topics of study and their relations to technics and technological questions, on the one hand, and to disciplinary concerns, on the other. Its central concern is in how, to what extent, and by which means the technical applicability of the field shaped the way scientists studied it, and which other factors shaped its research. I, therefore, examine the technological relevance of the various topics of study, and the connections between the questions studied by physicists and those useful for particular design. To explore the factors that drew scientists to study piezoelectricity and to the particular questions and methods they chose, I look at their

---

[2] Historians have claimed that postwar physics was shaped by societal interests in improving technics, especially those of the military, to an unprecedented degree (e.g. Paul Forman, 'Behind Quantum Electronics: National Security as Basis for Physical Research in the United States, 1940–1960', *Historical Studies in the Physical and Biological Sciences* 18 (1987): 149–229), and they continue to examine the way Cold War policy shaped research, e.g. Naomi Oreskes and John Krige, eds, *Science and Technology in the Global Cold War* (Cambridge, MA: MIT Press, 2014). For a short discussion of this historiography see Shaul Katzir, 'Introduction: Physics, Technology, and Technics during the Interwar Period', *Science in Context* 31 (2018): 251–61. For a similar claim about interwar geophysics see Anduaga, *Geophysics, Realism, and Industry*.

career paths and the connections between their disciplinary, institutional, and personal interests and their studies in the field.

To discern the effect of technological interest, I examine the main research topics of interwar piezoelectricity and analyse their connection to disciplinary questions and their relevance to the development of technics of societal interests. My concern is in the practice of researchers rather than in their rhetoric. Although the forces that affected research were general, the historian can track their effect and the mechanism by which they functioned only through their impact on particular individual scientists and research projects. It is through the details of the multilayered interactions between researchers and the various disciplinary, institutional, economic, social, and cultural forces that we can evaluate their ability to direct research in science. To this end, one needs to examine the content of the experimental and theoretical activities and the researchers' career paths, to connect the kind of research they carried out with their institutional settings. For this aim, I examine here dozens of scientific studies, previously neglected in the historical literature, and a corresponding number of biographical sources regarding the scientists who were involved. Viewing their earlier research and intellectual and institutional connections I assess the main factors that led individual scientists to their choice of research topics, and thereby the factors that directed research in the field at large.

It would, of course, be beyond the historian's ability to identify the specific reasons that led each individual to each particular topic. Still, their common paths can indicate the main factors that directed their research. Moreover, the examination of individual paths helps discern the various channels by which technological interests affected the field. Some of these channels were more concrete and easier to find (e.g. a connection to a corporation with an interest in related devices); some were more elusive. The knowledge that the field might be commercially useful attracted researchers to it, also when they did not have direct ties with organizations with stakes in the technics. Through the examination of these particular channels in the case of piezoelectricity, this chapter identifies forces and processes that directed research toward topics of interests of technics users in also other scientific fields.

That technological interests directed physics, however, did not make it a part of engineering or an indissociable 'technoscience'. The research at the focus of the current examination was regarded as physics, rather than engineering or technology. It was scientific research distinguished from, although connected with, technological research or a 'utilitarian regime'. As discussed in this book's Introduction, the latter aims at providing knowledge useful for designing and using technics, the former at general knowledge of the phenomena beyond its use in particular technics. For example, research into variations in the value of the piezoelectric coefficient of quartz with temperature was part of physics, while research into particular cuts whose resonance frequency would be stable under changes of temperature was engineering. As explained earlier, I regard the separation between the two as instructive for our understanding of the different kinds of research, work, publications, and careers of practitioners, and of the interactions between science, technology, and technics. That the borderline between the two kinds of research was sometimes blurred does not negate the distinction between them and its significance for the practice of the researchers.

To establish my claim that technological interests shaped physics and not only engineering, I examine studies in the former discipline, using two criteria for classification as either physics or engineering: first, by observing the way contemporaries perceived the different endeavours, e.g. in the journal *Science Abstracts*, which separated physics from engineering in the reports (a classification not necessarily identical to the way a particular author thought about his work), and second by examining the content of the research according to the above-mentioned thematic lines. The two criteria usually led to the same conclusions. When they differed, I followed my thematic criterion with a more thorough examination. Note that according to this distinction both basic and applied research are considered part of science distinct from engineering.[3] Neither contemporaries nor I classified individual studies by authors' institutional affiliation. Institutional setting affected the kind of research carried out in the laboratories, in ways examined in this chapter. Yet, individual researchers at state and industrial laboratories and at engineering schools carried out scientific studies, often along engineering research, as did some members of academic departments of physics. Within the realm of physics examined here, technological interests directed research to questions pertinent for technical development that were still legitimate from a disciplinary logic. Otherwise, the studies were regarded as external to physics. Thereby, the interests of technics users shaped the discipline and changed physics not in overt contradiction to its logic but by reordering its priorities.

[3] After it became a basis for useful devices, almost any study of piezoelectricity could have been connected to technical improvements, and in this sense, it was not 'pure research', for its own sake.

9.1 From Pure to Applicable Research: A Quantitative Effect

As expected, most attention following the technical employment of piezoelectricity was directed to its practical uses. Yet scientists and engineers not only exploited it, but also studied its properties at various levels. Their interest in the effect manifested in the sharp rise in the number of related publications. A 1928 bibliography of piezoelectricity records 208 items for the decade 1919–28, compared to less than 30 for each of the previous decades.[4] About half of the items in the list examined general properties of piezoelectric crystals and resonators. The other half dealt with piezoelectric instruments and applications.[5] General digests of physics show a similar picture. The British journal *Science Abstracts* aimed at covering all publications in physics that appeared in major European languages journals. Papers in the discipline whose *main* subject was piezoelectricity numbered only a handful before 1927, which was still higher than their number before the war. The number of papers rose sharply to 32 in the 1932 volume and to an interwar record of 49 in 1935 (see Figure 9.1), much faster than the general rise in papers in physics.[6] These papers were written by more than 150 scientists. Scientists working in Germany and the USA were dominant among them (about 45 from each), but there were also quite a few from the USSR, France, and Japan and a handful from Britain and other countries. Table 9.1 presents the main centres of research and topics studied in them.

Moreover, the content of research in the field suggests that it was shaped by those interested in improving technics based on piezoelectricity. At the centre of the research stood the vibrator and resonator, which were essential for sonar and frequency control. This was an entirely new research topic. As mentioned, before WWI scientists had examined piezoelectricity only under static or semi-static conditions. They began to study vibrations in the third phase

[4]For the years 1880 to1899, I identified 59 items on piezoelectricity: Shaul Katzir, *The Beginnings of Piezoelectricity: A Study in Mundane Physics* (Dordrecht: Springer, 2006), 6. Since interest in the field declined, the number for the first two decades of the twentieth century was probably lower. The ISI Web of Knowledge citation index gives 97 results for 'piezoelectr*' in the topic for the years 1918–28, 475 for the years 1929–39 (these include also articles in engineering), while only 11 for 1904–14. Physics in general also grew during these years but much slower. The topic 'physics' had 564 hits for the year 1910, compared with 880 for 1924 and 929 for 1935.

[5]Walter G. Cady, 'Bibliography of Piezo-Electricity', *IRE Proceedings* 15 (1928): 521–35.

[6]These reports appeared in *Science Abstracts* 'Section A: Physics'. The journal reported on papers a few months after their appearance, so the numbers of publications in each volume do not coincide exactly with the number of publication in that year.

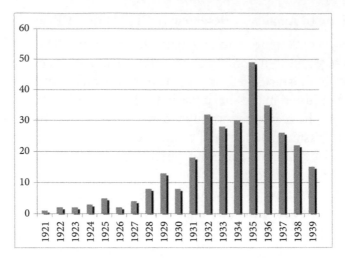

Figure 9.1 Annual number of papers on piezoelectricity, as mentioned in the volumes of *Science Abstracts*, Section A: Physics.

of research, in the scheme suggested here, namely, after Langevin had suggested using vibrating crystals as transducers for sonar (Ch. 3). This study led Cady to discover piezoelectric resonance and consequently its application for frequency standards and control in the fourth phase of research. Thus, their study became potentially useful for technical developments in the fifth phase. In particular, the research provided guidance for producing resonators more suitable for the needs of wireless communication and telephony. Yet, as a novel phenomenon, piezoelectric resonance also posed puzzles for physics, questions worth answering according to the disciplinary logic, like how do they behave under various physical conditions? How can the abrupt electric resonance be accounted for by knowledge and theories of static piezoelectricity and elasticity? What is the mechanism that leads to the phenomenon and how does it fit the known structure of crystals? These kinds of questions provided an additional, scientific, reason to study resonance.

9.2 Researchers and Themes in the Early Studies of the Resonator

While most of the scientists engaged in the research and development of sonar left the study of the vibrators with the termination of their work on the project, Walter Cady embarked on an extensive study of the piezoelectric resonator with the help of his students at Wesleyan University. He examined the resonator's effect on the electric circuit in the laboratory, and explored ways to account for the observations in terms of electric and elastic theories. By early 1921 Cady began publicizing his findings and inventions. In October of that year he sent an inclusive report about his scientific and technological work on the piezoelectric resonator. His long paper included a description of the fundamental phenomena of the piezoelectric resonator as observed by its effect on the electric circuit, a theoretical account of these findings, a discussion of the experimental settings needed for their observations, and their applications for frequency measurement and control. Cady's publications addressed the technical audience interested in applying the new findings, and scientists interested in extending and securing knowledge about these findings.[7]

In reporting about his finding, Cady described his laboratory setting of a regular RLC circuit to which he added a piezoelectric resonator (Chs. 3 and 4) and the abrupt change in the electric current in a circuit when the quartz crystal approaches resonance frequency. He explained the change by piezoelectric and elastic theory. The rise in the electric current, he inferred, followed the alteration in the electric displacement inside the crystal, itself caused by the piezoelectric reaction of the resonator. In resonance, the elastic elongation of the crystal ('strain' in a more technical term) is larger than that induced by converse piezoelectricity. In this case, and only then, the excess in elongation produces electric polarization and displacement inside the crystal in an amount determined by the direct piezoelectric effect. To account quantitatively for the phenomena, Cady relied on Voigt's phenomenological theory of piezoelectricity, on the mathematical theory of elasticity as presented by Horace Lamb, and on techniques from telephone engineering. This use of mathematical, physical, and engineering techniques reflects his interests as well as his resources.

Since this was the first theoretical account of a piezoelectric resonator, Cady chose the simple case of lengthwise vibrations of a thin quartz rod induced by piezoelectricity in the transverse direction (a strain in the y direction due to electric field in the x direction; see Figure 2.5), and employed an approximation of the exact elastic theory for that case. The use of this relatively simple case made it possible to reduce the complex elastic theory of crystalline vibration into a solvable expression. Yet, the case still required a few approximations, which I discuss below, and an extensive solution, which Cady and the

[7] Walter G. Cady, 'The Piezo-Electric Resonator', *IRE Proceedings* 10 (1922): 83–114; Walter G. Cady, 'Note on the Theory of Longitudinal Vibrations of Viscous Rods Having Internal Losses', *Physical Review* 15 (1920): 146–7; Walter G. Cady, 'The Piezo-Electric Resonator', *Physical Review* 17 (1921): 531.

editors deemed to warrant a separate publication.[8] The potentially eighteen equations of piezoelectricity (the five in the case of quartz) are reduced in the assumed case to one relation between the electricity in the x direction and the strain in the elongation in the y direction, with one coefficient. These simplifications notwithstanding, the resulting equation is still very complicated, so Cady resorted to a graphical method for its solution. He relied on a 'resonance circle' of the mechanical variables that he developed similarly to 'motional impedance', which A. E. Kennelly and others had developed to study the telephone receiver. The case of lengthwise vibrations under transverse electric voltage was not only a simple case from a theoretical and computational point of view but also the one used in Cady's experiments and in his applications of the resonator. Thus, the case suggested practical, empirical, and theoretical advantages. This combination fitted Cady's research on these three levels. It also characterizes much of the early research on the piezoelectric resonator beyond Wesleyan laboratory.

In the years that followed, a few scientists mainly from state and industrial laboratories joined Cady in studying the behaviour of piezoelectric crystals, while engineers and inventors developed methods for their employment. The first who joined the study were three of Cady's MA students (his university did not grant PhDs). W. F. Powers measured the changes in the frequency of a lengthwise vibrator, like that described by his professor, due to alteration in temperature. He published his results in 1924. Tadashi Fujimoto, Cady's former assistant, made the resonator the subject of his 1927 PhD dissertation at Ohio State University. He obtained the piezoelectric coefficients by empirical study of the resonator, applying and examining Cady's theory. Working in Cady's laboratory, J. R. Harrison excited resonance at frequencies considerably lower than expected by exploiting flexural vibrations.[9] Cady's former student and from 1921 colleague Karl S. Van Dyke[10] joined the research of his erstwhile professor with a 1925 theory of the resonator, which is discussed in the subsection 'Dynamic Theory and Equivalent Circuit Accounts

---

[8]Walter G. Cady, 'Theory of Longitudinal Vibrations of Viscous Rods', *Physical Review* 19 (1922): 1–6.

[9]From 1920 to 1921, Powers was Assistant Professor of Physics (Cady to Arnold, 30.4.1920, AT&T Archive, loc: 80 03 02 07). On early piezoelectric publication see Walter G. Cady, 'Bibliography of Piezo-Electricity', *IRE Proceedings* 15 (1928): 521–35. On Fujimoto see *Alumni Record of Wesleyan University, Middletown, Conn*, 5th edn (New Haven, CT: Tuttle, Morehouse & Taylor, 1921), 847.

[10]'Contributors', IRE Proceedings 30 (1942), 528.

of the Piezo-resonator'. As discussed in Chapter 4, many of the other early students and developers of the resonator also had encounters with Cady and his devices, through his efforts of spreading the high-frequency piezoelectric technics.

With a single exception, the first who embarked on scientific study of the resonator shared Cady's concern in its practical usage for telecommunication, although not all the researchers were directly engaged in improving technics. As mentioned in Chapter 7, David Dye, at the NPL, began experimenting with piezoelectric resonators for frequency measurements methods following Cady's visit in 1923. In addition to research and development on piezoelectric technics, Dye embarked, like Cady, on an empirical and theoretical study of the resonator. The results of this study had some bearings on Dye's design of the resonators for frequency standards. Yet, his research went beyond his direct technical needs; it was directed at gathering general knowledge about the resonator and at a theoretical account of the observed phenomena. The English researcher expanded the realm examined by Cady by achieving a larger spectrum of frequencies (with different quartz plates), allowing for an air gap between the quartz and the electrodes, changing the settings of the electronic circuit, and performing the experiment at different temperatures.[11]

The use of the piezoelectric resonator for frequency standards also attracted the attention of Erich Giebe and Adolf Scheibe, Dye's colleagues at the Physikalisch-Technische Reichsanstalt (PTR), at the laboratory for alternating current standards (*Noramlen*) and high frequencies in the electricity and magnetism department. In 1925 they invented a method for displaying the resonance frequency of piezoelectric crystals useful as a frequency standard. Their 'luminous crystals' produced a luminous discharge in a container filled with rarefied gas due to secondary electric tension induced by the piezoelectric effect at resonance. In contriving the technics, Giebe and Scheibe fulfilled one of the missions of their institute, namely providing some means for exact laboratory measurements. They continued with a further institutional mission, that of expanding knowledge relevant to practical technics, with an extensive study of resonance phenomena in quartz. In particular, they examined different modes of elastic vibrations, due to different cuts, and various crystallographic directions of the electric field (for example, by placing the electrodes

---

[11]D. W. Dye, 'The Piezo-Electric Quartz Resonator and Its Equivalent Electrical Circuit', *Proceedings of the Physical Society of London* 38 (1926): 399–458. His theory is discussed in the subsection 'Dynamic Theory and Equivalent Circuit Accounts of the Piezo-resonator'. See also the discussion of Dye in Chapter 7.

at various locations on the crystal). They also employed inhomogeneous electric fields. Like most of the early students of the resonator, they advanced a theoretical explanation for their findings, relying directly on Voigt's phenomenological theory and on the theory of elasticity. Their results extended the knowledge about the properties of piezoelectric resonance and its relations to the structure of crystals. At the same time these results, which were published in a physics journal, also suggested practical rules for obtaining resonance at frequencies higher and lower than those previously attained.[12] They further made their crystals illuminate at these resonance frequencies useful for telecommunication. They described these aspects in journals for technical physics and engineering.[13] This combination of scientific and technological benefits fitted the aims of the PTR.[14]

The resonator drew also the attention of their colleague Alexander Meissner at Telefunken, the German major radio technology company. A prominent researcher, Meissner had been carrying out scientific and technological research and development at the company since he had joined it in 1907, two years before receiving a PhD in electrical engineering. Famous among his many achievements was the co-invention of the amplifying triode circuit (Figure 2.2). Meissner first employed piezoelectric vibrators for a new microphone in 1923, in a manner resembling Nicolson's invention (Section 4.3). Yet, it seems that he and his colleagues began research on the resonator only in 1925, probably with the progress of radio technics and the resulting interest of Telefunken in frequency standards and control (sharing the interest of corporations discussed in Chapters 4–7). Their research led both to patents and to journal articles. Meissner and his colleagues experimented with various circuits for detecting resonance and for frequency control, and studied physical influences on the frequency and characteristics of the resonance. For example, Kurt Heegner examined how mechanical and electrical condition can dampen

[12] E. Giebe and A. Scheibe, 'Sichtbarmachung von hochfrequenten Longitudinal schwingungen piezoelektrischer Kristallstäbe', *Zeitschrift für Physik* 33 (1925): 335–44; E. Giebe and A. Scheibe, 'Piezoelektrische Erregung von Dehnungs-, Biegungs- und Drillungsschwingungen bei Quarzstäben', *Zeitschrift für Physik* 46 (1928): 607–52.

[13] E. Giebe, 'Leuchtende piezoelektrische Resonatoren als Hochfrequenznormale', *Zeitschrift für technische Physik* 7 (1926): 235; Giebe and Scheibe, 'Piezoelektrische Kristalle als Frequenznormale', *Elektrische Nachrichten-Technik* 5 (1928): 65–82.

[14] David Cahan, *An Institute for an Empire: The Physikalisch-Technische Reichsanstalt, 1871–1918* (Cambridge, UK: Cambridge University Press, 1989); Ulrich Kern, *Forschung und Präzisionsmessung. Die Physikalisch-Technische Reichsanstaltzwischen 1918 und 1948* (Weinheim: VCH, 1994).

the resonance curve, i.e. making the crystal resonate at a wider spectrum of wavelengths albeit with a weaker electric effect.[15] Damping was useful in order to have devices more tolerant to common deviations of contemporary instruments, and the subject of a patent.

Meissner extended the research to properties of vibrating quartz crystals beyond those directly pertinent to their applications. An observation of an unexpected mechanical rotation of a quartz plate 'cut in a way not usually used for oscillations' led him into a new line of study. He realized that the rotation originated in the production of an asymmetric air blast of uncommon strength by a crystal vibrating in the direction of the optical axis. Usually the motion of a vibrator produces sound waves, but due to its strength and dimensions, Meissner explained, the plate generated an air current, a phenomenon known from earlier studies in acoustics. He attributed its origins to the same properties responsible for the rotation of polarized light along a quartz's optical axis. To demonstrate the effect Meissner built a small crystal motor, which rotates when the quartz is excited by an external high-frequency source. This motor had an instructive rather than a practical use; still it shows Meissner's interest in producing practical devices from his findings, a common trait among industrial researchers.[16] Moreover, he patented a method for detecting resonance by the violent air currents it produces, which was useful in principle, although of no real practical value.[17]

In 1927 Edgar-Pierre Tawil reported on a similar observation of air blast in a study that followed his research on the effect of vibrations on the optical properties of resonating quartz bars. An independent French researcher with

---

[15] Heegner had earned a PhD in 'technical physics' in 1920. Meissner referred a few times to Heegner's unpublished results, and the latter assigned patents to Telefunken from which he received substantial royalties during WWII, but he probably was not a regular employee of the company. Samuel Patterson, 'Kurt Heegner—Biographical Notes', Mathematisches Forschungsinstitut Oberwolfach, Report no. 24 (2008): 1354–6.

[16] Reich, 'Irving Langmuir and the Pursuit of Science and Technology in the Corporate Environment'.

[17] Between 1923 and 1929, Meissner filed about ten different patents related to piezoelectricity (see Espacenet database) among them: 'Arrangement for converting acoustic energy into electrical energy', US1633186, filed: 19240529; 'Wave measurement', US1779259 filed 19261120; 'Indicating means for high-frequency oscillations', US1783297, filed 19270311; Alexander Meissner, 'Über piezo-elektrische Kristalle bei Hochfrequenz', Zeitschrift für Technische Physik 7 (1926): 585–92; Ingrid Ahrens, 'Meißner, Alexander', in Neue Deutsche Biographie, vol. 16)1990): 695–97; Paul Vigoureux, Quartz Resonators and Oscillators (London: H.M. Stationery Office, 1931), 196–97; Cady, Piezoelectricity, 440–41.

his own means probably of Syrian origins, Tawil was attracted to the possible use of piezoelectric quartz resonators to modulate light for the nascent television technics. He began physics research in 1925 with a related study of double refraction in quartz resonators. Soon he extended his empirical study to other topics of piezoelectricity, including vibrating crystals and the electric effect of torsion. Like Meissner, Tawil combined scientific research of the phenomenon with practical suggestions for its use, publishing his findings in scientific journals and filing patents on his inventions. Unlike Meissner, who contributed to a wide array of fields, Tawil concentrated his intellectual efforts on piezoelectricity. He received the 1931 Henri Becquerel Prize of the French Academy of Science for his findings in the field.[18]

Beyond the disciplinary interest in examining the new effect as such, Meissner saw an opportunity to probe the structure of quartz. He identified lines in the quartz crystals that produce stronger sound (the source of the air blast), and assumed that they form surfaces of the greatest molecular density. Thus, he inferred from the macroscopic sound production the crystalline atomic structure. On one hand, he employed this conclusion to suggest a more efficient crystal cut along these surfaces. On the other hand, he combined this insight with common assumptions about the structure of quartz to propose a speculative atomic structure of quartz and mechanism for the appearance of (static) piezoelectricity (his theory is discussed in the 9.4 subsection 'Piezoelectricity and the Structure of Matter').[19]

Taking for convenience 1927 as the end of the early study of piezoelectric vibrations, since the number of publications on the phenomenon expanded that year, two more researchers of the first phase should be mentioned. François Bedeau, a lecturer of physics at the Parisian Faculty of Science, advanced a synthetic theoretical exposition on the knowledge about the resonator (which he expanded in 1928). Like quite many physicists, Bedeau had a long interest in wireless communication on which he also taught, an interest that brought him also to the study of the piezoelectric resonator. Another physicist who reached

---

[18]Tawil published a dozen articles connected to piezoelectricity until 1936, and filed a similar number of patents during a longer period. Pierre Edgar Tawil, 'Procédé et dispositifs pour moduler l'intensité d'une source lumineuse', French patent FR601732, filed 7 August 1925. The British version of the patent states that Tawil was a French citizen. André de Wissant in *L'impartial* (19.7.1927), 1, describes Tawil as 'riche savant syrien'; on the prize 'Prix et subventions attribues en 1931', *Comptes rendus* 193 (1931): 1237–96, on 1286. Papers and patents are according to the above-mentioned databases.

[19]Alexander Meissner, 'Piezo-electric Oscillator', US patent, 1,875,087, filed 7 July 1928 (in Germany on 19 June 1927).

the study of the resonator from his interest in radio was Earle M. Terry, a professor in Wisconsin, known for his research on wireless communication. Terry examined physical factors that affect the constancy of the resonance frequency, a study directly related to practical usage.[20] Like the other early students of the resonator, Bedeau and Terry sought preliminary data and empirical rules useful for practical technics and for scientific understanding of the phenomena.

9.3 Further Studies of the Resonator

Dynamic Theory and Equivalent Circuit Accounts of the Piezo-resonator

The study of the resonator can be divided analytically into theoretical and empirical examinations. As illustrated by the examples mentioned earlier, these kinds of research were closely related, often being carried out by the same individuals. Cady had examined ways to elucidate and account for electric resonance phenomena already in the process of analysing his early observations (Ch. 4), and advanced a theory in 1922. Major contributors to the early study of the field joined him and suggested more elaborate resonator theories. Van Dyke published his theory in 1925, Dye and van der Pol published theirs in 1926, and Bedeau in 1927 suggested a theory based on those of Van Dyke and Dye. The German theoretician Max von Laue proposed his theory in 1925. Among the early proponents of resonator theories, only Laue did not engage in a more general study of the effect and its technological employment. The others developed their theories in close connection to experimentation in their own laboratories.

The theories suggested to account for the resonance phenomena can be divided into two kinds. The first accounts for the behaviour of the resonator by applying piezoelectric, electric, and elastic theories to quartz crystals under high-frequency electric voltage and pressure. Since this kind of analysis considers mechanical and electric forces in the crystal, I've coined it 'dynamic theory'. The second kind modelled the quartz resonator by an electric system that behaves like the resonator under defined conditions not too far from resonance frequency. Those systems were often called equivalent circuits or networks, since the resonator was replaced by capacity (represented

---

[20]'Bedeau, Jean François', in *Dictionnaire biographique français* contemporain, 2nd edn (Paris: Pharos 1954–5); Earle M. Terry, 'Factors Affecting the Constancy of Quartz Piezo-Electric Oscillators', *Physical Review* 29 (1927): 366; L. R. Ingersoll, 'Earle Melvin Terry— 1879–1929', *Science* 69, no. 1797 (6 July 1929): 592.

by a condenser), induction (represented by a coil), and resistance (represented by a resistor) arranged in a particular order as elements of an electronic circuit, which produced an equivalent effect. Since the resonator was almost always inserted within an electronic circuit, one could analyse its behaviour by examining its equivalent electric system using the well-known tools of electricity. Probably first suggested by Hermann Helmholtz in 1853, and developed largely by scientists, equivalent circuits were very common among electrical engineers. Still their simplicity and the analogy that they suggested were also appreciated by physicists, as they made it possible to describe vibrating electromechanical systems in simpler purely electrical terms.[21]

Equivalent systems helped elucidate the behaviour near resonance and its dependence on external physical influences, since they offered a model of known components (coils and condensers), whose behaviour was well understood by contemporary physicists and engineers. It helped identify the main properties of the resonator taken as a whole, i.e. without considering its internal constitutes and structure, and also without tracing its behaviour to any underline mechanism. These properties made the theory more efficient and simple to use than the dynamic theory in the analysis and design of experiments and devices. Thus, often an equivalent network answered better the needs of technology, and of research designed directly to help future practical design. Yet, in some cases (like the production of resonators more stable under temperature variation, discussed below and in the next subsection 'Experimental study of the resonator'), equivalent systems suggested no help and the researchers employed dynamic theory. Dynamic theory also provided the required link between the magnitudes of the modelled electric components (i.e. the values of the resistance, capacitors, and coils) and those of the real quartz resonator (i.e. the values of the piezoelectric, dielectric, and elastic coefficients). In this sense, dynamic theories are basic to equivalent system theories.

Still the dynamic theories by themselves do not show that one can describe the electric behaviour of the resonators by an equivalent electric system

[21] D. H. Johnson, 'Origins of the Equivalent Circuit Concept: The Voltage-Source Equivalent', *Proceedings of the IEEE* 91 (2003): 636–40; Kline, *Steinmetz: Engineer and Socialist*, 112–13. Bromberg suggests that maser and laser engineers applied the method of equivalent circuits, which was not useful for many physicists, perhaps because of their lack of experience with its application. (Joan L. Bromberg, 'Engineering Knowledge in the Laser Field', *Technology and Culture* 27 (1986): 798–818). A systematic examination is necessary for learning when and in which fields physicists became as familiar with the method as were Cady and Van Dyke.

and certainly not what the structure of such a system should be. As told in Chapter 4, the answer to this question eluded Cady, who had failed to contrive a system that would model the resonator under all the conditions and at all the values relevant to the problem. Thus, in his 1922 publication he used equivalent values of the capacity and resistance (in two different arrangements) of the resonator, but their values depended on the external frequency. Probably due to this imperfect form, Cady did not explicate the use of these values as an equivalent network. Three years later, his colleague Van Dyke solved the problem and formulated an equivalent network that presented the electric behaviour of the resonator, without a need to change its values under changing electric conditions. He succeeded in that by adding self-inductance to the capacity and resistance used by Cady, although the resonator does not show any magnetic effect. Clearly, the model was not designed to present the process inside the crystal, but its effect on the circuit instead. Van Dyke's equivalent network consisted of resistance, induction, and capacity in series all in parallel to another capacitor (Figure 9.2), whose values are determined by the properties of the resonator.[22]

Simultaneously with Van Dyke, Dye at the NPL also devised an equivalent network. Such a network not only allowed theoretical predictions of the resonator's behaviour, but also helped to analyse the effect of changing frequency, and surrounding temperature, and the existence of air gaps of varied widths between the crystal and the electrodes. It was, thus, closely related to experimental practice. Dye employed the same electric components as Van Dyke had done but he first devised a different network (where capacity, resistance, and induction are in parallel to each other and in series

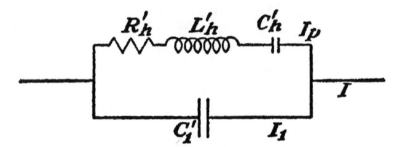

Figure 9.2 Electrical equivalent of a piezo resonator (without an airgap) near resonance frequency in the form suggested by Van Dyke. *Source*: Cady, *Piezoelectricity*, 298.

[22] Cady, *Piezoelectricity*, 333–5. Cady, 'The Piezo-Electric Resonator' (1922), 102–3.

to further capacitors). By early 1926 he independently reached the electric arrangement found by Van Dyke. Van der Pol in Philips was the third independent 'discoverer' of a useful equivalent network to model the behaviour of the piezo-resonator. His circuit was of a less general kind (lacking the capacity in parallel). Van der Pol developed the equivalent network to facilitate analysis of resonator circuits used for frequency control, which was of technical interest to his corporation. While van der Pol enjoyed relative freedom to carry out in-depth extensive scientific research on questions merely related to the interests of Philips (as discussed in Ch. 7), in this case he seems to have engaged in narrower utilitarian research on the properties of a technical system. Apparently, the equivalent network provided van der Pol with a way to analyse and understand the electric behaviour of the resonator in familiar terms. He dwelt neither on its derivation nor on its relation to the piezoelectric and elastic properties of quartz, which were important for Van Dyke and Dye. He was satisfied with the claim that the resonator's electric behaviour is like that of resistance, inductance, and capacity in series.[23]

In developing his model, Dye, who unlike van der Pol made quartz resonance the centre of his research, relied on a 1915 mathematical theorem by Stephan Butterworth. Butterworth showed that 'when set in motion by the interaction of charged bodies on an electrostatic field [a vibrating system of one degree of freedom] behaves as a series combination of inductance, resistance and capacity'. Butterworth also pointed at corresponding systems for higher degrees of freedom. A lecturer of physics with a gift for mathematics, Butterworth first developed an electric equivalent system for a specific problem (an electric vibration of a new kind of Galvanometer) in 1914. He then generalized his results for any vibrating electromechanical system. Dye probably knew Butterworth personally as the latter worked at the NPL from 1919 to 1921, and then moved to the nearby Admiralty Research Laboratory. Still, Butterworth's theorem was obscure enough that Dye learnt about it only after he had begun working on equivalent systems. Across the ocean, Cady and Van Dyke had not known the theorem before Dye introduced it into the research of piezoelectricity.[24] Apparently, in other fields, like electroacoustics,

---

[23] Balthasar Van der Pol, 'Het gebruik van piëzo-electrische kwarts-kristallen in de draadlooze telegrafie en telefonie', in *Gedenkboek ter herinnering aan het tienjarig bestaan van de Nederlandsche vereeniging voor radiotelegrafie, 1916–1926* (Zutphen: Nauta, 1926), 293–6; and an abstract in *Jahrbuch der drahtlosen Telegraphie und Telephonie*, 28 (1926): 194.

[24] Dye, 'The Piezo-Electric Quartz Resonator and Its Equivalent Electrical Circuit'; S. Butterworth, 'On Electrically-Maintained Vibrations', *Proceedings of the Physical Society of London*

the theorem was unknown, although equivalent circuits were useful.[25] Butterworth's theorem showed that one can contrive an electric equivalent system and its basic structure. Still one had to identify the specific properties of the electromechanical system with their equivalent components. Dye did that for the system studied by Van Dyke and extended it for the useful case of an air gap between the crystal and the electrodes.

Dye did not develop expressions for the coefficients of the equivalent system in terms of the piezoelectric constants (or coefficients). Instead, he relied directly on Butterworth's theorem to state the value of the equivalent coefficients in terms of a coefficient proportional to the piezoelectric constant. Van Dyke, on the other hand, expressed the equivalent coefficients in terms of the piezoelectric one, using the dynamic theory. Unlike the equivalent network, which fits all piezoelectric resonators (with proper values for the parameters), the dynamic theory is specific to the kind of vibration. A dynamic theory accounts for the electric and elastic properties of the resonating crystals by considering the piezoelectric and consequent elastic effects of the alternating electric voltage or (in rare cases) mechanical oscillations exercised on the crystal rod. Due to the complexity of the involved effects, the equations of the piezo-resonator depend nontrivially on the kind of crystal, the directions of its cut, the directions of the alternating electric field, and the dimensions of the plate. Each account of the resonance is based on the crystals' piezoelectric and elastic properties and, thus, is considered a specific family of cuts and mode of vibrations. In particular, theories were divided between those that explained lengthwise vibrations, i.e. mechanical oscillation along the length of a thin crystal, and those that accounted for the perpendicular thickness vibrations. As most treatments were of quartz, lengthwise vibrations were usually perpendicular to the direction of the electric field and parallel to the electrodes and thus free from external mechanical constraints.[26]

As mentioned earlier, since Cady's early resonators vibrated lengthwise, and as this is a simple case, he had first advanced a theory for this mode of vibration in 1922. In 1925, Van Dyke employed Cady's finding in identifying the

27 (1915): 410–24; S. Butterworth, 'On a Null Method of Testing Vibration Galvanometers', *Proceedings of the Physical Society of London* 26, no. 1 (1914): 264–73; A. B. Wood, 'Stephan Butterworth, OBE: An Appreciation', *Journal of the Royal Naval Scientific Service* 1 (1945): 96–8; about Dye's research during 1925 see *Report of the NPL for 1925*, 97.

[25] Roland Wittje, *The Age of Electroacoustics: Transforming Science and Sound* (Cambridge, MA: MIT Press, 2016), 20–1.

[26] Cady, *Piezoelectricity*, 284–332.

variables of the equivalent network. He elaborated the treatment and in 1928 suggested a dynamic deduction for both lengthwise and thickness vibrations. This more general case, however, was limited to quartz cut perpendicular to one of the electric axes and the optical axis ('Curie cut', see Figures 2.3 and 2.5).[27] Among others, Van Dyke's account did not cover most cuts and modes of vibrations used by Giebe and Scheibe for their frequency-standards resonators. In 1928, three years after they had begun using flexural and torsional vibrations they presented a theoretical explanation of the major properties of these vibrators, like the directions of maximal effect, and the variables that influence their resonance frequency. The theory was based on the piezoelectric and elastic phenomenological theories. Yet, although dynamic, the theory remained semi-qualitative, without elaborate, exact mathematical expressions of the vibrations. Five years later, Giebe, now with his co-worker E. Blechschmidt, presented a quantitative theory for such vibrations, but confined it to elongations of tubes and plates. As in many other theories of the resonator, the PTR researchers elaborated mathematically the equations of elasticity, ignoring those of piezoelectricity. They assumed that piezoelectricity excites the vibrations, but did not regard it significant in examining the exact manner of the resulting vibrations. In 1940, the two researchers suggested a quantitative theory also for torsional vibrations.[28]

Practical use of specific cuts often preceded theory; dynamic theory was not necessary for their application. This was the case with thickness vibrations, which Pierce introduced in a quartz cut perpendicular to a Curie cut (x-cut) in 1923. Still, a few researchers thought that quantitative dynamic theory would be helpful for shaping the resonator for their technical needs. This view seems to have motivated Issaku Koga, whose work on the frequency divider was discussed in Chapter 7. In 1932, as a professor for electrical engineering at the Tokyo Institute of Technology, he suggested a thickness vibration theory for all crystals and all cuts, accounting also for boundary reflections of elastic waves. The theory was part of a larger research project on characteristics of different cuts and modes of vibrating quartz plates,

---

[27] Cady, 'The Piezo-Electric Resonator' (1922); K. S. Van Dyke, 'The Electric Network of a Piezo-Electric Resonator', *Physical Review* 25 (1925): 895; K. S. Van Dyke, 'The Piezo-Electric Resonator and Its Equivalent Network', *IRE Proceedings* 16 (1928): 742–64.

[28] Giebe and Scheibe, 'Piezoelektrische Erregung von Dehnungs'; E. Giebe and E. Blechschmidt, 'Experimentelle und theoretische Untersuchungen über Dehnungseigenschwingungen von Stäben und Rohren (I u. II)', *Annalen der Physik* 18 (1933): 417–56, 457–85; Cady, *Piezoelectricity*, 450.

closely connected to their technical use. Using the theory and experiments with various quartz cuts, Koga found a crystal cut whose resonance frequency is practically independent of temperature, a highly useful property for exact frequency measurement and control. He reported on the cut in 1933 and two years later incorporated it in a new quartz clock.[29] Technological interests, thus, played a central role in directing Koga's research, even as his empirical and theoretical study of thickness vibrations went beyond the needs of technology.

An intent to provide rules for technical improvements seems to be only one among several motivations to develop the dynamic theories. Their authors were also concerned with disciplinary questions of completeness, and agreement with observations. Cady extended the theory for a resonator with a gap and damping, and showed how to derive the equivalent electric constants of thickness resonators in 1936. Other contributors with an institutional interest in technology, B. van Dijl from the Radio-Laboratory of the Dutch state-owned telecommunication provider and Rudolf Bechmann from Telefunken suggested, respectively, another mathematical approach and a fuller account of elastic boundary effects in 1936 and 1940. Academic researchers free from such institutional obligations also contributed. Ernest Baumgardt, who worked at Paul Langevin's laboratory mainly on ultrasonics,[30] explained a divergence between thickness theory and experiment with an assumed additional effect of electric charge on the elasticity of quartz in 1938. In 1942 A. W. Lawson from Pennsylvania University advanced a more complete mathematical account based on Cady's assumptions.[31] Most theoretical studies of the resonator, however, did not elaborate on piezoelectricity but on the elastic vibrations and their dependence on the elastic characteristics of the crystal.[32]

[29] *Progress in Radio in Japan* (Japanese National Committee for URSI: Tokyo, 1963): 8–9; for more information on Koga see Chapter 7.

[30] Yves Le Grand, 'Obituary, Ernest Baumgardt (1904–1969)', *Vision Research* 9 (1969): 1315–17.

[31] Cady, *Piezoelectricity*, 306–7.

[32] The elaborate discussions dedicated to lengthwise vibrations were devoted completely to elasticity. At Bell Labs W. P. Mason elaborated elastic theory to account for flexure vibration of a quartz bar in 1935. Two years later Ernst Lonn explained some discrepancies between theory and experiments in resonance through an elastic theory, Ernst Lonn, 'Zur Theorie der Schwingungen von Kristallplatten (Bemerkungen zu Arbeiten von Petržílka)', *Annalen der Physik* 30 (1937): 420–32.

Rigour and Approximations in Dynamic Theory

While the use of the resonator in physics and technology was central to the development of dynamic theories by most contributors to the field, the theoretician Laue was mainly concerned with the challenge it posed to the phenomenological elastic and piezoelectric theory of crystals. Current theory, as expressed in Voigt's 1910 *Lehrbuch der Kristallphysik*, did not describe such vibrations. While Cady later, if not at the time, described his 1922 resonator theory as 'an approximate theory of a new device', in 1925 Laue aimed for an exact account in the tradition of Kirchhoff and Voigt, who was one of his main teachers. He found Cady's theory lacking in completeness and rigour. Laue did not elaborate on the extant theory, but suggested his own assumptions, slightly different from Cady's. His theory was more detailed and rigorous in four aspects. First, it considers effects of the second order neglected by Cady. Second, Laue regarded effects overlooked by Cady, such as vibrations excited by forces not at the end of the bar, which the Wesleyan physicist ignored since he neglected the effect of friction. Third unlike Cady, Laue considered the piezoelectric effect of shearing stress (e_{14}). Fourth, Laue expressed the elastic frictional factor, regarded by Cady as given, in terms of generally known and more basic elastic coefficients of quartz.[33]

Cady admitted that his initial theory was not as exact as Laue's, but claimed that its elaborations in later publications and in more complete form in his 1946 textbook made the theory equivalent to Laue's. This was at least the case regarding the first two above-mentioned aspects. Still, the American physicist saw the two latter aspects superfluous. He claimed that for the vibrations the contribution of piezoelectricity by torsion is less significant than other effects ignored by both theories by assumptions like the uniformity of the electric field and the negligence of effects due to the width of the bar. Although the frictional factor of the resonator could be reduced to the frictional coefficients of the crystal, Cady did not see a need to spell out the connection as long as the precision of resonance measurements would not allow their determination. In line with the common view of American scientists, these reasons are of a pragmatic kind as they refer to possible measurable consequences of the theoretical assumptions. Note that Cady did not refer to technical consequences. Cady's preference seems to reflect also his attitude as an experimentalist, while Laue was a theoretician. As a theoretician Laue was more interested than Cady in

---

[33] M. v Laue, 'Piezoelektrisch erzwungene Schwingungen von Quarzstäben', *Zeitschrift für Physik* 34 (1925): 347–61; Cady, *Piezoelectricity*, 305–6, quote on 306; Armin Hermann, 'Laue, Max Von', in *Dictionary of Scientific Biography* (Detroit: Charles Scribner's Sons, 1973).

the completeness of the theory, its reference to all kind of influences, and the ability to reduce it to more basic magnitudes, even if such an endeavour did not provide immediate tools for the experimentalist. Theoreticians like Laue might have found that the additional necessary mathematical work was an advantage of the theory—an opportunity to employ their skills—rather than a disadvantage. Two young Soviet theoreticians, Sokoloff in 1928 and Lissütin in 1930, employed their mathematical competencies to elaborate Laue's theory for more complicated cases. Unlike Laue, Sokoloff and Lissütin worked at institutions with a declared interest in frequency control technology.[34] Bechmann at Telefunken, an organization with a clear interest in technology, also employed Laue's method in his work on thickness vibrations. Yet, neither their preference to Laue's approach nor Cady's 'pragmatic attitude' seem to be connected to technical use of the resonator.

Laue, like Cady and most other workers on piezoelectric theory at the time, was satisfied with a phenomenological account for the resonator. Dynamic theories explained the phenomena on the macroscopic–phenomenological laws of elasticity and piezoelectricity. Laue related the theory to more basic variables of the theory, but these were still phenomenological. No attempt had been made to connect the resonator to Born's lattice dynamic theory (Ch. 1). Born's theory was isolated from other trends in the research on piezoelectricity. Although Meissner did connect his atomic theory to the study of resonance, his theory, like other hypotheses about the molecular origins of piezoelectricity, was not relevant for the study of resonators. The separation between the micro-theory and research on the resonator is exemplified in the research of Reginald E. Gibbs, which is discussed in the subsection 'Piezoelectricity and the Structure of Matter'. In two collaborative works Gibbs studied in one the crystalline atomic structure in relation to piezoelectricity and in the other variations of resonance frequency with temperature, without making any connection between the two.

Experimental Study of the Resonator

Theory, in piezoelectricity, was mainly important for its use in experimentation and in developing devices, as experiments constituted the vast majority

---

[34]Sergei J. Sokoloff (written also Sokolov) worked at Leningrad's Central Radio Laboratory, A. Lissütin at the city's 'central chamber for measurement and weight'. Sergei J. Sokoloff, 'Schwingungen piezoelektrischer Quarzstäbe, hervorgerufen durch ein ungleichförmig verteiltes Feld', *Zeitschrift für Physik* 50 (1928): 385–94; A. Lissütin, 'Die Schwingungen der Quarzlamelle', *Zeitschrift für Physik* 59 (1930): 265–73.

of studies in the field. Researchers employed dynamic and, more frequently, equivalent network theories as well as elastic and static piezoelectric theory in planning and analysing their experiments. As is common, some experiments related more closely to theoretical developments, while others employed only a few theoretical relations and concepts, like equivalent capacity and inductance, or none at all. Experimental questions rarely originated in the theory of the resonators, and when they did, that was in studies of developers of theories or their associates. Dye, Cady, Giebe, Blechschmidt, and Koga tested some of their theoretical predictions in the laboratory. I found only one case where 'outside' researchers examined particular theoretical consequences, namely the 1933 experiment of S. Nishikawa, Y. Sakisaka, and I. Sumoto, who confirmed Koga's predictions for the modes of thickness vibrations through X-ray diffraction.[35] Fujimoto, Cady's former student, tested the dynamic theory by using it to determine piezoelectric coefficients whose values were known through other methods. More often, experimentalists relied on theory in planning experiments that originated in other concerns. For example, Andreeff et al. were interested in the changes of the piezoelectric coefficient with temperature. For that goal they designed their experiment according to the dynamic theory of resonators, which showed which variables should be measured for an experimental determination of the coefficient. This, of course, was true also for later 'dynamic' determinations of the coefficients (on which see section 9.3). Only a few of the other experimentalists, like Václav Petržílka, announced their relationships to an extant dynamic theory.[36]

Although dynamic theory was developed to explore crystal vibrations, most experimental studies of the characteristics of these oscillations did not employ it. Vibrations are an elastic effect, and elasticity attracted the most experimental attention, identifying and measuring the modes of vibrations, nodal points, amplitudes, and elastic moduli. Experimental planning and analysis of such questions often did not call for an elaborated discussion of the piezoelectric effect beyond pointing at the directions and approximate magnitudes of their resultant mechanical force. Thus, it did not require a dynamic theory. When they employed theoretical means, experimentalists were often satisfied in developing elastic theory for the particular crystal rods or plates under study and their modes of vibrations. For example, for his 1930 examination

[35] S. Nishikawa, Y. Sakisaka, and I. Sumoto, 'An X-Ray Examination of the Harmonic Thickness Vibration of Piezoelectric Quartz Plates', *Physical Review* 43 (1933): 363–64.

[36] In 1935 Petržílka studied longitudinal vibrations of quartz plates in relation to Lissütin's dynamic theory.

of transitions between flexural and transverse vibrations of quartz plates, the Frankfurter physicist Heinz Doerffler developed elastic equations of these vibrations, without regard of their particular piezoelectric source.[37] A year later, Harald Straubel in Jena used the known thermal variation of a quartz elastic modulus to calculate the dependence of specific resonators' frequency on temperature. In the first part of the same research, he relied on empirical means, without resorting to theory, to identify 'natural directions' of oscillations and pointed at their correspondence with Young's modulus. In this search, he followed Meissner's earlier identification of these preferable directions by the production of sound waves. Like Straubel, other experimentalists sometimes did not employ mathematical theory in exploring various modes of piezoelectric vibrations.[38]

Dynamic theory seemed to be more of interest to physicists, like Doerffler, than to developers of practical devices, with a few exceptions like Koga. The latter tended to employ theory only when deemed it useful for design. While I did not carry out a systematic study of engineering research, the cases of Lack's and Marrison's findings of temperature-stable quartz resonators (Ch. 6) suggest that, indeed, mathematical theory played a lesser role in the research of engineers. As mentioned, they reasoned from earlier empirical findings on modes of vibrations in quartz and how different modes depend on temperature instead of deriving preferable modes from a rigorous theory. Still six years later, W. P. Mason, their colleague at Bell Labs, developed an elastic theory for flexural vibration, probably with the aim of assisting in the design of further resonators in a more analytical manner. Experimental studies about modes of vibrations seemed to be more useful in developing efficient resonators for various practical ends, like stability and lower resonance frequency.

Still, Lack and Marrison enjoyed earlier mathematical and empirical findings on the modes of vibrations in quartz. A concern about practical cuts directed a few of the early researchers in the field, as seen for example in the work of Giebe and Scheibe. Indeed, practitioners did not miss that the mapping of various cuts and directions of oscillations suggested ways of reaching

---

[37]Heinz Doerffler, 'Biegungs- und Transversalschwingungen piezoelektrisch angeregter Quarzplatten', *Zeitschrift für Physik* 63 (1930): 30–53.

[38]Experiments on modes of vibrations that used laws of elasticity include those of Skellet (1930), Giebe and Scheibe (1931), Wright and Stuart (1931), Osterberg and Cookson (aimed at theory, 1935), Tykoncinski-Tykociner and Woodruff (1937), and Petržílka and A. Žáček (1938). Experiments that did not employ theory include those of Wachsmuth and Auer (1928), Koga (1930), and Bücks and Müller (1933).

cuts more efficiently or usefully for various technical ends. Straubel, in the above-mentioned research, urged practitioners to cut quartz according to its 'natural directions' of oscillations. While others were perhaps less explicit, they usually mentioned the practical implications of their research.

Physicists' interest in the question waned in the early 1930s, perhaps because there seemed to be little scientific interest after the main modes were mapped. The properties of different cuts became a subject of engineering research and development, aimed at designing new products. Scientists and engineers in industry (Telefunken, Zeiss, AT&T) and in engineering departments (Koga in Tokyo) employed quantitative resonance theory, data, and experiment to develop cuts useful to their needs. Among others they found the above-mentioned cuts of practically zero temperature coefficient, cuts of low- and of high-frequency resonance.[39] A few physicists, prominently Harold Osterberg in Wisconsin and Petržílka and Žáček in Prague, studied connected questions in the later 1930s. They concentrated on what seemed as open issues, because the phenomena in the laboratory and in technical use lacked a satisfactory theoretical account, because of disagreements about their nature and explanation, or because specific kind of cuts had not been studied previously. Osterberg, thus, accounted for Lack's finding of a crystal cut that practically does not change its resonance frequency with the temperature, by analysing the temperature coefficient's dependence on the vibration modes in quartz, a study that attracted also the attention of Bell Labs researchers. Petržílka and Žáček studied torsional vibrations following a controversy over the ability of Voigt's phenomenological theory to account for the static effect and how to account for the vibrations observed in different cuts. Earlier the two Czech physicists studied wedge-shaped quartz resonators.[40]

In addition to the study of modes of oscillations in quartz, tourmaline, and Rochelle salt, researchers studied the effect of external physical magnitudes on the resonator as a unit. In these studies they paid little attention to the process inside the crystal and to its particular elastic and piezoelectric coefficients. Yet, as they experimented with particular crystal specimens whose behaviours could have depended on such factors as vibration modes, they often discussed the vibrations they used, their main properties, and earlier

[39]On the research of quartz cut, see Cady, *Piezoelectricity*, 451–61.

[40]Harold Osterberg, 'The Temperature Coefficients of Shear and Longitudinal Modes of Vibration', *Review of Scientific Instruments* 7 (1936): 339–41; Václav Petržílka and A. Žáček, 'Radial and Torsional Vibrations of Annular Quartz Plates', *Philosophy Magazine* 25 (1938): 164–75.

studies of the vibrartion modes. Experimentalists explored changes in the crystals' electric and some mechanical properties and their frequency. For these aims they sometimes employed equivalent network formalism. Following the early studies at Wesleyan University and the NPL, physicists studied the effect of temperature, mechanical stress, and the magnitude and frequency of the electrical voltage on the resonator. A few studied the effect of air gaps on the resonance frequency and the influence of the surrounding medium. Others examined related questions like the effect of decrement of the vibrations on their frequency, and the production of heat by vibrations.[41] The conditions studied were similar to those that affected resonators when they were used for frequency standards and control devices, which required the stability of resonance frequency. Technological interests, thus, seem to be a decisive force in directing scientific research on these topics. Physicists showed interest in studying the effect of these physical conditions on frequency from the mid-1920s to the early 1930s, but did not pursue it later. As with the study of vibrational modes, they probably deemed the current knowledge of the main effects as satisfactory. During the 1930s, they turned to related questions, like the microscopic displacement of vibrators through the application of optical and X-ray methods, which are discussed in 9.5 subsection 'The Use of Other Effects to Study Piezoelectricity'.

Paths to the Study of the Piezoelectric Resonator

Most of the experimentalists who examined the resonator shared a concern with its technical application. After 1928, their concern usually did not follow from a personal engagement with designing devices, but from a general aim and usually a traceable motivation to provide useful knowledge for the developers of piezoelectric technics. Often, it was this concern with applications that led the researchers to piezoelectricity, or formed a major motivation for its study. To be more exact, they shared a general concern of improving the resonator for various uses in frequency standards and control. The profile of these

---

[41] Among the experiments on the effects of physical influences on the resonance frequency were the above-mentioned one of Powers, Dye, Heegner, and Terry and those of Strout (1928), Brown and Harris (1931), Petržílka (1932), and Gibbs and Thate (1932). The effect of air gaps was studied by Dye and Hehlgans (1928), and Grossmann and Wien (1931). The engineer Mario Boella examined the effect of decrement (1930); de Gramont and Beretzki showed that resonators are heated by their own motion (1932). In addition to *Science Abstracts* I also consulted R. E. Gibbs and V. N. Thatte, 'The Temperature Variation of the Frequency of Piezoelectric Oscillations of Quartz', *Philosophical Magazine* 14 (1932): 682–94.

researchers was quite similar to that of the early students of the resonator, with an increase in the share of university researchers. In the following I survey the main students of the resonator during the period by their institutional affiliation and trace their plausible routes to this research.

Notwithstanding the rise in the academic research, the research in national and industrial laboratories continued. Blechschmidt joined Giebe and Scheibe at the PTR; Barret and Howe at the NRL continued research begun earlier by Dawson (discussed in section 9.4); Hund, Wright, and Stuart began scientific research of the resonator connected to its possible employment by the American BoS. Bechmann joined Meissner at Telefunken and continued the research after the latter left for AEG. At Bell Labs, Mason and Skellett supplemented the more technological research and development of Marrison, Lack, and their associates. Academic researchers also joined their colleagues. S. Leroy Brown at the University of Texas, Austin, and Osterberg in Wisconsin joined their colleagues Terry in Wisconsin and Cady in Wesleyan (see Table 9.1 for the main schools). German professors and their students joined their American colleagues: Wien, Grossmann, Hehlgans, and Straubel in Jena; and Wachsmuth, Auer, and Doerffler in Frankfurt. Brown showed a longstanding interest in high-frequency oscillations. Max Wien and his associates in Jena directed much of their research in 'technical physics' to the needs of industry, especially regarding high-frequency electric oscillations, Wien's expertise. In particular, they cultivated strong connections with the science-based company Zeiss, the main funder of the natural science faculty at the university, and the founder of its institute for technical physics.[42] Although primarily an optics manufacturer, Zeiss engaged also in the production of crystal cuts for frequency control, probably due to its expertise in cutting and grinding. The company was closely involved in the research of Harald Straubel, who had assigned a few patents in the field to Zeiss while he worked at the university before he became its employee in 1935. Straubel continued to cooperate with the university while a Zeiss employee. The tight relationships benefited also from the high position that Haralad's father, Rudolf, a physicist himself, occupied in the company.[43]

---

[42]Stefan Gerber, Jürgen John, and Rüdiger Stutz, *Traditionen, Brüche, Wandlungen: die Universität Jena 1850–1995* (Köln: Böhlau Verlag, 2009).

[43]Zeiss had a long tradition of scientific research: Stuart Feffer, '*Microscopes to Munitions: Ernst Abbe, Carl Zeiss, and the Transformation of Technical Optics, 1850–1914*' (Berkeley: University of California Press, 1994). Straubel first patent, 'Scheibenförmiger piezoelektrischer Oszillator (bzw. Resonator)' DE530582, filed 3 December 1930, was assigned to Zeiss. He thanks

Japan provides another example of the involvement of government laboratories and academic institutes in the research. Koga began his work on piezoelectricity at Tokyo's Electro-technical Research Institute while studying at the electrical engineering department of Tokyo's Imperial University, and continued at the Tokyo Institute of Technology. In parallel to Dye and Giebe and Scheibe, Shogo Namba worked on Japan's radio frequency standards at the Electro-technical Laboratory of the Ministry of Communication. In 1929 he explored resonators' vibration modes. Nishikawa and his associates worked at Riken, an academic institute of physical and chemical research, and did not have a direct interest in frequency control. Yet, one of the main goals of the institute was to provide a scientific basis for fostering technology, as Nishikawa did with the resonator. Moreover, Tsunetaro Kujirai, Koga's supervisor, was a group leader in physics at Riken, suggesting an example for the close link between physics and the better-established electrical engineering in Japan.[44]

Another academic site for research was the University of Prague, where physics professor Žáček, an expert on radio and the measurement of high frequencies, prompted his assistant Petržílka to examine relationships between optic and piezoelectric properties in vibrating quartz in 1931. The next year

the company not only for material help but also for help in overcoming experimental difficulties. Harald Straubel, 'Fundamental Crystal Control for Ultra-High Frequencies', QST, April 1932, 10–13. A mention of his work in Zeiss and support for research at the university in H. E. R. Becker, 'Die Debye-Sears Beugungserscheinung und die Energiebilanz bei Erzeugung von Ultraschallwellen', *Annalen der Physik* 25 (1936): 384. On resonators produced by Zeiss see Arthur O. Bauer, 'Some Aspects of Precision Time Measurements, Controlled by Means of Piezoelectric-Vibrators, as Deployed in Germany prior to 1950', unpublished manuscript (6.2.2000), 19, available at http://www.cdvandt.org/time%20symposium%20webversion.pdf (accessed 18 January 2015). Norbert Günther's work at the university's Institute for Applied Optics was closer to the core activities of the company, examining the effect of mechanical and electric forces on double refraction in quartz using quartz crystals on loan from Zeiss. Norbert Günther, 'Untersuchung der Wirkung mechanischer und elektrischer Kraftfelder auf die Doppelbrechung des Quarzes', *Annalen der Physik* 13 (1932): 783–801.

[44]'Contributors to this issue', *IRE proceedings* 21 (1933), 326; Y. Namba, 'The Establishment of the Japanese Radio-Frequency Standard', *Proceedings of the IRE* 18 (1930): 1017–27; I. Nitta, 'Shoji Nishikawa (1884–1952)', in P. P. Ewald (ed.), *50 Years of X-Ray Diffraction* (Utrecht: N.V.A. Oosthoek, 1962), 328–334. On Kujirai and the connection between physics and electrical engineering see Kenji Ito, '"Electron Theory" and the Emergence of Atomic Physics in Japan', *Science in Context* 31 (2018): 293–320. Samuel K. Coleman, 'Riken from 1945 to 1948: The Reorganization of Japan's Physical and Chemical Research Institute under the American Occupation', *Technology and Culture* 31 (1990): 229–31. On Nishikawa's path to piezoelectricity see the subsection 'The Use of Other Effects to Study Piezoelectricity'.

Petržílka moved closer to answering technological interests with research on high-frequency vibrations of tourmaline crystals at the Heinrich Hertz Institute for vibration research in Berlin, an institute with an explicit interest in frequency control. At this point Petržílka acquired expertise and professional interest in the study of resonators. Thus, when he returned to Prague as a lecturer in physics, he continued with its study, establishing a local centre for studying resonators by engaging Žáček and younger physicists L. Zachoval, František Khol, and Franz Krista. The technologically oriented Hertz Institute, thus, pushed Petržílka to study areas of its interest, also after ceasing to pay his salary. Yet Petržílka himself found other areas more promising for further research and in 1938 began studying nuclear physics, a move hampered by the closure of Czech universities by the German occupation. Under these conditions, he co-authored an introductory book on piezoelectricity and its applications in Czech in 1940.[45]

Still, some of the participants in its research did not show an explicit interest in the technical employment of resonators. Richard Wachsmuth, Frankfurt's experimental physics professor, came to the subject due to his concern in elastic vibrations. Technical applications seemed less prominent in his group's work. Yet, the technical relevance probably helped him attain support for the research of his student Doerffler from the Notgemeinschaft der Deutschen Wissenschaft (NG), the main German research fund, and the Helmholtz-Gesellschaft, which promoted applied research. The higher prospects of receiving financial support for a technologically relevant study might have directed the student to this research rather than to another work on elasticity.[46]

[45] Petržílka published on the piezoelectric resonator alone and with Zachoval in 1934 and 1935 and with Žáček in 1935 and 1938. Khol published alone in 1938 and 1939 and thanked both Petržílka and Žáček (*Zeitschrift für Physik* 108 (1938): 225–31); Krista published in 1939. Petržílka's research probably encouraged also that of Dolejšek in Prague (on which see the subsection 'The Use of Other Effects to Study Piezoelectricity'). Václav Petržílka, 'Turmalinresonatoren bei kurzen und ultrakurzen Wellen', *Annalen der Physik* 15 (1932): 72–88; The editor, 'Sixty-Fifth Birthday of Professor Václav Petržílka', *Czechoslovak Journal of Physics B* 20 (1970): 369–74; 'The 70th birthday of Prof. Dr. August Žáček', *Czechoslovak Journal of Physics* 6 (1956): 204–5; Václav Petržílka and Josef B. Slavík, *Piezoelektřina a její použití v technické praxi* [*Piezoelectricity and Its Use in Technical Practice*] (Praha: Jednota českých matematiků a fysiků, 1940), 116. I thank Jan Kotůlek for the information about the book.

[46] Doerffler, 'Biegungs- und Transversalschwingungen'. Jochen Kirchhoff, 'Wissenschaftsförderung und forschungspolitische Prioritäten der Notgemeinschaft der Deutschen Wissenschaft 1920–1932' (Dissertation, LMU München, 2003), 84. 'Gepris-historisch', https://gepris-historisch.dfg.de/fall/145855? (accessed 17 April 2022).

Across the Atlantic, the resonator attracted the attention of Harold Osterberg, who examined its vibrations by an interferometer in his 1931 dissertation at the University of Wisconsin. Osterberg made piezoelectricity his major research area, publishing alone and with students about a dozen papers on the resonator. Apparently, the combination of a new subfield with open questions and their relevance for technology attracted the young physicist, who sought his own niche. He explored questions of some generality, yet many of his studies were relevant to practical applications. Among these were an interferometer examination of quartz cuts that were used in commercial devices (in 1933), the stability of high-frequency tourmaline resonators (with John W. Cookson in 1934), and the above-mentioned research on zero temperature coefficients. The latter is an example of a study of more general and theoretical character inspired by technological research. Ernst Lonn's 1937 mathematical solution of the vibrations of crystal plates that was stimulated by findings of Petržílka suggests another example. In the case of Osterberg the connection to technics was not only thematic. Bell Labs lent him quartz plates.[47] His inclination to technology-oriented research would lead him to leave the academy for the optical industry in 1939.

Clearly, technical prospects motivated much of the research on the piezoelectric resonator. Yet, the participation of scientists without direct personal or institutional links to the technology suggests that disciplinary concern and sheer curiosity also played a role. I have traced already some of the scientific concerns that researchers had; it is more difficult to trace the effect of the researcher's own curiosity—a private state of mind with little objective remains. Yet, one should not infer from our methodological shortcomings that curiosity was inconsequential. There are a few indications that curiosity motivated scientists and engineers alike, if not all and not always. Their published papers and notebooks show that their research often went beyond apparent scientific or practical questions. Another indication is the enthusiasm conveyed in some passages of their published material. In addition, researchers explicitly referred to their inner interest in the subjects they studied, usually retrospectively when they reflected on their work. Although curiosity sometimes led physicists to issues outside the central disciplinary and technical concerns, it often did not conflict with other motivations. Scientists found intellectual interest also in questions derived from technology. The curiosity and disciplinary concerns in such phenomena like piezoelectric resonance made them more attractive for

[47] H. Osterberg, 'A Multiple Interferometer for Analyzing the Vibrations of a Quartz Plate', *Physical Review* 43 (1933): 829. Osterberg, 'The Temperature Coefficients of Vibration'.

scientific research, thus facilitating physicists' move to questions of interest for users of technics.

9.4 Studies beyond the Resonator

Although the resonator attracted most attention, the technical application of piezoelectricity led also to an expansion of research on other properties of the phenomenon. It did this through three main paths. First, the effort to elucidate resonance triggered the examination of related questions regarding piezoelectricity in static conditions. These were often topics that had already been studied in the pre-applicable phase. Studies of the magnitudes of the piezoelectric coefficient (the ratio between strain and electric polarization in particular crystals and directions) and their variations under changing physical conditions form the prominent example. Second, through its usage and the discovery of piezoelectric resonance, the phenomenon became better known among physicists. Thus, researchers were mindful of the phenomenon and considered its study when they worked on connected questions, or when they were looking for a proper niche for their research. This was especially the case with those who had encountered the phenomenon in their earlier work. Third, the resonator and its theory suggested new methods for measuring and detecting piezoelectric effects. As is common in disciplinary research, physicists seized the new methods to measure magnitudes determined by other means. In this section I review the main studies about piezoelectric phenomena beyond the resonator, as they evolved through these three paths and consequently the inner dynamics of the disciplinary research that often followed the first studies. I examine also other factors like the continuation of former research questions and relations to developments in understanding solids. Studies related to interactions with other physical phenomenon are discussed in section 9.5.

To begin with the first path, a few physicists examined the dependence of piezoelectric coefficients on temperature. In 1927 Leo Dawson from the US NRL examined static piezoelectricity in quartz. Among others he observed a sharp decrease in the value of quartz's main coefficient (d_{11}) with high temperature. Joseph Valasek carried out similar research on Rochelle salt, a crystal which he intensively studied due to its peculiar behaviour.[48] A few researchers from the Leningrad Physico-Technical Institute (LFTI) further examined the relationship between temperature and the magnitude of the effect. In 1929 A.

[48]Later in this section I discuss briefly the research on Rochelle salt, which led to the identification of ferroelectricity.

Andreeff, V. Fréedericksz, and I. Kazarnowsky measured temperature coefficients of quartz resonators using a dynamic method, namely, measuring the coefficient through the behaviour of resonators. They did not find the decrease observed by Dawson. In a separate publication, R. D. Schulwas-Sorokina, one of the few female researchers in the field, explained the divergence by criticizing Dawson's method, claiming that the static method is inadequate due to electric conductivity at high temperatures (caused by thermal vibrations), making the coefficient seem lower than it is. While her paper was on its way to print, the Swiss research Albert Perrier joined the discussion and showed that he had reached results similar to those of the Leningrad group and different from Dawson's with a static method already in 1916.[49] Physicists continued measuring the changes in the piezoelectric coefficient with temperature during the 1930s, by the static and dynamic methods. Their results varied but tended to disagree with Dawson's.[50]

Perrier's research on the temperature dependence of piezoelectricity preceded the technical interest in piezoelectricity. He first examined the effect in 1916, following his longstanding study of the thermodynamics of phase changes in solids, and of the fact that the change in the structure of quartz (from α to β quartz) at 579°C causes the disappearance of its piezoelectricity.[51] Yet, Perrier returned to the question due to increasing interest in the field, which followed its practical application. As mentioned in Chapter 4, the NRL carried out research and development of crystal frequency control for use in short-wave navy communication.[52] In this context of understanding vibrating crystals, Dawson examined the properties of static piezoelectricity, showing only superficial familiarity with current knowledge. Vselolod Fréedericksz at the Leningrad group (or Frederiks in modern transliteration) had much more thorough knowledge of and experience with piezoelectricity. Between

[49] Leo H. Dawson, 'Piezoelectricity of Crystal Quartz', *Physical Review* 29 (1927): 532–41; R. D. Schulwas-Sorokina, 'Is It Possible to Determine the Piezoelectric Constant at High Temperature by The Statical Method?', *Physical Review* 34, 11 (1929): 1448–50; Albert Perrier, 'Zur Temperaturabhängigkeit der Piezoelektrizität', *Zeitschrift für Physik* 58 (1929): 805–10.

[50] Experiments were conducted by Fréedericksz with G. Michailow (1932), G. Michailow (1932), Van Dyke (1935), Pitt and McKinley (1935), A. Langevin (1935–6), and Clay and Karper (1937). Cady, *Piezoelectricity*, 221–3.

[51] Anon, 'Albert Perrier', in *Université Lausanne- rapport annuel 1961–1962* (Lausanne: Imprimerie Vaudoise, 1963), 5–6.

[52] The laboratory studied also other aspects of radio, most famously 'skip zones' Chen-Pang Yeang, *Probing the Sky with Radio Waves: From Wireless Technology to the Development of Atmospheric Science* (Chicago: University of Chicago Press, 2013), 171–5.

1910 and 1918 Fréedericksz stayed in Göttingen, serving at the institute as an assistant to the doyen of piezoelectricity, Woldemar Voigt, with whom he wrote on the electric effect of torsion and bending. While this can explain his renewed interest in the field, there are reasons to assume that piezoelectricity's applicability contributed to his return. He employed the new methods for the resonator, and his research had possible implications for understanding its behaviour, which depends on piezoelectric coefficients. In particular, one might hope to gain some insight into the variation of quartz's resonance frequency with temperature, which aroused much concern among practitioners (see Ch. 6). Michailow, Fréedericksz's research partner, further explored these implications for Rochelle salt, which has a critical point close to room temperature, but was less interesting from a practical point of view. Lastly, as an applicable field the study of piezoelectricity, including its basic properties, seemed fitting to the LFTI's aim of attaining knowledge pertaining to technical development.[53]

Examining the variation of the effect with temperature, the Leningrad group also measured the absolute value of quartz main piezoelectric coefficient (d_{11}), a quite common practice at the time. At least fourteen different determinations of its value were carried out between 1927 and 1939, versus nine in the thirty-five years before WWI. Some of these determinations were taken while examining variations in its value due to external condition, as in the case of high temperature, or under very high pressure. The latter question had also a practical side, since piezoelectricity was used also to measure high pressures, for example in explosions (Ch. 8). The exact value was important also for using the effect in other measurements like that of electric capacity. Still, much of the attention paid to the coefficient's exact value and its variations originated out of interest in the resonator and its dynamic theory. That effort of determining piezoelectric coefficients were in most cases of quartz, the only crystal actually used for frequency control, points to this connection. Quartz was also the most useful crystal in laboratory examination and drew the most attention at the pre-applied phase of research. Still, while it was the single most determined crystal before WWI, its study predominated

[53]Vladimir Vizgin and Viktor Frenkel, 'Vsevolod Frederiks, Pioneer of Relativism and Liquid Crystal Physics', in Yuri Balashov and Vladimir Vizgin (eds), *Einstein Studies in Russia* (Boston: Birkhäuser, 2002), 149–80. Michailow published papers about temperature's influence on the frequency of piezoelectric oscillations in Rochelle salt in 1936; Paul R Josephson, *Physics and Politics in Revolutionary Russia* (Berkeley, CA: University of California Press, 1991), 104–40.

the interwar period, drawing more measurements than all other crystals combined. A casual remark made by Cady in his comprehensive 1946 textbook *Piezoelectricity* reveals the importance of practical applicability in the study of specific crystals. He dismissed the need to have quantitative information about the coefficients of crystals of class 12, by pointing out that the possible candidate for examination 'cannot be recommended for piezoelectric applications'.[54]

The resonator provided not only theoretical motivation for measuring the coefficients but also new means for carrying out these measurements. Physicists often employ new methods to measure magnitudes known by other means. The theory of the resonators made it possible to determine the piezoelectric coefficient through frequency measurements by a few experimental methods, and several physicists seized on the technique. Some of them like Fujimoto and Van Dyke had worked on resonators; others like Fréedericksz and his colleagues showed a clear interest in them. For still others like Angelika Székely from Graz University and her student Berta Nussbaumer this was the only research related to piezoelectricity, which probably attracted Székely's attention due to her work on wireless communication. Despite the novelty attraction, most physicists preferred the older static methods.[55]

A number of researchers on questions not directly related to piezoelectric applications were triggered by their, or their teachers', encounter with the phenomenon in technological research, as Cady was in his exploration of resonance immediately after the end of the First World War. Another important case is that of Joseph Valasek in Minneapolis, who embarked on the study of piezoelectricity in Rochelle salt following the suggestion of William F. G. Swann, his PhD supervisor. During the war, Swann worked on submarine detection and thereby had access to confidential reports by Joseph A. Anderson and Cady that pointed out a curious behaviour of Rochelle salt crystals (discussed in Ch. 3). He put Valasek on the task of exploring the crystal's nonlinear electric polarization and its hysteresis by the exertion of pressure and electric voltage.[56] This kind of spontaneous electric polarization in crystals,

---

[54]Cady, *Piezoelectricity*, 200–31, quote on 209. Cady does not mention Székely's 1932 measurement of the quartz coefficient.

[55]Cady, *Piezoelectricity*, 387–92. Székely and Nussbaumer published their results in 1932.

[56]The reports are mentioned in Joseph Valasek, 'Piezo-Electric and Allied Phenomena in Rochelle Salt', *Physical Review* 17 (1921): 475–81; J. A. Anderson, 'Behavior of Rochelle Salt Crystals', a report in two parts 13.3 and 22.4.1918, and W. G. Cady, 'Report on Experiments with Rochelle Salt', 8.5.1918, in American National Archives, Records of the

known first as Seignette electricity and later as ferroelectricity, due to its analogy to ferromagnetism, became Valasek's main research subject. A few joined the research in the 1920s. Their number increased significantly from the early 1930s, as further research left many questions about the phenomena unanswered. By the middle of the decade, ferroelectricity became a distinct vibrant subject of research, with its own experts, collection of central phenomena, and a few competing theories for their explanation.[57] The discovery of ferroelectricity was, thus, another outcome of the phase of research and development regarding sonar, the third in the schematic sequence of the research on piezoelectricity. Like Cady's discovery of electric resonance, Valasek built on previous findings gained in the technological research and opened a new subject of study within physics.

Paul Langevin's use of piezoelectricity in the same phase also promoted research in the field. As an assistant in his laboratory, René Lucas employed a new method inspired from the work on sonar for detecting piezoelectricity in crystals in 1924. Langevin began publishing on piezoelectricity in the mid-1930s, inspiring also a few junior collaborators, including his son and son-in-law.[58] It seems that Langevin's experience influenced also students at the nearby laboratory of Charles Fabry at the Faculty of Science, who was well familiar with Langevin's work, as he had introduced it in the USA in June 1917 (Ch. 3). For his 1927 dissertation Ny Tsi-Ze (Yan Jici) examined static piezoelectricity in quartz, using a crystal specimen cut as an ultrasonic transducer and enjoying assistance from a commercial company that used it (*Société de Condensations et d'Applications Mécanique*). His study included an

National Academy of Science, Box 72 #2 and #4; Roger H. Stuewer, An interview with Dr. Joseph Valasek (8.5.1969), https://www.aip.org/history-programs/niels-bohr-library/oral-histories/2345 (accessed 16 November 2014). L. E. Cross and R. E. Newnham, 'History of Ferroelectrics', in W. D. Kingery (ed.), *Ceramics and Civilization, Volume III. High-Technology Ceramics: Past, Present, and Future* (Westerville, OH: American Ceramic Society, 1986), 289–305, on 292–3. Jan Fousek, 'Joseph Valasek and the Discovery of Ferroelectricity', in *Proc. 9th IEEE International Symposium on Applications of Ferroelectrics* (Piscataway, NJ, 1994), 1–5.

[57] Since ferroelectricity became a distinct subfield with its own questions and dynamics, I do not elaborate on it here, although it attracted a considerable portion of the research on piezoelectricity from more than twenty contributors.

[58] René Lucas, 'Sur la piézoélectricité et la dissymétrie moléculaire', *Comptes rendus* 178 (1924): 1890–2. Andre Langevin examined the effect of temperature on the piezoelectric coefficient of quartz, partly with A. M. Moulin in 1936–7. His brother-in-law Jacques Solomon published with Paul Langevin. As mentioned, Baumgardt studied piezoelectricity at Langevin's laboratory in 1938.

examination of the effect under very high voltage (finding that the linearity breaks down), a quantitative comparison of the longitudinal and transverse converse effects in quartz (i.e. the mechanical deformation due to electric tension along the electrical axis, x, and the perpendicular axis in the hexagonal plane, y), and the effect of converse piezoelectricity on the dielectric coefficient of quartz and on its optical behaviour (electro-optics).[59] These questions continued pre-war concerns, even if Ny was original in putting the converse rather than the direct effect at the focus of its work. For example, Nachtikal in 1899 examined the linearity of the direct effect under high pressure, Ny examined the case of high voltage. These questions, however, did not attract much attention during the interbellum. Noteworthy exception is Norbert Günther's theoretical and experimental demonstration of the interdependence of the elastic and optical coefficients of quartz by piezoelectricity and the Kerr effect (change of refractive index due to electric field) in his 1932 dissertation in Jena.[60] An interest of their laboratories in optics probably contributed to the choice of topic of these two dissertations.[61]

Ny's own experience with piezoelectricity explains his further research in the field. In 1934 as a director of the Institute of Physics in Beijing, he returned to piezoelectricity. With Tsien Ling-Chao (Qian Lingzhao) he examined electrification by torsion in a hollow quartz cylinder, in an attempt to elucidate differences between the findings of Röntgen from 1889 and those of Tawil from 1928 and their relationship to the phenomenological theory. This created a small controversy, mostly in the French literature, and further research on torsion in quartz by the involved researchers and their collaborators: Fang Sun-Hung in China and R. E. Gibbs with whom Tsien studied in England, and by Paul Langevin and Jacques Solomon in Paris. As with other questions related to quartz's piezoelectric properties, the issue of torsion also had practical implications, which probably contributed to the interest in the question. In

---

[59] Ny Tsi Ze, 'Étude expérimentale des déformations et des changements de propriétés optiques du quartz sous l'influence du champ électrique', *Journal de Physique et le Radium* 9 (1928): 13–37.

[60] Already Jacques and Pierre Curie in 1881 and then Röntgen and Joffé in 1913 had examined the relationship between the direct longitudinal and transverse effects in quartz. In the early 1890s, following earlier experiments by Röntgen, Kundt, and Czermak, Pockels developed a theory and carried out experiments on piezo-optics and electro-optics. Katzir, *Beginnings of Piezoelectricity*, 41–7, 150–3, 200–4; Cady, *Piezoelectricity*, 217–21, 717–24; Günther, 'Wirkung mechanischer und elektrischer Kraftfelder'.

[61] Fabry was an expert in optics. On the connection between University of Jena and the optical company Zeiss, see above p. 280.

their first publication Ny and Tsien pointed out that their results suggested the possibility of vibrating crystals by torsion. Shortly after, they employed them to explain an earlier failure in producing resonators from particular cuts. In a few of their subsequent studies they examined such vibrators.[62]

Piezoelectricity and the Structure of Matter

Only a few studies of piezoelectricity in the 1920s substantially continued research projects that began before the application of the phenomenon, which is not surprising given the number of earlier studies. One was Perrier's abovementioned exploration of the phenomenon at high temperatures. Another was Anna Beckman's 1922 study of piezoelectricity at very low temperatures, which continued her 1913 joint research on the same subject with Kamerlingh Onnes, the doyen of low-temperature physics.[63] Two additional and related research projects had begun shortly before piezoelectricity was applied, yet like the research on the resonator they presented a novel direction in the study of the phenomenon. In this case, the novelty originated in new developments in physics rather than in new findings about the phenomenon and its applications. One project concerned the consequences of Born's lattice dynamic theory for piezoelectricity, discussed in Chapter 1. Born, one of his collaborators, and one of his students elaborated on the subject.[64] Despite its novelty the project was not extended further at the time, probably since it did not seem to yield verifiable results. The second subject was the atomic structure of quartz and its relationship to the crystal's piezoelectricity. In 1914 in one of the earlier uses of X-ray diffraction techniques, William Henry Bragg examined the atomic structure of quartz. Motivated by a 1922 suggestion of an alternative structure by Maurice Huggins, Bragg returned to the subject, collaborating now with his student Reginald E. Gibbs, who continued studying the subject alone at University College London. Bragg and Gibbs inferred from the knowledge of piezoelectricity that the oxygen and silicon atoms that constitute quartz must be separated, and employed this conclusion in constructing a lattice structure that fitted their X-ray observations. Although piezoelectricity

---

[62]Tsi-Zé Ny and Ling-Chao Tsien, 'Sur le développement d'électricité par torsion dans les cristaux de quartz', *Comptes rendus* 198 (1934): 1395–6; Tsi-Zé Ny and Ling-Chao Tsien, 'Oscillations with Hollow Quartz Cylinders Cut along the Optical Axis', *Nature* 134 (11 August 1934): 214–15.

[63]Cady, 'Bibliography of Piezo-Electricity'.

[64]Born published on the theory with his assistant Elisabeth Bormann in 1920 (when he was in Berlin) and alone in 1921 and 1923 (when he was in Göttingen). In 1925, his student Gustav Heckmann corrected a formal error in Born's theory.

was rather peripheral in his X-ray diffraction research, Gibbs returned to its study with collaborators in the 1930s, now without connection to X-ray techniques. Apparently, his acquaintance with the phenomenon encouraged its later study.[65] The question of the exact atomic lattice structure and its connection to the piezoelectric properties of the crystal continued to attract attention in reports and summaries of quartz and piezoelectricity, but received only a few remarks in research papers.

While Bragg and Gibbs relied on modern methods of probing into the structure of materials using radiation with a wavelength of the atomic scale, Alexander Meissner inferred the structure of quartz by a more traditional macroscopic observation of its electroacoustic behaviour. As mentioned earlier, he reached the subject from his study of the resonator. Intrigued by his success to identify surfaces of higher molecular density in quartz, Meissner combined this conclusion with the common assumptions and knowledge about the hexagonal structure of the crystal to suggest a model that did not take the later theory of Bragg and Gibbs' into account.[66] In this model Meissner inferred the location of the silicon atoms form the direction of the air blast that he observed. Since they are heavier and thus more significant for the density of quartz, he merely coupled the oxygen atoms to them. Within a few months, Meissner revised his theory to include also the results of X-ray interference experiments. Now doublets of oxygen atoms got an equal role to that of the silicon atom, forming each a corner of a three-dimensional structure, whose projection is a hexagon on the plane perpendicular to the optical axis. This structure allowed Meissner to suggest an explanation of static piezoelectricity. According to the proposed structure a pressure, or pulling, causes a displacement of the atoms vis-à-vis each other, deforming the hexagons and resulting in a net electric charge. The Telefunken scientist showed how these displacements explain the appearance of positive and negative charges on different electrodes due to pressing, or pulling, in the electrical axis (E_1 in Figure 9.3) and perpendicular to it.[67]

As Meissner pointed out, his theory resembles the 1893 suggestions of Lord Kelvin (William Thomson) and Voigt. Yet, Kelvin, who proposed similar hexagons, assumed that these are independent molecules, an assumption discredited by the findings of X-ray diffraction. Voigt proposed a general model that did not relate to the actual lattice structure of quartz. Like the suggestions of his predecessors, Meissner's theory was also speculative but in line with contemporary understanding of atomic lattice structure. The new view about the

[65] Vigoureux, *Quartz Resonators and Oscillators*, 177–96; William Bragg and R. E. Gibbs, 'The Structure of Alpha and Beta Quartz', *Proceedings of the Royal Society of London. Series*

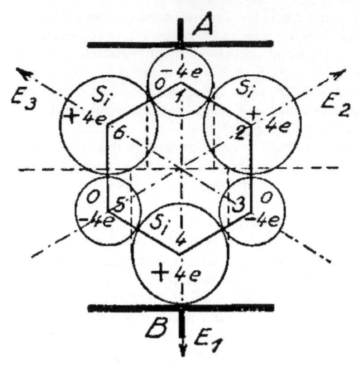

Figure 9.3   Generation of electric polarization by pressure in the direction of an electric (x) axis of quartz according to Meissner. *Source*: 'Über piezo-elektrische Kristalle bei Hochfrequenz II', 76.

role of individual atoms led Meissner to an original explanation of pyroelectricity. While earlier theories assumed that heat produces electricity through thermal expansion, like Boguslawski in 1914 (Ch. 1), he explained the effect by thermal motion of the individual atoms (but unlike the Russian physicists, he did not impose any quantum condition). Inferring from their behaviour as pure elements, he assumed that the oxygen atoms in the quartz retain their heat energy, and thus displacement, longer than the silicon atoms. The difference in the time of returning from displacement accounts, according to Meissner, for

A 109 (1925): 405–27. In 1932 Gibbs examined the changes of resonance frequency with V. N. Thate, and in 1936 studied torsion in quartz with L. C. Tsien.

[66] At the time, X-ray diffraction experiments did not provide information to determine atomistic structure unambiguously. André Authier, *Early Days of X-Ray Crystallography* (Oxford: Oxford University Press, 2013), 172.

[67] Alexander Meissner, 'Kristalle bei Hochfrequenz'; Alexander Meissner, 'Über piezo-elektrische Kristalle bei Hochfrequenz II', *Zeitschrift für technische Physik* 8 (1927): 74–7.

the electric effect of heat. With this hypothesis, Meissner explained the appearance of pyroelectricity in quartz despite its central symmetry, which forbids its appearance according to Voigt's theory. As mentioned in Chapter 1, in the 1910s the induction of pyroelectricity in quartz was a subject of dispute between Voigt and Röntgen, who observed such an effect. Although he did not refer to this discussion, Meissner entered here a known disciplinary question. Further, his theory indirectly supported the existence of 'true' pyroelectricity that is not a result of thermal expansion, which Voigt's theory allowed in crystals devoid of central symmetry, like tourmaline, but whose existence was in doubt due to the lack of experimental support. Although fitting the new findings, Meissner's theory, like those of his predecessors suggested only a qualitative account of the phenomenon. Moreover, like Kelvin's, his theory did not account for the piezoelectric effect of torsion.[68]

Meissner's atomic theory of piezo- and pyroelectricity could not help in designing more useful resonators or circuits, not even indirectly. In this sense, it was a kind of 'pure research'. Moreover, it was connected to disciplinary concerns that originated in the pre-applied phase of research, and to current questions about atomistic structure. Still it was strongly connected to applied and even technological research, beginning with attempts to improve the means for frequency standard and control, continuing in a study of the vibrating crystal, a discovery of unknown effect, and ending with an inference about the structure of matter. Meissner's research well exemplifies this connection between the different levels of research, where potential utility directed parts of the research, while others followed their curiosity about observations, the disciplinary logic of following empirical results, and answering open questions about natural effects, albeit often with an open eye to possible applications of the findings to new devices (even if not always practical).

9.5 Interactions between Piezoelectric and Other Fields of Physical Research

Like Meissner's theory, part of the interest in the piezoelectric vibrator and its underling static effect emerged from disciplinary research and questions in physics and crystallography. One such source of interest was the above-mentioned concern in understanding piezoelectric resonance itself. Another source of interest originated in the use of the resonator in the study of other physical effects and in the employment of other effects in its study. Most important among these were the study of acoustics, especially ultrasonics, and

[68] Ibid., Katzir, *Beginnings of Piezoelectricity*, 130–6.

the exploration of crystalline structure. The use of a piezoelectric vibrator in ultrasonic experiments led a few physicists to study the effect beyond the experimental device, namely piezoelectricity. Piezoelectricity found a few additional laboratorial uses, which increased interest in the phenomenon. In the other direction, physicists exploited their expertise in X-ray diffraction, and to a smaller extent in light interferometry for a finer exploration of the resonator. In a related concern, scientists became interested in the interactions between piezoelectricity and radiation (from visible light to γ rays). For a number of scientists, these research fields provided an entry point to the study of piezoelectricity. Yet, usually they carried out only a few studies on the subject.

Laboratory Usage of Piezoelectricity

The study of piezoelectric vibrations began within the research on ultrasonic transducers for submarine echo detection. As discussed in Chapter 3, Cady's discovery of the sharp change in the electric properties of piezoelectric resonators originated from this study. After the war, ultrasonics became a subject of scientific as well as technological research. A few of the scientists involved in its research during the war, like Paul Langevin and Robert Boyle, continued with its study, and others joined them. Partly following these new developments in piezoelectricity, experimentalists contrived new methods for studying ultrasonics using the piezoelectric resonator. During the 1920s this study did not stimulate much research into piezoelectricity. In the 1930s, however, with the introduction of more sophisticated methods in ultrasonics, piezoelectricity attracted the attention of a few researchers.

One of the first was Pan Tcheng Kao. In 1930–1, while he was working on a dissertation about the velocity of the propagation of ultrasonic waves at Charles Fabry's laboratory in Paris, he published a few articles on piezoelectricity in quartz plates and their vibrations. He returned to the vibrator in 1935. Pan joined a few of the researchers associated with Fabry who had studied piezoelectricity, among them Bedeau, Tawil, and Ny (discussed earlier).[69] Simultaneously with Pan, Ernst Grossmann employed the quartz vibrator in his dissertation on sound waves. With his teacher Max Wien in Jena, he used sound waves to examine the effect of air gaps on the frequency of resonators. As mentioned earlier, this research was part of a larger interest in frequency control technics at Wien's laboratory.

---

[69] On the field Cady, *Piezoelectricity*, 678–85, 695–8. Kao presented his works to *Comptes rendus* and *Revue d'Optique*; he published his dissertation in 1931.

Two groups in Breslau connected interests in acoustics and piezoelectricity: one at the laboratory of W. Waetzmann at the Technische Hochschule, and the other at the university. Waetzmann and his assistants and students studied acoustics using new piezoelectric technics. This research led Hans Müller to an acoustic examination of quartz resonators, with two different collaborators in 1932 and 1933.[70] A year later, Ludwig Bergmann and Clemens Schaefer at the university's Institute of Experimental Physics followed an effect of optical diffraction grating by ultrasonic waves in liquid, independently discovered by students of Langevin and a group at MIT. The Breslau physicists found patterns analogous to those gained by X-ray diffraction (Laue's patterns) from the light diffracted by piezoelectric crystal, which was the source of the supersonic vibrations. They used the method to probe into the modes of vibration and structure of the oscillating crystals, in addition to examining the targets of the ultrasonic waves that they produced. Extending their research, they collaborated with Erwin Fues, a professor of theoretical physics, and his student Hanfried Ludloff on a theory about the observations. In 1935 Friedrich Klauer from their laboratory reported on analogous results for X-ray diffraction in oscillating crystals. Bergmann at the university conducted research on modes of vibrations in quartz crystals by applying optical–acoustic methods and used light diffraction for exact wave measurements. His main interest, however, remained ultrasound, and he authored the first book dedicated to the phenomenon in 1937. Following his engagement with piezoelectricity, Schaeffer also advised Ernst Lonn to examine the above-mentioned discrepancy between Petržílka's results and resonance theory.[71] Thus, scientific use of the piezoelectric resonator led quite a few researchers to study it, even, as in Lonn's case, to questions quite remote from its laboratory usage. Use of the phenomenon in the laboratory as in practical technics stimulated further study.

[70] Hans Müller from Breslau should not be confused with his more famous Swiss-German namesake (usually written as Mueller), who worked among others on ferroelectricity. Karl Bücks and Hans Müller, 'Über einige Beobachtungen an schwingenden Piezoquarzen und ihrem Schallfeld', *Zeitschrift für Physik* 84 (1933): 75–86. This research also received support from the Helmholtz-Gesellschaft. In the second study, with T. Kraefft, Müller examined the Doppler effect between two vibrators.

[71] Clemens Schaefer and Ludwig Bergmann, 'Über neue Beugungserscheinungen an schwingenden Kristallen', *Die Naturwissenschaften* 22 (1934): 865–90; Friedrich Klauer, 'Röntgen-Laue-Diagramme an piezoelektrisch schwingenden Kristallen', *Physikalische Zeitschrift* 36 (1935): 208–11; Lonn, 'Zur Theorie der Schwingungen von Kristallplatten'.

Egon Hiedemann a Privatdozent at the University of Cologne, applied the same methods from ultrasonics in liquids, his expertise, for two related experiments. In 1934 with two associates, he studied nodal points of vibrating quartz crystal. With another collaborator, he showed that the vibrations of a quartz crystal immersed in liquid are strongly damped. For these works he received support from the German *NG*. Like Bergmann, Hiedemann kept his focus on ultrasonics, authoring a book on the subject in 1939. Two years after Hiedemann's study, H. E. R. Becker, a student of Wien in Jena, gained similar results for the damping of quartz in other liquids in his doctorate research. In the second part of the work, Becker linked the dissipation of energy from the resonator to the effect of ultrasonic waves on light. He, thus, linked two local interests in ultrasonics and frequency control shared by physicists at the university and at Zeiss. The company supported his research with material and the assistance of Straubel, then already a company employee.[72] A few other researchers examined piezoelectricity following their work on acoustics. Some examined the resonator's general properties. Beyond such a study, Arnold Pitt and D. W. R. McKinley, two physics students at Toronto, followed observations they made in measuring the velocity of sound at low temperature to study variation of the piezoelectric effect in quartz at a wide range of temperatures.[73]

Piezoelectricity provided also methods for exploring contested or unknown material structure. Its existence in particular directions indicates a lack of corresponing symmetry in those directions. Such studied had begun already in 1924, a year before Giebe and Scheibe contrived an effective method for testing the mere existence of the effect in many crystals. In this 'powder method' they put many—at least fifty—small crystal pieces (of 0.1 to 5 mm in length) between the condenser's plates in a circuit for detecting resonance of the kind suggested by Cady. The electric effect of resonance is strong enough that a few of the small crystal pieces laying close to a direction sensitive to the alternating electric voltage were sufficient to induce an effect detectable by a sensitive telephone headset.[74] Here, the two PTR researchers employed knowledge

[72] Egon Hiedermann, H. R. Asbach, and K. H. Hoesch, 'Die Sichtbarmachung der stehenden Ultraschallwellen in durchsichtigen festen Körpern', *Zeitschrift für Physik* 90 (1934): 322–6; H. E. R. Becker, 'Die Rückwirkung einer umgebenden Flüssigkeit auf die Schwingungen einer Quarzplatte', *Annalen der Physik* 25 (1936): 359–72; Becker, 'Debye-Sears Beugungserscheinung'.

[73] In addition to those mentioned, those reaching the subject from acoustics included Speight (1935) and his co-author Jatkar (1939).

[74] E. Giebe and A. Scheibe, 'Eine einfache Methode zum qualitativen Nachweis der Piezoelektrizität von Kristallen', *Zeitschrift für Physik* 33 (1925) 760–6.

acquired from their applied research, i.e. the resonance effect explored for improving frequency standards, for scientific goals, i.e. the exploring the structure of 'non-useful' crystals. Following their lead, about a dozen scientists employed the laboratory method to examine dozens and by 1943 hundreds of minerals. Whereas Giebe and Scheibe reached their method from the study of crystal resonance, most other users of the method employed it due to their interest in the structure of crystals and other materials. Most of these researchers, like a few others who utilized piezoelectricity to probe into material structure, studied physical chemistry, crystallography, and mineralogy. They employed piezoelectricity following increasing familiarity with it and the possibilities offered by Giebe and Scheibe's method.[75]

Once known, piezoelectricity found further laboratory uses. For example, quite a few researchers used resonators composed of piezoelectric crystal and another solid to measure the latter's elastic moduli.[76] Other employed the effect to measure electric capacity (G. A. Kjandsky, 1928; Clay, 1932) and the specific charge of electrons (Kirchner, 1932). It found also surprising uses such as in confirming special relativity in an experiment analogous to Michelson-Morely's (A. B. Wood, G. A. Tomlinson, and L. Essen, 1937). The former studies seem to originate from a concern about the subject matter rather than about the piezoelectric method, the latter at least partly from an experience with piezoelectric vibrators for ultrasonic and frequency standards, respectively, by Wood at the British Admiralty and Essen who continued Dye's work at the NPL. Wood had also joined David A. Keys in developing a piezoelectric method for measuring pressure during explosions, following a wartime suggestion from J. J. Thomson (mentioned in Ch. 8).[77]

These technical and laboratory applications led also in this case to further study of piezoelectricity. In 1937, following his use of its effect to measure capacity, Jacob Clay, professor of experimental physics in Amsterdam, determined the piezoelectric coefficient of quartz under high pressure and at various temperatures, needed for validating the accuracy of the capacity measurement. The pressure-measuring method led Keys, who had just left the

[75]These studies include those of R. Lucas (1924); Hettich with Schleede (1927, 1928) and with Schneider (1928); S. B. Elings and P. Terpstra (1928); H. Mark and K. Weissenberg (1928); W. A. Wooster (1929); L. Egartner et al. (1932); H. Seifert (1932); G. Greenwood and D. Tomboulian (1932, 1935, 1937); B. Gossner and N. Heff (1934); J. Engl and I. P. Leventer (1937); and W. L. Bond (1943). Cady, *Piezoelectricity*, 231–3.

[76]Cady, *Piezoelectricity*, 486.

[77]A. B. Wood, 'Admiralty Experimental Station, Shandon (Gareloch) Dunbartonshire', *Journal of the Royal Naval Scientific Service* 20 (July 1965): 42–78, on 52.

group to teach physics at McGill University, to study possible variability in the piezoelectric coefficient of tourmaline, the crystal his group found useful for the method. The method required that it maintain its value when measured in relatively long intervals of changing pressure (isothermal process) and under abrupt changes in pressure (adiabatic process). In 1923, Keys found that the differences between them is very small and thus negligible for any measurement. Other groups also developed piezoelectric methods for measuring an explosion's pressure. The American work led John C. Karcher at the Bureau of Standards to examine the linearity of the relation between pressure and electricity in quartz under very high pressures of up to about 3500 atm in 1922. A decade later Paul Bernard in Paris showed interests in the invariability of the piezoelectric effect in quartz and in its use to measure explosive pressures. He examined the exact reversibility of piezoelectricity and the absence of hysteresis before reporting about his pressure measurements. In 1934 André Langevin (Paul's son) with different collaborators employed piezoelectricity for measuring pressure in blood vessels and explosions at his father's laboratory. Following its use, he examined the variation of quartz piezoelectric coefficient with temperature (joining the above-mentioned discussion) and in 1940 determined its absolute value. Determining pressure had many laboratory uses in science and technology, leading quite a few researchers to employ piezoelectricity for that aim. Most of them used piezoelectricity due to an interest in the measured magnitude, and did not study the effect itself.[78]

The Use of Other Effects to Study Piezoelectricity

The power of X-ray diffraction to probe the inner structure of piezoelectric crystals drew a handful of the many experts on those methods to their research. As mentioned, Bragg and Gibbs had already employed X-ray diffraction to probe the structure of quartz, the paradigmatic piezoelectric crystal in 1922. In the 1930s physicists used the method to finely examine piezoelectric

[78]Jacob Clay and J. G. Karper, 'The Piezo-Electric Constant of Quarz [Sic]', *Physica* 4 (1937): 311–15; David A. Keys, 'On the Adiabatic and Isothermal Piezo-Electric Constants of Tourmaline', *Philosophical Magazine* 46 (1923): 999–1001; J. C. Karcher, 'A Piezoelectric Method for the Instantaneous Measurement of High Pressures', *Scientific Papers of the Bureau of Standards* 445 (1922): 257–64; Cady, *Piezoelectricity*, 220, 691–5. Bernard published his results in three separate articles in 1934 and 1935. Seidl, who is mentioned in the subsection 'The Use of Other Effects to Study Piezoelectricity', seems as a unique example in Cady's list of a researcher whose interest in piezoelectricity led her to its application in mechanical measurements. A few others had experience with piezoelectric technics but not with its research.

vibrations and their structure, complementing and sometimes exceeding the resolution reached via other methods. In 1931, for example, Gerald W. Fox and Percy H. Carr determined the amplitude of vibration of quartz lattice atoms due to piezoelectric resonance, by measuring the intensity of X-rays reflected from the crystal. With another junior collaborator, Fox extended the research to tourmaline and Rochelle salt. Fox had just become a professor at Iowa State College, after receiving his PhD and teaching in Michigan for three years, and was looking for new research subjects. His former teacher and an X-ray expert, James M. Cork, probably suggested to him using x-ray diffraction to study piezoelectric resonators. Shortly after, Cork joined his former student carrying out experiments to explain a surprising observation in Fox and Carr's experiment regarding the intensity of X-ray reflection.

Cork alone further compared oscillating and static crystals by their Laue's X-ray diffraction patterns with indecisive results, and in 1936 suggested an X-ray study of vibrating crystals to his student Carl V. Bertsch. Fox also returned to such studies in 1935 and 1938. From 1932, however, he extended his experimental study of piezoelectricity (always jointly with a junior associate) to other methods and issues like possible causes of breakage and properties of the static effect, which suggest an interest in questions related to technology. X-ray methods, thus, led him to study piezoelectricity, but he extended its study and earned expertise in the field also due to his search for a research project (following the disciplinary logic) and its relevance to technics.[79]

A similar combination of experience in X-ray methods and an interest in practical usage of piezoelectric crystals probably led Václav Dolejšek and Miroslav Jahoda to study the variation in X-ray diffraction patterns of quartz due to the effect of static piezoelectricity on the lattice, in 1938. Dolejšek was the director of the Institute of Spectroscopy at Charles University in Prague, working mainly on X-ray spectroscopy. This research led him to study crystals used for grating X-rays for scientific and technical aims. His institute had an interest in the latter, as it hosted the Physical Research Institute of Škoda, a major manufacture of a wide range of metallic machines and armaments.

[79] Moving to Iowa, Fox discontinued the research he had done in Michigan. Percy H. Carr and Lester T. Earls, 'Gerald W. Fox', *Physics Today* 27, no. 4 (April 1974): 135. Fox and Cork submitted their joint article two months after the publication of Fox and Carr's paper. Gerald Fox and James Cork, 'The Regular Reflection of X-Rays from Quartz Crystals Oscillating Piezoelectrically', *Physical Review* 38 (1931): 1420–3. In a 1933 publication with Morris Underwood, Fox concluded that tourmaline plates are 'inferior' to quartz plates in the power output of their oscillations.

Piezoelectricity probably attracted his attention also due to the activity of Petržílka's group in the same physics department.[80]

Expertise in X-ray diffraction technics also led K. S. Knol in Groningen and G. E. M. Jauncey in St. Louis to study piezoelectricity. With W. A. Bruce, Jauncey joined Fox's above-mentioned study of the effect of piezoelectric vibrations on diffraction patterns, and exploited it to infer about the lattice structure of quartz. Knol, however, did not use X-rays in his study of piezoelectricity. With Coster and Prins and alone he determined the piezoelectric coefficient of zinc blende (sphalerite) via a static method. They carried out this work in order to measure the effect of electric polarity on the intensity of the crystal's X-ray reflection. That zinc blende has a simple lattice structure was probably why they studied the crystal. That it was the first crystal for which a value of a piezoelectric coefficient was theoretically calculated, via Born and Bormann's 1920 method, provided another reason to measure its value. The predicted value was an order of magnitude higher than a 1913 measurement. Knol et al. had reasons to assume that the earlier determination was too low. Indeed, Knol found a higher value, which was still five times lower than the theoretical prediction. Regarding the simplicity of the theory, Cady commented that 'to have arrived at the same order of magnitude at all was an achievement'.[81]

As the work of Knol suggests, the use of X-ray diffraction was often connected to an interest in crystal structures, an interest which provided another motivation for studying piezoelectricity. Shoji Nishikawa, who is mentioned in the subsection 'Paths to the Study of the Piezoelectric Resonator', had studied crystal structure by X-ray analysis since working in W. H. Bragg's laboratory in 1920. A decade later he recruited his students to study, first, the effect of vibrations on the intensity of X-rays reflected from quartz and, second, modes of vibrations of quartz plates by X-ray. As mentioned, Nishikawa also had an institutional incentive to examine technically promising issues.[82] Malcolm

[80] V. Dolejšek and M. Jahoda, 'Sur les variations du réseau des cristaux piézoélectriques produites par une tension électriquestatique', *Comptes rendus* 206 (1938): 113–15; Marie Neprašová and Miroslav Rozsíval, 'In Memory of Professor Dr. Václav Dolejšek', *Czechoslovak Journal of Physics* 5 (1955): 115–16.

[81] The Groningen group carried out the experiment in 1930 and Knol suggested a more exact one in 1932. He also measured quartz's coefficient at the same time, but this seems mainly to verify the accuracy of his method, as the value for quartz was better established. Cady, *Piezoelectricity*, 7, 229–30 (quotation), 742–3. Jauncey and Bruce published their results in 1938.

[82] Nitta, 'Shoji Nishikawa'.

Colby in Texas also studied X-ray reflection from vibrating quartz in 1932. Two years later he carried out an X-ray diffraction analysis of lengthwise vibrations in quartz, and an optical observation to measure strain ratio in quartz under static electric voltage. Colby had been interested in crystal structure since writing his 1929 dissertation in Chicago. Yet, he had also an interest in the applications of piezoelectricity. Before going to Chicago, he studied at the University of Texas and in 1926–7 published on frequency measurements, with his former teacher, and later colleague, Brown. As mentioned earlier, Brown reported on his study of the piezoelectric resonator in 1931.

For those already interested in piezoelectricity, X-ray diffraction and other modern observational techniques suggested further tools in its study. The NRL (which had clear interest in frequency control) employed the expertise with X-ray technics that Barret acquired in his work at the metallurgy department for a similar study of vibrating quartz in 1932. Schaeffer and Bergmann directed Klauer to perform an X-ray experiment analogous to their acoustic-optical one. Osterberg had become interested in piezoelectricity early in his studies, and then employed an interferometer method in his 1931 dissertation research on the piezoelectric vibrator. Others who used the methods in the field do not seem to have a special expertise in it.[83] K. Eichhorn applied another new optical method of the stroboscope to study the strain in quartz rods piezoelectrically excited into flexural vibrations.

Physicists working on strong electromagnetic radiation found in piezoelectricity one among many fields for examining its effect. For his 1928 dissertation at the Vienna Institute for Radium, Research J. Laimböck examined the influence of X- and γ-rays on the piezoelectric coefficients. His more experienced colleague at the institute and an assistant at the university, Franziska Seidl, followed Laimböck with a similar a study. In 1935 she returned to the question with E. Huber, examining also the effect of radiation on conductivity. Neither of these experiments found conclusive evidence for an effect. Seidl had used X-rays to study crystals before, yet apparently her encounter with piezoelectricity led her to a few additional studies of the effect itself. In the most interesting of these studies, she examined the changes in the conductivity of quartz due to the electric polarity produced by the piezoelectric effect. For this goal she also

---

[83] In 1927 Osterberg wrote his BA thesis (jointly with Jacob W. Moelk) on 'The Influence of Temperature on the Piezo-electric Effect in Quartz' (viewable through Google Books'). Interferometry was used also by G. Wataghin and G. Sacerdote in Turin (1931) and Samuel H. Cortez at Indiana University in 1934. Sacerdote had an interest in high frequencies, which explains his study of vibrating crystals.

measured the piezoelectric coefficient of the crystal she used.[84] While the Viennese experts in radiology examined the effect of radiation on the piezoelectric properties, in 1926–7 Tawil in Paris and R. Moens and Jules-Émile Verschaffelt in Ghent examined the effect of the piezoelectrically induced vibrations on the optical properties of quartz. Yet, unlike the scientific motivation of their Viennese colleagues theirs seems to lie in the practical potential of the effect for nascent television technology.[85] Still, as mentioned earlier in the chapter, the research and development of the technology led Tawil also to a more general study of piezoelectricity.

9.6 The Sites and Kinds of Piezoelectric Research

The relevance of piezoelectricity to studies within physics and crystallography widened the circle of individuals engaged in its study. It attracted also researchers who had worked on the resonator in academic and government laboratories. Most of those who joined the study due to that interest worked in academic research institutes, making universities the most important site of its scientific–disciplinary study. In the German-speaking world they included Hiedemann and his collaborators in Cologne, Müller and Bergmann in Breslau and their associates at the university and the Technishce Hochschule, and Seidl and Laimböck in Vienna. In the United States they included Fox's group in Iowa, with its connection to Cork in Michigan, and Colby and his associates in Austin. Paul Langevin and Fabry in Paris served as the nuclei of two additional groups of students, although unlike the groups discussed above, it seems that experience with piezoelectric technics was central in drawing some of their associates to the study. Sonar and its use were central to Langevin's group, and an important context for Ny's early study. Technical relevance and prospects drew Bedeau, Tawil, and de Gramont, all connected to Fabry, to the field.[86] Questions regarding static piezoelectricity also attracted a number

---

[84]Franziska Seidl, 'Elektrische Leitfähigkeit von belasteten Piezoquarzen', *Zeitschrift für Physik* 75 (1932): 488–503; 'Franziska Seidl' http://lise.univie.ac.at/physikerinnen/historisch/franziska-seidl.htm#_ftn5 (accessed 28 January 2015).

[85]Like Tawil, the Ghent group suggested application of the effect for television. Neither of these researchers (Verschaffelt was a physics professor) seems to have had a prior connection to the study of optics or piezoelectricity.

[86]Armand de Gramont, a wealthy aristocrat with an aerodynamics laboratory and an interest also in optics, was a central figure in the establishment and management of the Institut d'Optique Théorique et Appliquée, which Fabry directed, in addition to his chair at the Faculty of Science and later also at the Polytechnique. Gérard Roblin, 'L'Institut d'Optique a 75 ans',

of academic researchers at many sites (Wesleyan, Minneapolis, Graz, Berlin-Göttingen, Lausanne, London, and Beijing). Researchers at the NRL and at Telefunken laboratories also engaged with the topic. An important centre for the static effect that was not prominent in the study of other questions was formed at the LFTI.

The applications of piezoelectricity transformed the geography of its research. The United States, which had not contributed to the study in the pre-applicable phase, became a major centre of study (see Table 9.1 for the research centres). This move coincided with a more general rise of American physics at the time. Still, piezoelectricity became known and used there only through World War I research. American scientists Cady and Valasek followed the knowledge gained in the war to new discoveries. Both exploited their initial lead to keep a central position in the subsequent study of their findings. Others employed expertise acquired in the war (initially often rudimentary) to follow Cady's findings and inventions into new methods and studies of the phenomenon. In accordance with a common view of American science, the technical prospects of the field helped attract quite a few additional physicists to piezoelectricity. Although British and French scientists enjoyed a similar experience with the use of piezoelectricity, and Cady introduced a few of them to crystal frequency control at an early stage, only a few of them engaged in studying the resonator. Paris became a centre for piezoelectric study, but its scientists rarely worked on the main issues examined by scientists in other places. Most of the researchers there were young, a high portion came from the developing world, mainly from China (to which they transfer their expertise), and usually they did not pursue the study of piezoelectricity for long.[87]

Although its scientists did not share an experience with piezoelectric vibrations in their war research, and despite a short delay in the transfer of frequency standards technics due to the post-war boycott, Germany maintained a leading role in the field. The old tradition of piezoelectric research in Germany was one reason for its continuous strength in the field. Yet, personal

Opto 118 (1996): 13–24. Beginning in 1930, de Gramont published with Mabboux, Beretzki, and alone on piezoelectric oscillators, comparing resonators at neighbouring frequencies, suggesting ways of keeping beats between two resonators independent of the temperature and ways of producing audial waves, useful electroacoustics, Armand de Gramont and Georeges Mabboux, 'Camprison de quartz piézoe-électriques oscillant á des fréquients voisines', Comptes rendus 190 (1930): 1394–5.

[87]The one exception in Tawil, but he was an outsider to the academic world with a clear technological aim in the field.

Table 9.1 Main research centres on piezoelectricity by geography
The major subjects of research are mentioned for each centre. Main figures in the research
groups appear in bold, names of government institutions appear in italics, and names of
industrial companies in small caps. The list does not include all sites of research. Initials
appear to prevent confusion and when the person is not mentioned in the text.

1) Britain
   a) *NPL*: **Dye**, Vigoureux, Essen: resonators (theory and experiment)
   b) **Gibbs** connected to Bragg (lattice structure) and Tsien
      (a connection to Ny's group), Thate (resonators)

2) France
   a) **Associated with Fabry**: Bedeau, Bernard, Yeou Ta (cotton), Tawil, Pan, Ny (later
      in Beijing), de Gramont (with Beretzki): static effects (torsion), effect on light
   b) **P. Langevin's** circle: A. and P. Langevin, Moulin, J. Solomon, Baumgardt,
      Guerbilsky, Lucas: static effects and vibrations

3) Germany
   a) Jena (connection to ZEISS): Günther, Grossmann, M. Wien, Hehlgans, Straubel,
      Becker: vibrations, effect on light
   b) Göttingen: Born, Heckmann, Laue, Gockel: theory, connection to Voigt
   c) Breslau: Uni: Schaefer, Bergmann, Fues, Ludloff, Lonn (?); TH: Müller, Kraefft
      & Bücks (students of Waetzmann): vibrating crystal connected to ultrasonics
   d) *PTR*: Giebe, Scheibe, Blechschmidt: resonators
   e) TELEFUNKEN: **Meissner** and Bechmann, Heegner: resonators, atomistic theory

4) USA
   a) Academic
      i) Wesleyan: **Cady**, Fujimoto (in Ohio), Hans Jaffe, Van Dyke, Harrison,
         Powers: resonators, static determinations
      ii) Michigan: Cork, Bertsch: X-ray studies
      iii) Iowa: **Fox** (connected to Cork) with Carr, Fraser, Hutton, Fink, Underwood,
         Fredrick: X-ray studies and resonators
      iv) Wisconsin: Terry (pre 1929), **Osterberg** with Cookson, Hestenes: vibrations
         and static measurements
   b) US government and industrial laboratories: resonators and static measurement
      i) *NRL*: Dawson, Barrett, and Howe
      ii) *BoS*: Hund and Wright, Wright and Stuart
      iii) BELL LABS: Mason, Skellett

5) USSR: Leningrad (*LFTI*): Andreeff, Fréedericksz and Kazarnowsky, and
   Michailow; Schulwas-Sorokina, M. V. Posnov, Lissütin, Sokoloff (but connection
   to Odessa), Kjandsky, Eremejew and Kurtschatow: effect of temperature, resonator
   theory

6) Others
  a) Japan: Tokyo: Koga, Shoyama (Tokyo Eng Uni), Namba & Matsumaura (*Electrotechnical Laboratory Tokyo*); Watanabe (Sendai): all in electrical engineering; Nishikawa, Sakisaka and Sumoto (Riken): resonators, X-ray studies
  b) Czechoslovakia: Prague: Dolejšek & Jahoda, Khol, **Petržílka**, Žáček, Zachoval: vibrations, X-ray studies
  c) China: Beijing: **Ny**, Tsien, Fang, Shang Keng-Yi: torsion, including vibrations

---

connections with a pre-war researcher existed only in the case of Born and Laue (both studied and worked with Voigt in Göttingen). Other reasons for the German position in the field were probably the general strength and diversity of its physics community, and the good connections between industrial and academic research in the country. German research enjoyed the contribution of industrial (Telefunken) and government (PTR) laboratories. Their contributions were especially important at the early stages when only few academic scientists examined connected questions.[88] Similar laboratories carried out research in the USA, although they tended to concentrate on narrower questions related to specific applications. National laboratories did play a role, albeit more limited, in Japan and the UK. In particular, the NPL performed a few studies comparable to those of the PTR, but it did not continue with similar research after Dye's untimely death. Commercial companies in other countries, including Britain, did not study general properties of piezoelectricity.[89] Ironically, the connection to the older German tradition of research in the field was more significant at the new LFTI, with Fréedericksz and the directorship of Joffé, who worked on piezoelectricity with Röntgen before the war. Another relatively important new centre was developed in Tokyo, due to interest in the applications and the transfer of knowledge on X-ray

[88]The relation between industry and academy shows that not all groups in Germany shared the view that science should be pure and idealistic.

[89]The one exception is van der Pol's development of an equivalent circuit at Philips, but it did not have any continuation. Although this is only one example, the minor British contribution suggests that it lagged behind both the USA (as accepted) and Germany. This fits Marsch's conclusions that German industry invested considerably more than the British in both research and development, correcting an earlier claim by Edgerton and Horrocks. Both did not elaborate on the character of research in the two countries. Ulrich Marsch, *Zwischen Wissenschaft und Wirtschaft: Industrieforschung in Deutschland und Grossbritannien 1880–1936* (Paderborn: Schöningh, 2000); D. E. H. Edgerton and S. M. Horrocks, 'British Industrial Research and Development before 1945', *Economic History Review* 47 (1994): 213–38.

diffraction. Researchers in most other industrial countries took part in the study of piezoelectricity research; among them, it is noteworthy to mention the Prague school.

A few individuals engaged in research on piezoelectricity, like Cady, Langevin, Fabry, Ny, Petržílka, and Fox, played a central role in directing their (usually junior) colleagues to its study. Fabry himself did not study piezoelectricity, but still induced other researchers to do so. They created small research centres that were often connected to institutional interest in technology. As the example of LFTI suggests, however, such an institutional interest did not necessarily imply direct research on the resonator, but led to relevant studies of static piezoelectricity. Moreover, sometimes scientists who began with questions related to technology widened their research to issues quite remote from those of design, or even without any apparent technical implications. These included also researchers working in industrial and government institutes (e.g. Giebe, Scheibe, and Meissner).[90] Still, piezoelectric research in national and industrial laboratories, like those of the NPL, NRL, PTR, and Telefunken, began with an explicit concern to improve practical devices. This was also the case at the engineering departments in Tokyo (research at other engineering schools was usually regarded as contributing to technology rather than to science, and so had at least an equal concern in practical methods). The concern with technology was explicit, but the research was less direct in the case of the LFTI.

At the early stage, research at the universities often originated from a particular interest of an individual like Cady, Langevin, and Terry in related technics. In subsequent years, particular extant connections to technical uses of piezoelectricity waned in importance. A more general inclination towards research relevant to technology seems to be the most important single factor in directing scientists to such studies of piezoelectricity. As discussed in the Introduction, the view beyond this inclination, namely that academic physics should help technology, was quite common at the time and received official endorsement in policy statements and in material support from foundations and

---

[90]The industrial and government researchers who examined nonapplicable issues in piezoelectricity worked only in Germany. However, the fact that industrial researchers in other fields, especially in the USA, carried out general research suggests that the national difference in this case might be rather accidental. Yet, it might indicate that the freedom of research enjoyed by Langmuir and a few other heroes of interwar American industrial research was an exception in these institutes. Such an assumption is supported by the limited freedom of the researchers who worked on the field in Bell Labs, as seen in Chapters 5 and 6.

states. Providing basic knowledge for technical progress was more explicit in a few places like the Riken Institute in Tokyo, the Optic Institute, which Fabry directed, and the chair for technical physics in Jena. Still, the idea was shared by scientists in regular physics departments. It was particularly strong at science funds. Quite a few studies in Germany and the USA enjoyed financial aid from the NG, the Helmholtz Gesellschaft, and the National Research Council (USA).[91] Commercial companies also assisted research relevant to their technics. Zeiss, AT&T, and the Société de Condensations et d'Applications mécanique provided funds, crystals, and expertise to academic physicists. At least in one case an institute aimed at fostering technology, the Hertz Institute for High-Frequency Research, assisted a university scholar, Petržílka, by providing temporary employment. Thereby it strengthened his involvement with questions related to the use of the resonator. Moreover, prospects of employment in such institutes or in industry (an option that Osterberg and Straubel realized) encouraged young academics, struggling to secure employment or sufficient income, to pursue technologically relevant questions. Hopes to gain material support from commercial and government agencies probably also encouraged physicists who reached piezoelectricity from studies of other fields of physics, like Fox and Colby, to explore questions connected to its technical uses. It was through these particular connections to technological interests and through the general expectation to support technology that the needs and aims of users of piezoelectric technics directed research in the field.

9.7 Conclusions

Almost any study of the resonator and many studies of static piezoelectricity were of potential value for future technical uses. Studies of the resonator that were regarded as part of physics were interesting from technological and scientific perspectives, as they revealed novel knowledge about the phenomena and knowledge useful for technical design. Studies of the resonator were connected to other issues of piezoelectricity, be they the properties of the static effect, crystalline structure, or ultrasonic effects. Studies aimed at scientific and technological goals were connected and sometimes even combined into one and the same project, and this made research pertinent to technics

[91] Charlotte Bigg, 'L'optique de Précision et La Première Guerre Mondiale', *Schweizerische Zeitschrift für Geschichte* 55 (2005): 34–45. Kirchhoff, 'Wissenschaftsförderung und forschungspolitische Prioritäten der Notgemeinschaft der Deutschen Wissenschaft'; Kevles, *The Physicists*, 117–54.

attractive to aspiring physicists. They could move from studying general properties of the resonators, the motion of atoms in them, and the changes in the
value of the piezoelectric coefficient near critical temperatures, to examining
special cuts and their usefulness for particular ends, or the variation of resonance frequency with temperature. When they pursued questions relevant to
technics, like the latter, they could still publish their results in physics journals and contribute to the discipline, and so advance their career, and open
prospects in both academic and industrial research. Physicists could continue
with the kind of disciplinary questions that they usually examined, mainly
the (quantitative) effect of varying different physical conditions on a particular magnitude, and maintain their self-image as scientists while answering
questions relevant to technology, even when the particular questions did not
originate from 'the logic of the discipline'. This double use of the studies facilitated the engagement of academic scientists with technologically relevant
questions, an engagement driven by particular connections to government,
public, and commercial institutions aimed at fostering technics (e.g. state
and industrial laboratories, scientific funds) and a general societal interest in
employing science for material benefits.

Beyond the concern for technology, two factors led scientists to study piezoelectricity in the fifth phase of its evolution—that of a field recognized as
technically useful: expertise in related research topics, and experience in its
use in the laboratory. While the concern for technology often reflected institutional interests (utilitarian logic), the two other routes originated in a
research programme of the scientists, or of his (and in rare cases her) senior
colleague, and in the disciplinary logic of exploring new effects. The applicability of piezoelectricity in the study of ultrasonics and crystal structure not
only made the phenomenon well known to a larger group of scientists but
also provided expertise in its use, which some exploited in its study. Research
of other issues, primarily from the physics of crystals and their structure, provided relevant expertise, which scientists exploited in the experimental (e.g.
X-ray) and theoretical (e.g. elastic theory) study of piezoelectricity. Their
questions followed disciplinary logic in issues like interactions between different phenomena, determining coefficients and their variations, and explaining
discrepancies between experiment and theory. Since researchers in a disciplinary regime are often actively searching for topics they can study, their
expertise and the opening of connected questions were critical in directing
further research.

That the improvement of practical applications was not the main or explicit goal of many studies pertinent to applications makes the effect of

technological interests on their choice of problems more difficult to trace. As the examples discussed here show, usually a combination of a few factors drew scientists to these studies. Yet, in many cases, a concern with improving technics, often connected to telecommunication, played a major role among them. Although it is difficult to determine the effect of particular technological interests in guiding the research of an individual scientist, their overall impact on the research of piezoelectricity is evident. The two central characteristics of the research—the concentration on the resonator and the emphasis on quartz—were pertinent to the practical applications of the effect. Moreover, many studies of the resonator dealt with issues important to the users and developers of frequency standards and control, like the determination of resonance frequency and its thermal stability. While these were interesting questions also from a disciplinary perspective, no comparable research was done on other equally valuable questions that did not have similar technical prospects. For example, scientists paid limited attention to issues like the relationships between piezoelectricity and radiation, its relationship to pyroelectricity (which, as mentioned in Ch. 1, attracted considerable attention in the 1910s), or those between the lattice and crystallographic structure of the crystals and their piezoelectric properties. A closer quantitative examination of more crystal species could have shed light on these relations; it might have provided also some clues about the strength of the piezoelectric effect, which is not dealt with by Voigt's phenomenological theory. In short, due to technological interests in piezoelectricity, physicists directed their studies to topics relevant to technics rather than to other topics.

The expectations that scientific research would extend and improve knowledge about piezoelectric-based technics, i.e. technology, rather than the possibilities that the technics offered for experimentation directed research in the field into particular topics. Neither piezoelectricity nor technics based on the effect were a 'research technology', i.e. general purpose laboratory technics useful for experiments on various scientific question.[92] Scientists often chose the field and particular topics in its study because crystal frequency control and sonar, the main technics based on the effect, were highly useful for particular organizations in the industry, the state, and the military. The central role

---

[92] Even when piezoelectricity was used in laboratory instruments, as in the case of research on ultrasonics it was confined to a specific area and did not became a basis for a versatile research technology in Shinn's sense. Terry Shinn, *Research-Technology and Cultural Change: Instrumentation, Genericity, Transversality* (Oxford: The Bradwell Press, 2008).

of technological interests in the case of piezoelectricity suggests that similar interests also shaped other scientific fields. Moreover, the channels and factors that allowed organizations with interests in these technics to direct research in piezoelectricity were probably exploited also for advancing other technologies. The channels and factors exposed here are therefore instructive for our understanding of the interaction of scientific research with economic, state, social, and cultural forces in the twentieth century.

Organizations aimed at advancing technology were constitutive in channelling research to related topics. These included, as we saw, state, industrial, and military laboratories, which employed some of the scientists who studied piezoelectricity, and supported others. They were more influential in the early stages of the research, as were a few individual scientists who either encountered the phenomena in WWI military technological research or regarded it relevant for their interests in wireless. Research funds provided material support that encouraged study to be pertinent to piezoelectric technics more in the latter part of the period. The early stage was critical for creating a basic body of knowledge required to identify the new phenomenon as fruitful and worthwhile for disciplinary study. That piezoelectricity had already been a mature scientific field connected the research to particular disciplinary questions, like the explanation of resonance by the phenomenological theory.[93] The early stage also saw the formation of a small group of scientists with expertise in the phenomena, and thus with stakes in its further study. Thereafter, the subject became attractive also for disciplinary study to scientists without a former encounter or a special concern for its applications. It also made the phenomenon better known, drawing additional scientists to identify its possible theoretical and experimental relevance to their research interests and expertise, e.g. in X-ray diffraction crystallography.

Several factors allowed users and developers of technics to exploit these channels in order to expand and shape research on piezoelectricity. One important factor was the close epistemic connection between scientific and technological issues. In piezoelectricity, as in many other contemporary fields, some questions of technology and physics were very close and even overlapped, a link that provided both those working on technics and on science an incentive to collaborate. The new and strengthened institutions that aimed to foster collaboration between science and technology emerged partly due to this connection and helped to reinforce it. These institutes, including state and

---

[93] It would be interesting to compare this research with that of phenomena that became useful with their discovery, like X-rays.

industrial laboratories, scientific foundations, and academic institutes, aimed at connecting physics and technology, like the chair for technical physics in Jena, the LFTI, and Riken played an important role in directing research to topics related to technics of practical interest. Historians should pay more attention to the last group, namely of academic research institutes, which combined research in technology and physics,[94] as it has not received due discussion in the secondary literature. Not less important, technological interests directed research related to technical improvements also at 'regular' university departments, not by obligation nor by direct material rewards, but by subtler and thus more effective means: by expanding the research in piezoelectricity in general; by suggesting potential material gains from related studies, and more plausibly by suggesting a way to attain recognition, since these topics attracted attention in a larger circle than purely disciplinary topics; lastly and probably most important by the general view that academic research should help answer questions pertinent to technics.

These factors did not depend on the particular properties of piezoelectricity. They, thus, present general strands in society, industry, and science that shaped relevant fields of physics according to technological interests. Moreover, since some of these factors were either new or more powerful in the interwar period than previously, it is plausible to conclude that their effect was stronger at the time than earlier, a conclusion supported by other studies.[95] Interwar physics was less 'pure' than often assumed. Its tighter connections to societal powers foreshadowed the strong influence of the state and industry on post-WWII

---

[94] Among others, a few of the institutes of the KWG belong to this group.

[95] On the connections between physics and technology in the interwar period see *Interactions of Interwar Physics: Technology, Instruments, and Other Sciences*, vol. 31, issue 3 (2018) of *Science in Context*, especially Katzir, 'Physics, Technology, and Technics'. Technological interests directed research to pertinent topics among other fields in acoustics and especially electroacoustics (Robert Thomas Beyer, *Sounds of Our Times: Two Hundred Years of Acoustics* (New York: American Inst. of Physics, 1999); Wittje, *The Age of Electroacoustics*), in physics of the Earth's crustal layer and atmospheric-electricity by oil, mining, and radio industries (Anduaga, *Geophysics, Realism, and Industry*), in thermal emanation of electrons (thermionics) and electronic technics by electric and telecommunication corporations, and in natural and artificial radioactivity and particle accelerators by the interests of medical and power electricity industries (J. L. Heilbron and Robert W. Seidel, *Lawrence and His Laboratory: A History of the Lawrence Berkeley Laboratory* (Berkeley: University of California Press, 1989)). The connection between industry and academic research in chemistry had already been well established before the First World War.

physics.[96] Moreover, the combination of direct, indirect, and often subtle ways by which research pertinent to technics became attractive for scientists, in this case, seems to be central to the success of societal interests to channel research in modern physics also at earlier and later periods.

[96]Forman, 'Behind Quantum Electronics: National Security as Basis for Physical Research'.

CONCLUDING REMARKS TO THE BOOK

Chapters 1–9 have related the story of the transformation of piezoelectricity from a small, quiet, and unpractical research subject to a vibrant field known and studied primarily for its technical applications. This transformation was possible thanks to a combination of epistemic and social factors, which led to the bidirectional transfer of knowledge between science and technology and to concentrations of research and development efforts in particular directions and topics. Knowledge of the phenomenon allowed it to be applied for practical technics, but use of such previously unpractical scientific knowledge depended on scientists turning from disciplinary to utilitarian research, a shift made only in special social circumstances. It was only due to their joining their nations' war effort during the First World War that Langevin and his colleagues sought out some means to detect submarines, and explored the possibility of using quartz crystals toward that end. In contriving and examining suitable technics they transferred experimental and theoretical knowledge about piezoelectricity from science to technology. These circumstances led not only scientists, but also inventors, like Chilowsky, to research on submarine detection, to which they carried knowledge from other technological fields.

Scientists were central also in the transfer of disciplinary knowledge of piezoelectric resonance to technics. In their research, inventions, and consulting work, Cady and Pierce applied findings about piezoelectric resonance to frequency standards and control and encouraged their implementation. At this later phase of applying scientific knowledge to technics (fourth in the sequence of phases sketched here), when the phenomenon had already been applied to sonar, and so had already been studied by engineers and scientists with interest in practical technics, the transfer of knowledge to technology did not require emergency conditions. More common peacetime forces, such as the interest of commercial and state users of telecommunication in measuring and controlling frequencies, sufficed to encourage the research and development of piezoelectric-based technics and their implementation. Scientists were instrumental also in transferring knowledge from technology to science, especially in the third phase of research. That they worked as research engineers during the war and as disciplinary scientists in its aftermath enabled Cady and

Sonar to Quartz Clock. Shaul Katzir, Oxford University Press. © Oxford University Press (2023).
DOI: 10.1093/oso/9780198878735.003.0011

Swann to make findings from technological research a basis for novel studies and discoveries in physics (i.e. piezoelectric resonance and ferroelectricity). Two things made scientists crucial to this knowledge transfer. On the one hand, their knowledge of scientific effects, like piezoelectricity, that were less known to engineers enabled them to use these effects for technical ends. On the other hand, when the scientists returned to disciplinary study, they were expected to produce novel knowledge for which they could exploit findings and methods that they encountered in the technological research. Exploration of inapplicable phenomena, however, fell outside the professional interests of their colleagues working within the utilitarian regime.

In moving between the disciplinary and utilitarian regimes physicists transferred also research methods and attitudes. Aiming at knowledge useful for improving submarine detection technics, they carried out empirical and theoretical studies, mostly quantitative, on the properties of the devices (e.g. Rutherford's hydrophone) and the phenomena on which the technics relied (e.g. underwater (ultra)sonic waves, piezoelectric crystals and their vibrations) wider in scope and further from the specific devices under use than was common in technological research. These studies advanced the development of hydrophones and sonar, and also led to the first observations of the peculiarities of piezoelectric resonance. With other war projects that involved scientists, they contributed to the introduction of a new kind of systematic research into phenomena related to the technics within technical research and development enterprises. The recent industrial research laboratory presented a prior attempt at such a combination, but smaller in scale and with less scientific research. Also there, through recruitment and visits, academic scientists were central in transferring knowledge and methods to technology.[1]

In the early crucial stages, the transfer of knowledge between different scientific and technical fields depended on personal experience and expertise and on personal connections, not that this dependence was a logical necessity. In most cases, one could have found the required information in scientific and technical literature and obtained the required ingredients. Nevertheless,

---

[1] I refer here to the research in industrial laboratories before 1915; see Leonard S. Reich, *The Making of American Industrial Research: Science and Business at GE and Bell, 1876–1926* (Cambridge, UK: Cambridge University Press, 1985). On other studies during the war and their effect see Shaul Katzir, '"In War or in Peace": The Technological Promise of Science Following the First World War', *Centaurus* 59 (2017): 223–37, and also the Introduction to this book. Other pre-World War I institutions for promoting technologically related research did not combine research and development at the same site.

in practice, prior familiarity with the field, or with those already involved in it, was almost indispensable. In exploring using the effect for submarine detection technics, Langevin and, to a lesser extent, Rutherford drew on their knowledge of piezoelectricity, which they had learnt in different ways from the Curies. Personal visits to Langevin and exchanges with him and with those who examined the technics directly (e.g. in the June 1917 Washington, DC conference) were central in the dissemination of the knowledge about the piezoelectric sonar to Allies' researchers. Cady's research on the piezoelectric vibrator as a receiver led him to study resonance phenomena more generally. This stage was the only one that logically depended on personal experience, as his research followed his own unpublished observations, which he had not yet shared with colleagues.[2] Cady was central in promoting the use of his devices and findings by showing them to many researchers, among others at Western Electric and the NBS and to Pierce. He further helped in the first installation of crystal frequency control and promoted the use and study of piezoelectricity in his journey to Europe, which brought researchers like Dye to the field. Moreover, early researchers of piezoelectric resonance were connected to Cady. Those who joined them enjoined an early encounter with the study of piezoelectricity or with its technical usage. Only in the mid-1920s did researchers without a direct personal connection to students of piezoelectricity begin joining the study of piezoelectric resonance.

The critical roles played by a few individuals indicate that contingency was significant in the development of piezoelectric technics. Large forces in the development of science and technology (discussed in the Introduction) and in the development of the strong modern nation-state led scientists to participate in the World War I quest for some means of underwater detection.[3] Yet, Langevin and Rutherford's familiarity with piezoelectricity was quite accidental, in the sense that it was not why they were recruited for the mission. It is unlikely that the piezoelectric sonar would have been invented at that period without Langevin's uncommon expertise with it. All further development of piezoelectric technics, crystal frequency control included, hung on its invention. Contingency also characterized Cady's early research with piezoelectricity. Through a chance meeting he was invited to the June 1917 interallied technical

[2] Cady did not report on his observations in the classified memos probably because they did not seem relevant to the effort to improve the receivers and were still far from an experimental finding.

[3] I leave political historians the question whether the war itself, or only its timing, where contingent. It was certainly external to the development of science and technology.

conference in Washington, DC. Fortuitous observation and personal decision to pursue them led him to discovering the phenomena connected to piezoelectric resonance. Yet, in this case, the contingency seems to be more about the particular person discovering piezoelectric resonance than about its discovery. Since Cady followed common scientific practice both in his technological research on piezoelectric transducers (which led to the unexpected observation) and in his disciplinary pursuit of the phenomena beyond them (which led to the discovery), another physicist in charge of the research on crystal receivers could have reached similar results.[4] No uncommon expertise or fortuitous circumstances seem necessary for contriving crystal frequency control and the quartz clock. Well-known contemporary radio technics provided most of the resources needed for their development. Research on piezoelectric crystals, encouraged by their use, provided supporting knowledge. Once the potential of frequency control for telecommunication was realized, its exploitation to meet growing needs seemed inevitable, due to the strong corporations and states that had high stakes in them.

The appearance of multiple inventions and discoveries in the fifth phase of the research provides further indication of the changing significance of particular individuals versus that of large forces in this history. The lack of multiples in the earlier phases of the research suggests the role of contingency in these cases, whereas their frequent occurrence in the later phases indicates that inventions followed general trends in science and technology. Multiple inventions such as piezoelectric filters (invented by Espenschid and Hansell), the Pierce–Miller oscillator, and, of course, the quartz clock point not only to a widespread interest in the application of piezoelectricity to meet telecommunication needs but also to the scientific and technical resources available for employment. These resonators allowed also for multiples in the study of the phenomenon, as with the simultaneous formulation of the resonator's equivalent circuit (by Van Dyke, Dye, and van der Pol). Similar cases of multiple inventions occurred with the valve-maintaining tuning-fork (Abraham and Bloch and Eccles and Jordan) and tuning-fork time measurements (Dye and Smith at the NPL and Horton's group at AT&T). These cases indicate that knowledge of the tuning-fork and the vacuum tube was widespread among those working on exact measurements and on a societal interest in their results. It was different, however, with sonar. The failure of Richardson, Fessenden,

---

[4]This seems plausible within the American war research, which drew quite a few physicists to this technical research and allowed for further specialization than the research in the European Powers did.

and Behm to contrive an ultrasonic sonar (a case of a 'missed' multiple invention) suggests the significance of Chilowsky's expertise, which was not shared by the 'missed' inventors. One may imagine how Fessenden's scheme would have been developed into Chilowsky's had the latter not delivered his. It is hard to contrive, however, a scenario in which the use of piezoelectricity for sonar would have been suggested except by one of the few physicists familiar with the effect. It was not a multiple invention, but rather a singleton.[5] Such were also the other two groundbreaking innovations in this history: the discovery of piezoelectric resonance and the invention of crystal frequency control technics.

That these singletons were also groundbreaking inventions was not accidental. To a considerable extent, the transformative power of these innovations stemmed from their unexpectedness, leading thereby to novel research and development enterprises. That sonar and crystal frequency control were unforeseen explains why they remained singletons although they solved problems of much technological interest. Without being familiar with the idea of ultrasonic echo-sounding and with piezoelectricity, one would not suspect that their combination would allow for submarine detection. Those looking for some means for frequency control had no reason to explore piezoelectricity. The First World War brought Cady to a unique position in which he was not only well informed with the latest technical advancements but also performed research into the phenomena beyond them. He further explored his early findings, which did not seem relevant for any central problem in technology or physics, thanks to his return to disciplinary research. This specific combination explains why he was the only person who discovered piezoelectric resonance. That the invention of crystal frequency control was not shared by others needs further explanation, since the technical potential for such a

[5] I adopt the terms from Merton's famous discussion of multiples discoveries in science. Still the use of the term 'singleton' here in deliberately different form Merton's. In contrast to his position that every discovery is a potential multiple, I identify here real cases of singletons, in the sense that they would not have been invented at the time had their particular inventor failed to do so. Similarly, I do not assume that the missed attempts would have led to the invention, as Merton and his followers claim. Important differences between scientific discoveries and technical inventions including some mentioned in the text notwithstanding, I regard here both, as most authors on the subject have done. See Robert K. Merton, 'Singletons and Multiplies in Science', in *The Sociology of Science: Theoretical and Empirical Investigation* (Chicago: University of Chicago Press, 1973), 343–70; David Lamb and Susan M. Easton, *Multiple Discovery: The Pattern of Scientific Progress* (Avebury, 1984).

method for electronic telecommunication was clear to many. Still, contriving such a method was a risky endeavour that required specific knowledge of piezoelectricity. In these circumstances large corporate and state laboratories, which contributed to all the multiples mentioned above, did not rush to enter the subject. It remained Cady's own until he demonstrated the feasibility of such a method.[6]

That crystal frequency control did not emerge from the industrial research laboratory illustrates the restrictions on work performed in these institutions. Researchers like those at AT&T's research branch were limited to the subject of their studies. Not only did they not pursue disciplinary research into the phenomenon, like the one that led Cady to the discovery of resonance, but they were restricted also in their technological study to known methods like those of electronically maintained tuning-fork frequency standards. Even that method, based on a well-known instrument, was the product of academic physicists and engineers mobilized to the war effort rather than that of a commercial laboratory. Notwithstanding the contribution of industrial laboratories to piezoelectric technics, the research performed there well supports Leonard Reich's general claim from 1985:

> with only a few exceptions, the results of industrial research have been advances of science and technology in preconceived directions because researchers and research directors usually had to foresee the types of results coming from projects in order to justify their support. By the very nature of preconception, the results were rarely revolutionary. Commercial and financial exigencies made most companies unwilling or unable to give many of their scientists and engineers research freedom needed to pursue revolutionary research. Rules, regulations, and red tape limited the researchers' initiative and the scope of their work, while the need to produce regular stream of results for corporate consumption forced many research directors to favor short-term low-risk projects over longer-term undertakings which were less certain of applicable results.[7]

Even the quartz clock, arguably the most unexpected piezoelectricity-based device to emerge from the industrial laboratory, was contrived through

---

[6]This is not to deny the plausible possibility that if Cady himself had not explored the technical usage of his discovery, others would have taken the enterprise later. Thus, a Merton could claim that the invention of crystal frequency control was a potential multiple. I doubt, however, the value of such a claim.

[7]Reich, *The Making of American Industrial Research*, 8.

research within extant fields (of frequency measurement and piezoelectric resonators), which suggested that the device is feasible. It was a typical product
of research within a field recognized as technically useful, which characterized the fifth phase of research in the scheme suggested in this book. As such
it well fitted industrial and state laboratories, which could invest their richer
resources in well-defined projects of technical interest, like those on frequency
meters. These institutions contributed also to improvements in piezoelectric
frequency control devices (which were contrived in the earlier phase), e.g. developing more stable piezo-resonators of wider use; to the invention of new
technics based on piezoelectric resonance, e.g. filters; and to the study of
resonance phenomenon, e.g. with equivalent network theory.

Industrial and government laboratories began studying piezoelectricity
only after it had proved to be technically useful, i.e. for sonar in the third
phase and generally in the fifth phase of research.[8] Industrial laboratories performed mostly technical research and development supplemented by some
applied research well connected to concrete technical aims. This was very clear
in the case of AT&T, the corporation with the highest stakes in the related
technics. Its intensive technical work in the field was supplemented by some
studies regarding the changing of resonance with temperature. Apparently,
the major technics developed by the corporation's laboratories, the quartz
clock included, did not require much original research into the phenomena,
i.e. scientific study. Two European industrial researchers, Meissner and van
der Pol, could follow an unexpected observation (the former) or explore a
general equation and possible behaviour shown by artefacts (the latter). They
enjoyed more freedom than their American colleagues, probably because they
personally had higher status in their laboratories rather than due to national
differences. Their studies led to unexpected scientific findings and, in the case
of van der Pol, to a practical device, even if eventually it was not used by his
employer.[9] More strictly defined research goals directed the outcomes of other
studies in these institutions. Thus, although the industrial laboratory became

---

[8] Nicolson's research on piezoelectric transducer for telephones is an exception, as he carried
it out while the central effort was on the development of transduces for sonar (third phase). Yet
since he had learnt about the technical usage of piezoelectricity from its application for sonar
he had already turned to the last phase of research in the scheme suggested, i.e. working in a
field known to be technically useful (for the use of crystals as transducers, which does not rely
on piezo-resonance, piezoelectricity did not go through the fourth phase).

[9] In these open-ended studies, van der Pol studied self-coupled triodes and consequently
nonlinear dynamics systems more generally, rather than piezoelectricity.

an important site for technological research and a major one for technical development, it did not turn into a central site for the disciplinary study of piezoelectricity, not even regarding resonance. The research and development practice of the industrial laboratory in this field indicates that it was not the home of science, despite contrary suggestions.[10] State research institutions for physics and technology provided a more welcoming home for scientific study (notably in Britain, Germany, and the USSR). Academia remained the main site for disciplinary research also when it was pertinent to applications.

Different institutions tended to support different kinds of research, leading to varying results. Universities, as expected, produced mostly disciplinary research, resulting in further knowledge about the phenomena. It was also marked by more flexibility in moving from one subject of study to another (e.g. from atomic structure to torsion in quartz, and from ultrasonic study and X-ray diffraction to the study of piezoelectricity). Yet a move from knowledge to application was rare within departments of physics (it was more common in departments of engineering). Cady provided an exception rather than a rule.[11] In this field, one does not see signs of the entrepreneurial university at the time.[12] State laboratories occupied a middle ground between the industrial laboratory and academia. In addition to their direct work on technics, these organizations aimed at the production of knowledge relevant for the employment and development of technics also beyond their walls. This approach led to deeper and more extended studies than those usually carried out in industry, like the development of equivalent network (NPL) and studies regarding the properties of piezo-resonators in different cuts and under varying physical conditions (NPL and PTR). That academic and industrial institutions also developed equivalent networks shows an overlap between the studies at these places. Notwithstanding, the scant discussion that van der Pol at Philips offered for his network reflects the lower priority for knowledge at the industrial laboratory. Academic researchers, on the other hand, carried out more general

[10]Relying on statistical claims that more scientists (presumable defined by education) worked in industry than at universities, Steven Shapin concluded that 'by mid-century the industrial, rather than the academic, scientist was closer to the institutional norm.' He, however, does not refer to their practice. Steven Shapin, *The Scientific Life: A Moral History of a Late Modern Vocation* (Chicago: University of Chicago Press, 2008), 110.

[11]Other academics studied and developed piezoelectric technics when they were mobilized to the war efforts. Pierce contributed to the technics but not to knowledge of the effect.

[12]Heilbron and Seidel do find signs for commercial entrepreneurship in J. L. Heilbron and Robert W. Seidel, *Lawrence and His Laboratory: A History of the Lawrence Berkeley Laboratory* (Berkeley: University of California Press, 1989).

research on question related to piezo-resonators and their mechanism, such as the effect of temperature of the phenomenon, and the underline static effect.

Transitions between the utilitarian regime aimed at applications and the disciplinary regime aimed at knowledge and understanding led to findings of groundbreaking technics and novel scientific phenomena. Arguably, Cady's discovery of piezoelectric resonance provides the clearest example of the benefits of a shift from technological research to a 'pure scientific' investigation, i.e. a study without a known technical goal. His utilitarian research on submarine detection was requisite for his observation of negative capacity in resonator circuits, yet his (and his colleagues') commitment to improving submarine detection blocked further research on these observations. Cady's return to disciplinary research as a university professor was necessary for turning these previously unexamined incidental observations into a discovery. Thus, the commitment of researchers to scientific or technical aims was not merely an issue of rhetoric, but it directed their research into different sets of questions and findings. Cady turned back to the utilitarian regime enjoying the expertise gained with the effect to invent crystal frequency standard and control.

Notwithstanding detours to the utilitarian regime as a consultant and patentee, after these important inventions, Cady's main endeavour was the disciplinary study of piezoelectricity. While Cady himself continued working on technics, the separation between those who studied the physics of piezoelectricity and those who developed technics based on it became clearer with the establishment of distinct enterprises of research in the field. Notwithstanding, research in the disciplinary regime followed questions of interest to those working in the utilitarian regime. Researchers who crossed the line between utilitarian and disciplinary regimes (making a transitory regime in Shinn's terms), who were crucial in the second and third phases of study, became rarer and less important in the fifth phase.

Propositional knowledge of science and technology, i.e. knowledge that could be and usually was explicated, played a central role in the development of piezoelectric technics at all stages. Not that know-how and experience did not have a role but they were secondary to the propositions and theories about natural effects and technical devices. The physics of piezoelectricity was crucial to the inventions of sonar and frequency control and continued to be helpful in improving these technics. For this reason, powerful forces (e.g. the industry and the state) encouraged studying questions pertinent to the technics. In this sense, the technics were historically and logically based on science, notwithstanding the components added in the technical development. Technological knowledge about triodes as developed for telecommunication was also indispensable for the invention of frequency control and of the quartz clock. Construction of the quartz clock relied on statements and rules about the behaviour of triode circuits, acquired in the technological study of them, and of piezoelectric resonators and oscillators, acquired in their study. The process also required expertise gained by direct experience with the technical components. Notwithstanding the crucial role of propositional knowledge and technical expertise, combining and modifying known components into working innovative technics like sonar and quartz clock required long methodological research and development process and ingenuity.

The centrality of propositional knowledge, which characterizes science, is connected in this case to the essential role of physics in the invention and development of piezoelectric technics. Beyond the acquaintance with effects discovered in scientific research, experimental and theoretical knowledge of particular properties of the phenomena and of experimental manipulation with them were crucial for Langevin's and Cady's inventions of sonar and frequency control. The case of sonar clearly exemplifies, however, that scientific resources were not enough, as the main resources had already been available for a few years before the invention. As is well known, technical developments required much work and thus societal interest. Pointing out the scientific origins of these technics, however, is far from an endorsement of a linear model of innovation. Even my schematic division of stages of research indicates a back-and-forth feedback mechanism between science, technology, and technics and a bidirectional rather than unidirectional casual influences among them. Technological research enhanced the study of the field, led to new observations, including those leading to the discovery of piezoelectric resonance, and back to disciplinary science. In the case of electronics, technology became a field of study on its own with elaborated mathematical theory and its own dynamics, which resembled disciplinary science. After the usefulness of piezoelectricity was recognized, its utilitarian and disciplinary studies continued to inform and influence each other. Technological interests and technical findings shaped scientific research no less than the findings and ideas of academic laboratories shaped technology.

BIBLIOGRAPHY

Archives consulted

AT&T Archives and History Center, Warren, NJ: notebooks of Marrison, Nicholson, and other researchers at AT&T, inner and external correspondences and memos.

Archives Center, National Museum of American History (ACNMAH)—Walter Guyton Cady Papers, 1903–74

Department of Manuscripts and University Archives, University Library, Cambridge (CUL), papers of Ernest Rutherford and Joseph J. Thomson

Harvard University Archives, the papers of George Washington Pierce

La Centre de resources historiques de l'École Supérieure de Physique et de Chimie Industrielles de la Ville de Paris. (ESPCI): papers of Paul Langevin.

National Archives of United Kingdom (UKNA): records of Admiralty, Naval Forces, Royal Marines, Coastguard, and related bodies: BIR and other reports

National Archives (United States of America) at College Park, College Park, MD, records of the national academy of science (USNA), London, Paris, and Washington, DC offices

Niels Bohr Library, American Institute of Physics, Cady's dossier and Elias Klein papers

Rhode Island Historical Society MSS 326 – Cady Family Papers (Walter Cady's diaries)

Articles, books, and dissertations

A. [Appleton], E. [Edward] V. 'David William Dye. 1887–1932'. *Obituary Notices of Fellows of the Royal Society (1932–1954)* 1, no. 1 (December 1932): 75–8.

Abraham, Henri, and Eugène Bloch. 'Amplificateurs pour courants continus et pour courants de très basse fréquence.' *Comptes rendus* 168 (1919): 1105–8.

Abraham, Henri, and Eugène Bloch. 'Entretien des oscillations d'un pendule ou d'un diapason avec un amplificateur à lampes'. *Journal de physique théorique et appliquée* 9 (1919): 225–33.

Abraham, Henri, and Eugène Bloch. 'Mesure en valeur absolue des périodes des oscillations électriques de haute fréquence'. *Journal de physique théorique et appliquée* 9 (1919): 211–22.

Ahrens, Ingrid. 'Meißner, Alexander'. *Neue Deutsche Biographie* 16 (1990): 695–7.

Aitken, Hugh G. J. *The Continuous Wave: Technology and American Radio, 1900–1932*. Princeton, NJ: Princeton University Press, 1985.

Alder, Ken. *The Measure of All Things: The Seven-Year Odyssey and Hidden Error That Transformed the World*. Reprint edition. Free Press, 2003.

Amoudry, Michel. *Le général Ferrié et la naissance des transmissions et de la radiodiffusion*. Grenoble: Presses universitaires de Grenoble, 1993.

Anduaga, Aitor. *Geophysics, Realism, and Industry: How Commercial Interests Shaped Geophysical Conceptions, 1900–1960*. Oxford: Oxford University Press, 2016.

Anon. 'A. M'L. Nicolson, Video Pioneer, 69', New York Times, 4 February 1950.

Anon. 'Albert Perrier'. In *Université Lausanne—Rapport Annuel* 1961–1962, 5–6. Lausanne: Imprimerie Vaudoise, 1963.

Appleton, Edward V. 'Automatic Synchronization of Triode Oscillators'. *Cambridge Philosophical Society Proceedings* 21 (1922): 231–48.

Arshadi, Roozbeh, and Richard S. C. Cobbold. 'A Pioneer in the Development of Modern Ultrasound: Robert William Boyle (1883–1955)'. *Ultrasound in Medicine and Biology* 33 (2007): 3–14.

Ashford, Oliver M. 'Richardson, Lewis Fry'. In *Complete Dictionary of Scientific Biography*, 24: 238–41. Detroit: Charles Scribner's Sons, 2008.

Aubin, David, and Catherine Goldstein (eds). *The War of Guns and Mathematics: Mathematical Practices and Communities in France and Its Western Allies around World War I*. Providence, RI: American Mathematical Society, 2014.

Authier, André. *Early Days of X-Ray Crystallography*. Oxford: Oxford University Press, 2013.

Bauer, Arthur O. 'Some Aspects of Precision Time Measurements, Controlled by Means of Piezo-electric-vibrators, as Deployed in Germany prior to 1950', unpublished manuscript (6.2.2000). http://www.cdvandt.org/time%20symposium%20webversion.pdf.

Becker, H. E. R. 'Die Debye-Sears Beugungserscheinung und die Energiebilanz bei Erzeugung von Ultraschallwellen'. *Annalen der Physik* 25 (1936): 373–84.

Becker, H. E. R. 'Die Rückwirkung einer umgebenden Flüssigkeit auf die Schwingungen einer Quarzplatte'. *Annalen der Physik* 25 (1936): 359–72.

Behm, Alexander. 'Einrichtung zur Messung von Meerestiefen und Entfernungen und Richtungen von Schiffen oder Hindernissen mit Hilfe reflektierter Schallwellen'. DE282009 (C), filed 22 July 1913, and issued 13 February 1915.

Bellaiche, Laurent. 'Piezoelectricity of Ferroelectric Perovskites from First Principles'. *Current Opinion in Solid State and Materials Science* 6 (2002): 19–25.

Bensaude-Vincent, Bernadette. *Langevin, 1872–1946: Science et Vigilance.* Paris: Belin, 1987.

Beyer, Robert Thomas. *Sounds of Our Times: Two Hundred Years of Acoustics.* New York: American Institute of Physics, 1999.

Bigg, Charlotte. 'L'optique de précision et la Première Guerre mondiale'. *Schweizerische Zeitschrift für Geschichte* 55 (2005): 34–45.

Boersma, F. Kees. 'Structural Ways to Embed a Research Laboratory into the Company: A Comparison between Philips and General Electric 1900–1940'. *History and Technology* 19 (2003): 109–26.

Boersma, Kees. *Inventing Structures for Industrial Research: A History of the Philips National Laboratory, 1914–1946.* Amsterdam: Aksant Academic Publishers, 2002.

Boguslawski, Sergei. 'Zur Theorie Der Dielektrika. Temperaturabhängigkeit Der Dielektrizitätskontante. Pyroelektrizität'. *Physikalische Zeitschrift* 15 (1914): 283–8.

Boguslawski, Sergei. 'Pyroelektrizität Auf Grund Der Quantentheorie'. *Physikalische Zeitschrift* 15 (1914): 569–72.

Boguslawski, Sergei. 'Zu Herrn W. Ackermanns Messungen Der Temperaturabhängigkeit Der Pyroelektrischen Erregung'. *Physikalische Zeitschrift* 15 (1914): 805–10.

Born, Max. *Dynamik der Kristallgitter.* Fortschritte der Mathematischen Wissenschaften in Monographien, Heft 4. Leipzig and Berlin: B.G. Teubner, 1915.

Born, Max, and Elisabeth Bormann. 'Zur Gittertheorie der Zinkblende'. *Annalen der Physik* 62 (1920): 218–46.

Boudia, Soraya. *Marie Curie et son laboratoire: Science et industrie de la radioactivité en France*. Paris: Editions des archives contemporaines, 2001.

Bowen, I. S. 'John A. Anderson, Astronomer and Physicist'. *Science* 131, no. 3401 (4 March 1960): 649–50.

Bragg, William, and R. E. Gibbs. 'The Structure of Alpha and Beta Quartz'. *Proceedings of the Royal Society of London. Series A* 109 (1925): 405–27.

Briscoe, Grace, and Winifred Leyshon. 'Reciprocal Contraction of Antagonistic Muscles in Peripheral Preparations, Using Flashing Neon-Lamp Circuit for Excitation of Nerve'. *Proceedings of the Royal Society of London. Series B* 105 (1929): 259–79.

Brittain, James E. 'Walter G. Cady and Piezoelectric Resonators'. *Proceedings of the IEEE* 80, no. 11 (1992).

Brittain, James E. 'Scanning the Past: Joseph Warren Horton'. *Proceedings of the IEEE* 82 (1994): 1470.

Brittain, James E. 'Harold D. Arnold: A Pioneer in Vacuum-Tube Electronics'. *Proceedings of the IEEE* 86 (1998): 1895–6.

Brittain, James E. 'Electrical Engineering Hall of Fame: Lloyd Espenschied'. *Proceedings of the IEEE* 95 (2007): 2259–62.

Bromberg, Joan L. 'Engineering Knowledge in the Laser Field'. *Technology and Culture* 27 (1986): 798–818.

Buckley, Oliver E. 'The Evolution of the Crystal Wave Filter'. *Journal of Applied Physics* 8 (1937): 40–7.

Bücks, Karl, and Hans Müller. 'Über einige Beobachtungen an schwingenden Piezoquarzen und ihrem Schallfeld'. *Zeitschrift für Physik* 84 (1933): 75–86.

Burns, Russell W. 'The Contributions of the Bell Telephone Laboratories to the Early Development of Television'. *History of Technology* 13 (1991): 181–213.

Burns, Russell W. *Television: An International History of the Formative Years*. London: Institution of Electrical Engineers, 1998.

Butterworth, Stephan. 'On a Null Method of Testing Vibration Galvanometers'. *Proceedings of the Physical Society of London* 26, no. 1 (1914): 264–73.

Butterworth, Stephan. 'On Electrically-Maintained Vibrations'. *Proceedings of the Physical Society of London* 27 (1915): 410–24.

Cady, Walter G. 'Note on the Theory of Longitudinal Vibrations of Viscous Rods Having Internal Losses'. *Physical Review* 15 (1920): 146–7.

Cady, Walter G. 'The Piezo-electric Resonator'. US 1,450,246, filed 28 February 1920, and issued 3 April 1923.

Cady, Walter G. 'Methods of Maintaining Electric Currents of Constant Frequency'. US 1,472,583, filed 28 May 1921, and issued 30 October 1923.

Cady, Walter G. 'New Methods for Maintaining Constant Frequency in High-Frequency Circuit'. *Physical Review* 18 (1921): 142–3.

Cady, Walter G. 'The Piezo-Electric Resonator'. *Physical Review* 17 (1921): 531.

Cady, Walter G. 'The Piezo-Electric Resonator'. *Proceedings of IRE* 10 (1922): 83–114.

Cady, Walter G. 'Theory of Longitudinal Vibrations of Viscous Rods'. *Physical Review* 19 (1922): 1–6.

Cady, Walter G. 'An International Comparison of Radio Wavelength Standards by Means of Piezo-Electric Resonators'. *Proceedings of IRE* 12 (1924): 805–16.

Cady, Walter G. 'Bibliography of Piezo-Electricity'. *Proceedings of IRE* 15 (1928): 521–35.

Cady, Walter G. *Piezoelectricity: An Introduction to the Theory and Applications of Electromechanical Phenomena in Crystals*. New York: McGraw-Hill, 1946.

Cady, Walter G. 'Piezoelectricity and Ultrasonics'. *Sound: Its Uses and Control* 2 (1963): 46–52.

Cahan, David. *An Institute for an Empire: The Physikalisch-Technische Reichsanstalt, 1871–1918*. Cambridge, UK: Cambridge University Press, 1989.

Campbell, John. *Rutherford: Scientist Supreme*. Christchurch, New Zealand: AAS Publications, 1999.

Carassa, Francesco. 'Francesco Vecchiacchi'. *Rendiconti del Seminario matematico e fisico di Milano* 27 (1957): xix–xxi.

Cardwell, Donald. *Wheels, Clocks, and Rockets: A History of Technology*. Reprint edition. New York: Norton, 2001.

Carr, Percy H., and Lester T. Earls. 'Gerald W. Fox'. *Physics Today* 27, no. 4 (April 1974): 135.

Cartwright, M. L. 'Balthazar Van Der Pol'. *Journal of the London Mathematical Society* 35 (1960): 367–76.

Chilowsky, Constantin, and Paul Langevin. 'Procédés et appareils pour la production de signaux sous-marins dirigés et pour la localisation á distance d'obstacles sous-marins'. FR502913, filed 29 May 1916, and issued March 1920.

Clapp, James K. '"Universal" Frequency Standardization from a Single Frequency Standard'. *Journal of the Optical Society of America* 15 (1927): 25–47.

Clapp, James K. 'A New Frequency Standard'. *General Radio Experimenter* 3, no. 11 (April 1929).

Clay, Jacob, and J. G. Karper. 'The Piezo-Electric Constant of Quarz [Sic]'. *Physica* 4 (1937): 311–15.

Cochrane, Rexmond Canning. *Measures for Progress: A History of the National Bureau of Standards*. New York: Arno Press, 1976.

Coleman, Samuel K. 'Riken from 1945 to 1948: The Reorganization of Japan's Physical and Chemical Research Institute under the American Occupation'. *Technology and Culture* 31, (1990): 228–50.

Colin, Victor. 'La téléphonie sans fil'. *Bulletin de la société internationale des électriciens*, 9 (1909): 427–50.

Collins, Harry M. *Changing Order: Replication and Induction in Scientific Practice*. London: SAGE, 1985.

Colpitts, E. H., and O. B. Blackwell. 'Carrier Current Telephony and Telegraphy'. *Transactions of the American Institute of Electrical Engineers* 40 (1921): 205–300.

Cross, L. E., and R. E. Newnham. 'History of Ferroelectrics'. In *Ceramics and Civilization,* Volume III: *High-Technology Ceramics: Past, Present, and Future*, 289–305. Westerville, OH: American Caeramic Society, 1986.

Crossley, A. 'Piezo-Electric Crystal-Controlled Transmitters'. *Proceedings of IRE* 15, no. 1 (1927): 9–36.

Curie, Jacques. 'Recherches sur le pouvoir inducteur spécifique et la conductibilité des corps cristallisés'. *Annales de chimie et de physique* 17 (1889): 385–434.

Curie, Jacques, and Pierre Curie. 'Sur un électromètre à bilame de quartz'. *Comptes rendus* 106 (1888): 1287–9.

Curie, Marie. *Traité de radioactivité*. Vol. 1. Paris: Gauthier-Villars, 1910.

Curie, Pierre. *Œuvres de Pierre Curie*. Paris: Gauthier-Villars, 1908.

Dawson, Leo H. 'Piezoelectricity of Crystal Quartz'. *Physical Review* 29 (1927): 532–41.

Dear, Peter. *The Intelligibility of Nature: How Science Makes Sense of the World*. Chicago: University of Chicago Press, 2006.

Dellinger, J. H. 'The Status of Frequency Standardization'. *Proceedings of IRE* 16 (1928): 579–90.

Dick, Steven J. *Sky and Ocean Joined: The U. S. Naval Observatory 1830–2000.* Cambridge, UK: Cambridge University Press, 2002.

Doerffler, Heinz. 'Biegungs- und Transversalschwingungen piezoelektrisch angeregter Quarzplatten'. *Zeitschrift für Physik* 63 (1930): 30–53.

Dolejšek, V., and M. Jahoda. 'Sur les variations du réseau des cristaux piézoélectriques produites par une tension électriquestatique'. *Comptes rendus* 206 (1938): 113–15.

Douglas, Susan J. *Inventing American Broadcasting, 1899–1922.* Baltimore: Johns Hopkins University Press, 1987.

Dye, David. W. 'The Valve-Maintained Tuning-Fork as a Precision Time-Standard'. *Proceedings of the Royal Society of London. Series A* 103 (1923): 240–60.

Dye, David. W. 'A Self-Contained Standard Harmonic Wave-Meter'. *Philosophical Transactions of the Royal Society of London. Series A* 224 (1924): 259–301.

Dye, David. W. 'The Piezo-Electric Quartz Resonator and Its Equivalent Electrical Circuit'. *Proceedings of the Physical Society of London* 38 (1926): 399–458.

Dye, David W., and L. Essen. 'The Valve Maintained Tuning Fork as a Primary Standard of Frequency'. *Proceedings of the Royal Society of London. Series A* 143, no. 849 (1934): 285–306.

Eccles, W. H. 'The Use of the Triode Valve in Maintaining the Vibration of a Tuning Fork'. *Proceedings of the Physical Society of London* 31 (1919): 269.

Eccles, W. H., and W. A. Leyshon. 'Some Thermionic Tube Circuits for Relaying and Measuring'. *Journal of the Institution of Electrical Engineers* 59 (1921): 433–6.

Eccles, W. H., and W. A. Leyshon. 'Some New Methods of Linking Mechanical and Electrical Vibrations'. *Proceedings of the Physical Society (London)* 40 (1928): 229–33.

Edgerton, David. *The Shock of the Old: Technology and Global History Since 1900.* Oxford: Oxford University Press, 2011.

Edgerton, David E. H., and S. M. Horrocks. 'British Industrial Research and Development before 1945'. *Economic History Review* 47 (1994): 213–38.

Editor. 'Sixty-Fifth Birthday of Professor Václav Petržílka'. *Czechoslovak Journal of Physics B* 20 (1970): 369–74.

Espenschied, Lloyd. 'Electrical Wave Filter'. US patent 1,795,204 (A), filed 3 January 1927, and issued 3 March 1931.

Espenschied, Lloyd. 'The Origin and Development of Radiotelephony'. *Proceedings of IRE* 25 (1937): 1101–23.

Espenschied, Lloyd. 'R. A. Heising, Former President of IRE, Dies at 76'. *Specturm, IEEE* 2 (1965): 222.

Essen, L. 'The Dye Quartz Ring Oscillator as a Standard of Frequency and Time'. *Proceedings of the Royal Society of London A: Mathematical, Physical and Engineering Sciences* 155 (1936): 498–519.

Essen, Ray. 'Greenwich Time: From Pendulum to Quartz 1'. *Horological Journal* 154 (2012): 198–201.

Essen, Ray. *Birth of Atomic Time: Includes The Memoirs of Louis Essen*. Peterborough: FastPrint Publishing, 2015.

Fagen, M. D. (ed.). *A History of Engineering and Science in the Bell System— The Early Years*. Murray Hill, NJ: Bell Laboratories, 1975.

Feffer, Stuart. 'Microscopes to Munitions: Ernst Abbe, Carl Zeiss, and the Transformation of Technical Optics, 1850–1914'. PhD, University of California, Berkeley, 1994.

Ferguson, Eugene S. 'The Mind's Eye: Nonverbal Thought in Technology'. *Science* 197, no. 4306 (1977): 827–36.

Ferguson, Eugene S. *Engineering and the Mind's Eye*. Cambridge, MA: MIT Press, 1992.

Fontanon, Claudine. 'L'obus Chiloswki et la soufflerie balistique de Paul Langevin: un épisode oublié de la mobilisation scientifique (1915–1919)'. In Dominique Pestre (ed.), *Deux siècles d'histoire de l'armement en France: de Gribeauval à la force de frappe*, 81–109. Paris: CNRS éditions, 2005.

Forman, Paul. 'Behind Quantum Electronics: National Security as Basis for Physical Research in the United States, 1940–1960'. *Historical Studies in the Physical and Biological Sciences* 18 (1987): 149–229.

Fousek, Jan. 'Joseph Valasek and the Discovery of Ferroelectricity'. *Proceedings 9th IEEE International Symposium on Applications of Ferroelectrics*, 1–5, 1994.

Fox, Gerald, and James Cork. 'The Regular Reflection of X-Rays from Quartz Crystals Oscillating Piezoelectrically'. *Physical Review* 38 (1931): 1420–3.

Friedel, Robert, and Paul Israel. *Edison's Electric Light: Biography of an Invention*. New Brunswick, NJ: Rutgers University Press, 1988.

Frost, Gary.L. 'Inventing Schemes and Strategies: The Making and Selling of the Fessenden Oscillator'. *Technology and Culture* 42 (2001): 462–88.

Galitzine, B. 'An Apparatus for the Direct Determination of Accelerations'. *Proceedings of the Royal Society of London A: Mathematical, Physical and Engineering Sciences* 95 (1919): 492–507.

Galitzine, B. 'Ueber das Dalton'sche Gesetz'. *Annalen der Physik* 41 (1890): 588–626.

Gebhard, Louis A. *Evolution of Naval Radio-Electronics and Contributions of the Naval Research Laboratory*. Washington, DC: Naval Research Lab, 1979.

Gerber, Stefan, Jürgen John, and Rüdiger Stutz. *Traditionen, Brüche, Wandlungen: die Universität Jena 1850–1995*. Köln, Weimar: Böhlau Verlag, 2009.

Gibbs, R. E., and V. N. Thatte. 'The Temperature Variation of the Frequency of Piezoelectric Oscillations of Quartz'. *Philosophical Magazine* 14 (1932): 682–94.

Giebe, E. 'Leuchtende piezoelektrische Resonatoren als Hochfrequenznormale'. *Zeitschrift für technische Physik* 7 (1926): 235.

Giebe, E., and E. Blechschmidt. 'Experimentelle und theoretische Untersuchungen über Dehnungseigenschwingungen von Stäben und Rohren (I u. II)'. *Annalen der Physik* 18 (1933): 417–56, 457–85.

Giebe, E., and A. Scheibe. 'Eine einfache Methode zum qualitativen Nachweis der Piezoelektrizität von Kristallen'. *Zeitschrift für Physik* 33 (1925): 760–6.

Giebe, E., and A. Scheibe. 'Sichtbarmachung von hochfrequenten Longitudinal schwingungen piezoelektrischer Kristallstäbe'. *Zeitschrift für physik* 33 (1925): 335–44.

Giebe, E., and A. Scheibe. 'Piezoelektrische Erregung von Dehnungs-, Biegungs- und Drillungsschwingungen bei Quarzstäben'. *Zeitschrift für Physik* 46 (1928): 607–52.

Giebe, E., and A. Scheibe 'Piezoelektrische Kristalle als Frequenznormale'. *Elektrische Nachrichten-Technik* 5 (1928): 65–82.

Ginoux, Jean-Marc, and Christophe Letellier. 'Van Der Pol and the History of Relaxation Oscillations: Toward the Emergence of a Concept'. *Chaos: An Interdisciplinary Journal of Nonlinear Science* 22, no. 2 (30 April 2012): 023120.

Godin, Benoît. 'The Linear Model of Innovation: The Historical Construction of an Analytical Framework'. *Science, Technology & Human Values* 31 (2006): 639–67.

Godin, Benoît. *Models of Innovation: The History of an Idea*. Cambridge, MA: MIT Press, 2017.

Gooday, Graeme. *The Morals of Measurement: Accuracy, Irony, and Trust in Late Victorian Electrical Practice*. Cambridge, UK: Cambridge University Press, 2004.

Greenspan, Nancy. *The End of the Certain World : The Life and Science of Max Born : The Nobel Physicist Who Ignited the Quantum Revolution*. New York: Basic Books, 2005.

Gruetzmacher, Johannes. 'Steuerungseinrichtung, insbesondere Wechselstromrelais'. DE613413 (C), filed 7 April 1932, and issued 16 April 1936.

Guerbilsky, Alexis. 'Lames piézo-èlectriques d'épaisseur non uniforme'. *J. Phys. Radium* 8 (1937): 165–8.

Guillet, A. 'Roue à denture harmonique, application à la construction d'un chronomètre de laboratoire à mouvement uniforme et continu'. *Comptes rendus* 160 (1915): 235–7.

Guillet, A., and V. Guillet. 'Nouveaux modes d'entretien des diapasons'. *Comptes rendus* 130 (1900): 1002–4.

Günther, Norbert. 'Untersuchung der Wirkung mechanischer und elektrischer Kraftfelder auf die Doppelbrechung des Quarzes'. *Annalen der Physik* 13 (1932): 783–801.

Hackmann, Willem D. *Seek and Strike: Sonar, Anti-Submarine Warfare and the Royal Navy, 1914–54*. London: Her Majesty's Stationery office, 1984.

Hackmann, Willem D. 'Sonar Research and Naval Warfare 1914–1954: A Case Study of a Twentieth-Century Establishment Science'. *Historical Studies in the Physical Sciences* 16 (1986): 83–110.

Hall, Karl. 'The Schooling of Lev Landau: The European Context of Postrevolutionary Soviet Theoretical Physics'. *Osiris* 23 (2008): 230–59.

Hansell, Clarence W. 'Filter'. US patent 2,005,083 (A), filed 7 July 1927, and issued 18 June 1935.

Hartcup, Guy. *The War of Invention: Scientific Developments, 1914–18*. London: Brassey's Defence Publishers, 1988.

Hartshorn, L. 'D. W. Dye, D.Sc., F.R.S'. Proceedings of the Physical Society 44 (1932): 608–10.

Heilbron, John L. *Ernest Rutherford and the Explosion of Atoms*. Oxford Portraits in Science. Oxford: Oxford University Press, 2003.

Heilbron, John L., and Robert W. Seidel. *Lawrence and His Laboratory: A History of the Lawrence Berkeley Laboratory*. Berkeley: University of California Press, 1989.

Heising, Raymond A. 'Alexander Mclean Nicolson 1880–1950'. *Bell Laboratories Record* 28 (May 1950): 221.

Heising, Raymond A. 'Introduction'. In Raymond Alphonsus Heising (ed.), *Quartz Crystals for Electrical Circuits, Their Design and Manufacture*, 1–9. New York: Van Nostrand, 1946.

Hentschel, Klaus. 'Gauss, Meyerstein and Hanoverian Metrology'. *Annals of Science* 64 (2007): 41–75.

Hermann, Armin. 'Laue, Max Von'. In *Dictionary of Scientific Biography*, 8: 50–3. Detroit: Charles Scribner's Sons, 1973.

Hessenbruch, Arne. 'Calibration and Work in the X-Ray Economy, 1896–1928'. *Social Studies of Science* 30 (2000): 397–420.

Hiedermann, Egon, H. R. Asbach, and K. H. Hoesch. 'Die Sichtbarmachung der stehenden Ultraschallwellen in durchsichtigen festen Körpern'. *Zeitschrift für Physik* 90 (1934): 322–6.

Hoddeson, Lillian. 'The Emergence of Basic Research in the Bell Telephone System, 1876–1915'. *Technology and Culture* 22 (1981): 512–44.

Homburg, Ernst. 'The Emergence of Research Laboratories in the Dyestuffs Industry, 1870–1900'. *British Journal for the History of Science* 25 (1992): 91–111.

Hong, Sungook. 'Historiographical Layers in the Relationship between Science and Technology'. *History and Technology* 15 (1999): 289–311.

Hong, Sungook. *Wireless: From Marconi's Black Box to the Audion*. Cambridge, MA: MIT Press, 2001.

Horton, Joseph W., and W. A. Marrison. 'Precision Determination of Frequency'. IRE Proceedings 16 (1928): 137–54.

Horton, Joseph W., Norman H. Ricker, and Warren A. Marrison. 'Frequency Measurement in Electrical Communication'. *Transactions of the American Institute of Electrical Engineers* 42 (1923): 730–41.

Horton, Joseph Warren. *Excursions in the Domain of Physics*, a typed manuscript, 1965, at AIP.

Hounshell, David A. 'The Evolution of Industrial Research in the United States'. In Richard S. Rosenbloom and William J. Spencer (eds), *Engines of Innovation: U.S. Industrial Research at the End of an Era*, 13–85. Boston, MA: Harvard Business School Press, 1996.

Hounshell, David A., and John K. Smith. *Science and Corporate Strategy: Du Pont R&D, 1902–1980*. Cambridge, UK: Cambridge University Press, 1988.

Howeth, Linwood S. *History of Communications-Electronics in the United States Navy*. Washington, DC: U.S. Government Printing Office, 1963.

Hughes, Jeff. 'William Kay, Samuel Devons and Memories of Practice in Rutherford's Manchester Laboratory'. *Notes and Records of the Royal Society* 62 (2008): 97–121.

Hughes, Thomas P. *American Genesis: A Century of Invention and Technological Enthusiasm, 1870–1970*. New York: Viking, 1989.

Hull, Lewis M., and James K. Clapp. 'A Convenient Method for Referring Secondary Frequency Standards to a Standard Time Interval'. *IRE Proceedings* 17 (1929): 252–71.

Hunt, Bruce J. 'The Ohm Is Where the Art Is: British Telegraph Engineers and the Development of Electrical Standards'. *Osiris* 9 (1994): 48–63.

Hunt, Frederick V. *Electroacoustics: The Analysis of Transduction, and Its Historical Background*. Cambridge, MA: Harvard University Press, 1954.

Ingersoll, L. R. 'Earle Melvin Terry—1879–1929'. *Science* 69, no. 1797 (6 July 1929): 592.

Israel, Giorgio. 'The Emergence of Biomathematics and the Case of Population Dynamics: A Revival of Mechanical Reductionism and Darwinism'. *Science in Context* 6 (1993): 469–509.

Israel, Giorgio. 'Technological Innovation and New Mathematics: Van Der Pol and the Birth of Nonlinear Dynamics'. In Ana Millán Gasca, Mario Lucertini, and Fernando Nicolò (eds), *Technological Concepts and Mathematical Models in the Evolution of Modern Engineering Systems*, 52–77. Basel: Birkhäuser, 2004.

Ito, Kenji. '"Electron Theory" and the Emergence of Atomic Physics in Japan'. *Science in Context* 31 (2018): 293–320.

Jackson, Myles W. *Harmonious Triads: Physicists, Musicians, and Instrument Makers in Ninteenth-Century Germany*. Cambridge, MA/London: MIT Press, 2008.

Jaffe, Hans. 'Professor Cady's Work in Crystal Physics'. 18th Annual Frequency Control Symposium: 5–11, 1964.

Jennings, O. E. 'Proceedings of the Baltimore Meeting of the American Association for the Advancement of Science'. *Science* 49 (1919): 11.

Joas, Christian, and Shaul Katzir. 'Analogy, Extension, and Novelty: Young Schrödinger on Electric Phenomena in Solids'. *Studies in History and Philosophy of Modern Physics* 42 (2011): 43–53.

Johnson, D. H. 'Origins of the Equivalent Circuit Concept: The Voltage-Source Equivalent'. *Proceedings of the IEEE* 91 (2003): 636–40.

Johnson, Jeffrey Allan. 'Chemical Warfare in the Great War'. *Minerva* 40 (2002): 93–106.

Jolliffe, Charles B., and Grace Hazen. 'Establishment of Radio Standards of Frequency by the Use of a Harmonic Amplifier'. *Bureau of Standards Scientific Paper* 530 (1926): 179–89.

Josephson, Paul R. *Physics and Politics in Revolutionary Russia*. California Studies in the History of Science Studies of the Harriman Institute. Berkeley: University of California Press, 1991.

Kaiser, Walter. 'What Drives Innovation in Technology?' *History of Technology* 21 (1999): 107–23.

Karcher, J. C. 'A Piezoelectric Method for the Instantaneous Measurement of High Pressures'. *Scientific Papers of the Bureau of Standards* 445 (1922): 257–64.

Katzir, Shaul. *The Beginnings of Piezoelectricity: A Study in Mundane Physics*. Dordrecht: Springer, 2006.

Katzir, Shaul. 'Hermann Aron's Electricity Meters: Physics and Invention in Late Nineteenth-Century Germany'. *Historical Studies in the Natural Sciences* 39 (2009): 444–81.

Katzir, Shaul. 'Who Knew Piezoelectricity? Rutherford and Langevin on Submarine Detection and the Invention of Sonar'. *Notes and Records of the Royal Society* 66 (2012): 141–57.

Katzir, Shaul. 'Scientific Practice for Technology: Hermann Aron's Development of the Storage Battery'. *History of Science* 51 (2013): 481–500.

Katzir, Shaul. 'Manchester at War: Bohr and Rutherford on Problems of Science, War and International Communication'. In F. Aaserud and H. Kraugh (eds), *One Hundred Years of the Bohr Atom: Proceedings from a Conference*, Scientia Danica. Series M: Mathematica et Physica 1, 495–510. Copenhagen: Danish Academy of science, 2015.

Katzir, Shaul. '"In War or in Peace": The Technological Promise of Science Following the First World War'. *Centaurus* 59 (2017): 223–37.

Katzir, Shaul. 'Technological Entrepreneurship from Patenting to Commercializing: A Survey of Late Nineteenth and Early Twentieth Century Physics Lecturers'. *History and Technology* 33 (2017): 109–25.

Katzir, Shaul. 'Time Standards for the Twentieth Century: Telecommunication, Physics, and the Quartz Clock'. *Journal of Modern History* 89 (2017): 119–50.

Katzir, Shaul. 'Introduction: Physics, Technology, and Technics during the Interwar Period'. *Science in Context* 31, no. 3 (September 2018): 251–61.

Kern, Ulrich. *Forschung und Präzisionsmessung. Die Physikalisch-Technische Reichsanstalt zwischen 1918 und 1948*. Weinheim: VCH, 1994.

Kevles, Daniel J. *The Physicists: The History of a Scientific Community in Modern America*, 2nd edn. Cambridge, MA: Harvard University Press, 1995.

Keys, David A. 'On the Adiabatic and Isothermal Piezo-Electric Constants of Tourmaline'. *Philosophical Magazine* 46 (1923): 999–1001.

Kilbon, Kenyon. 'Pioneering in Electronics: A Short History of the Origins and Growth of RCA Laboratories, Radio Corporation of America, 1919 to 1964', 1964. http://www.davidsarnoff.org/kil.html.

Kinsman, R. G. 'A History of Crystal Filters'. Proceedings of the 1998 IEEE International Frequency Control Symposium (Cat. No.98CH36165), 563–70, 1998.

Kirchhoff, Jochen. 'Wissenschaftsförderung und forschungspolitische Prioritäten der Notgemeinschaft der Deutschen Wissenschaft 1920–1932'. Dissertation, LMU München, 2003.

Klauer, Friedrich. 'Röntgen-Laue-Diagramme an piezoelektrisch schwingenden Kristallen'. *Physikalische Zeitschrift* 36 (1935): 208–11.

Klein, Elias. *Notes on Underwater Sound Research and Applications before 1939*. Washington, DC: Office of Naval Research. Department of the Navy, 1967.

Kline, Ronald R. *Steinmetz: Engineer and Socialist*. Baltimore: Johns Hopkins University Press, 1992.

Kline, Ronald R., and Thomas C. Lassman. 'Competing Research Traditions in American Industry: Uncertain Alliances between Engineering and Science at Westinghouse Electric, 1886–1935'. *Enterprise and Society* 6 (2005): 601–45.

Kloot, William van der. 'Lawrence Bragg's Role in the Development of Sound-Ranging in World War I'. *Notes and Records of the Royal Society* 59 (2005): 273–84.

Kloot, William van der. *Great Scientists Wage the Great War*. Oxford: Fonthill, 2014.

Klyukin, Igor I., and E. N. Šoškov. *Konstantin Vasil'evič Šilovskij: 1880–1958*. Lenigrad: Nauka, 1984.

Koga, Isaac. 'A New Frequency Transformer or Frequency Changer'. *IRE Proceedings* 15 (1927): 669–78.

Koga, Isaac. 'Frequency Demultiplication and the Origin of Frequency Shift Keying System'. *Journal of the Institute of Electronics and Communication Engineers of Japan* 56 (1973): 1335–40.

Kohler, Robert E. *Lords of the Fly: Drosophila Genetics and the Experimental Life*. Chicago: University of Chicago Press, 1994.

Korn, Arthur. *Elektrische Fernphotographie und Ähnliches*, 2nd edn. Leipzig: S. Hirzel, 1907.

Kragh, Helge. *Quantum Generations: A History of Physics in the Twentieth Century*. Princeton, NJ: Princeton University Press, 1999.

Kragh, Helge. 'The Lorenz-Lorentz Formula: Origin and Early History'. *Substantia* 2, no. 2 (September 2018): 7–18.

Ku, Ja Hyon. 'Uses and Forms of Instruments: Resonator and Tuning Fork in Rayleigh's Acoustical Experiments'. *Annals of Science* 66 (2009): 371–95.

Lack, Fredrick R. 'Observations on Modes of Vibration and Temperature Co-efficients of Quartz Crystal Plates'. *IRE Proceedings* 17 (1929): 1123–41.

Lamb, David, and Susan M. Easton. *Multiple Discovery: The Pattern of Scientific Progress*. Farnham, UK: Avebury, 1984.

Lang, Sidney B. 'A Conversation with Professor W. G. Cady'. *Ferroelectrics* 9 (1975): 141–9.

Lang, Sidney B. 'Walter Guyton Cady'. *Ferroelectrics* 9 (1975): 139–40.

Langevin, Paul. 'Procédé et appareils d'émission et de réception des ondes élastiques sous-marines à l'aide des propriétés piézo-électriques du quartz'. FR505703, filed 17 September 1917, issued 5 August 1920.

Langevin, Paul. 'The Employment of Ultra-Sonic Waves for Echo Sounding'. *Hydrographic Review* 2 (1924): 57–91.

Laue, M. v. 'Piezoelektrisch erzwungene Schwingungen von Quarzstäben'. *Zeitschrift für Physik* 34 (1925): 347–61.

Lavet, Marius. 'Propriétés des organes électromagnétiques convenant aux petits moteurs chronométriques a diapason'. *Annales Françaises de chronométrie* 15 (1961): 183–96.

Layton, Edwin T. 'Through the Looking Glass, or News from Lake Mirror Image'. *Technology and Culture* 28 (1987): 594–607.

Lécuyer, Christophe, and Takahiro Ueyama. 'The Logics of Materials Innovation: The Case of Gallium Nitride and Blue Light Emitting Diodes'. *Historical Studies in the Natural Sciences* 43 (2013): 243–80.

Le Grand, Yves. 'Obituary, Ernest Baumgardt (1904–1969)'. *Vision Research* 9 (1969): 1315–17.

Lelong, Benoit. 'Paul Langevin et la détection sous-marine, 1914–1929. Un physicien acteur de l'innovation industrielle et militaire'. *Épistémologiques* 2 (2002): 205–32.

Leyshon, W. A. 'Forced Oscillations in Self-Maintained Oscillating Circuits.' *Philosophical Magazine* 46 (1923): 686–98.

Leyshon, W. A. 'On the Control of the Frequency of Flashing of a Neon Tube by a Maintained Mechanical Vibrator'. *Philosophical Magazine* 4 (1927): 305–24.

Lissütin, A. 'Die Schwingungen der Quarzlamelle'. *Zeitschrift für Physik* 59 (1930): 265–73.

Lonn, Ernst. 'Zur Theorie der Schwingungen von Kristallplatten (Bemerkungen zu Arbeiten von Petržílka)'. *Annalen der Physik* 30 (1937): 420–32.

Lucas, René. 'Sur la piézoélectricité et la dissymétrie moléculaire'. *Comptes rendus* 178 (1924): 1890–2.

MacDougall, Robert. 'Long Lines: AT&T's Long-Distance Network as an Organizational and Political Strategy'. *Business History Review* 80 (2006): 297–327.

McGahey, Christopher Shawn. 'Harnessing Nature's Timekeeper: A History of the Piezoelectric Quartz Crystal Technological Community (1880–1959)'. Ph.D. thesis, Georgia Institute of Technology, 2009.

MacLeod, Roy M., and E. Kay Andrews. 'Scientific Advice in the War at Sea, 1915–1917: The Board of Invention and Research'. *Journal of Contemporary History* 6 (1971): 3–40.

Marcovich, Anne, and Terry Shinn. 'Regimes of Science Production and Diffusion: Towards a Transverse Organization of Knowledge'. *Scientiae Studia* 10, no. SPE (2012): 33–64.

Marrison, Warren A. 'Some Facts about Frequency Measurement'. *Bell Laboratories Record* 6 (1928): 385–8.

Marrison, Warren A. 'A High Precision Standard of Frequency'. *IRE Proceedings* 17 (1929): 1101–22.

Marrison, Warren A. 'The Crystal Clock'. *Proceedings of the National Academy of Sciences of the United States of America* 16 (1930): 496–507.

Marrison, Warren A. 'The Evolution of the Quartz Crystal Clock'. *Bell System Technical Journal* 27 (1948): 510–88.

Marsch, Ulrich. *Zwischen Wissenschaft Und Wirtschaft: Industrieforschung in Deutschland Und Grossbritannien 1880–1936*. Paderborn: Schöningh, 2000.

Mason, Warren P. 'Low Temperature Coefficient Quartz Crystals'. *Bell System Technical Journal* 19 (1940): 74–93.

Mason, Warren P. 'Quartz Crystal Applications'. In Raymond Alphonsus Heising (ed.), *Quartz Crystals for Electrical Circuits: Their Design and Manufacture*, 11–56. New York: Van Nostrand, 1946.

Maurain, Charles, and A. Pacaud. *La Faculté des sciences de l'Université de Paris de 1906 à 1940*. Paris: Presses universitaires de France, 1940.

Mehra, Jagdish, and Helmut Rechenberg. *The Historical Development of Quantum Theory*. Vol. 1, Part 1: *The Quantum Theory of Planck, Einstein, Bohr, and Sommerfeld: Its Foundation and the Rise of Its Difficulties, 1900–1925*. New York: Springer-Verlag, 1982.

Meissner, Alexander. 'Über piezo-elektrische Kristalle bei Hochfrequenz'. *Zeitschrift für technische Physik* 7 (1926): 585–92.

Meissner, Alexander. 'Piezo-Electric Crystals at Radio Frequencies'. *IRE Proceedings* 15 (1927): 281–96.

Meissner, Alexander. 'Über piezo-elektrische Kristalle bei Hochfrequenz II'. *Zeitschrift für technische Physik* 8 (1927): 74–7.

Merton, Robert K. 'Singletons and Multiplies in Science'. In *The Sociology of Science: Theoretical and Empirical Investigation*, 343–70. Chicago: University of Chicago Press, 1973.

Meyer, Stefan, and Egon R. von Schweidler. *Radioaktivität*. Naturwissenschaft und Technik in Lehre und Forschung. Leipzig: Teubner, 1916.

'Michigan Men in Service'. *The Michigan Alumnus* 24 (March 1918): 357.

Miller, David Philip. 'The Political Economy of Discovery Stories: The Case of Dr Irving Langmuir and General Electric'. *Annals of Science* 68 (2011): 27–60.

Miller, John M. 'Effective Resistance and Inductance of Iron and Bimetallic Wires'. *Bulletin of the Bureau of Standards* 12 (1915): 207–67.

Miller, John M. 'Quartz Crystal Oscillators'. *Monthly Radio and Sound Report (US Navy)*, June 1925, 53–64.

Miller, John M. 'Piezo-electric Oscillation Generator'. US patent 1,756,000 (A), filed 10 September 1925, and issued 22 April 1930.

Millman, S (ed.). *A History of Engineering and Science in the Bell System: Physical Sciences (1925–1980)*. New York: Bell Telephone Laboratories, 1983.

Mitchell, Daniel Jon. 'Measurement in French Experimental Physics from Regnault to Lippmann. Rhetoric and Theoretical Practice'. *Annals of Science* 69 (2012): 453–82.

Molinié, Philippe, and Soraya Boudia. 'Mastering Picocoulombs in the 1890s: The Curies' Quartz-Electrometer Instrumentation, and How It Shaped Early Radioactivity History'. *Journal of Electrostatics* 67 (2009): 524–30.

Moseley, Russell. 'The Origins and Early Years of the National Physical Laboratory: A Chapter in the Pre-History of British Science Policy'. *Minerva* 16 (1978): 222–50.

Mueller, Milton. *Universal Service: Competition, Interconnection, and Monopoly in the Making of the American Telephone System*. Cambridge, MA: American Enterprise Institute, 1997.

Mumford, Lewis. *Technics and Civilization*. London: Routledge, 1946.

Namba, Y. 'The Establishment of the Japanese Radio-Frequency Standard'. *IRE Proceedings* 18 (1930): 1017–27.

Nebeker, Frederik. *Dawn of the Electronic Age: Electrical Technologies in the Shaping of the Modern World, 1914 to 1945*. Piscataway, NJ: Wiley-IEEE Press, 2009.

Neprašová, Marie, and Miroslav Rozsíval. 'In Memory of Professor Dr. Václav Dolejšek'. *Czechoslovak Journal of Physics* 5 (1955): 115–16.

Neto, Climério Paulo da Silva and Alexei Kojevnikov. 'Convergence in Cold War Physics: Coinventing the Maser in the Postwar Soviet Union'. *Berichte zur Wissenschaftsgeschichte* 42 (2019): 396.

Nicolson, Alexander M. 'Generating and Transmitting Electric Currents'. US patent 2,212,845 (A), filed 13 April 1923, and issued 27 August 1940.

Nicolson, Alexander Mclean. 'The Piezo Electric Effect in the Composite Rochelle Salt Crystal'. Proceedings of the American Institute of Electrical Engineers 38 (1919): 1315–33.

Nicolson, Alexander Mclean. Piezophony. US patent 1,495,429, filed 10 April 1918, and issued 27 May 1924.

Nishikawa, S., Y. Sakisaka, and I. Sumoto. 'An X-Ray Examination of the Harmonic Thickness Vibration of Piezoelectric Quartz Plates'. *Physical Review* 43 (1933): 363–4.

Nitta, I. 'Shoji Nishikawa (1884–1952)'. In P. P. Ewald (ed.), *50 Years of X-Ray Diffraction*, 328–334. Utrecht: N.V.A. Oosthoek, 1962.Ny, Tsi-Zé. 'Étude expérimentale des déformations et des changements de propriétés optiques du quartz sous l'influence du champ électrique'. *Journal de Physique et le Radium* 9 (1928): 13–37.

Ny, Tsi-Zé, and Ling-Chao Tsien. 'Oscillations with Hollow Quartz Cylinders Cut along the Optical Axis'. *Nature* 134 (1934): 214–15.

Ny, Tsi-Zé, and Ling-Chao Tsien. 'Sur le développement d'électricité par torsion dans les cristaux de quartz'. *Comptes rendus* 198 (1934): 1395–6.

O'Connell, Joseph. 'Metrology: The Creation of Universality by the Circulation of Particulars'. *Social Studies of Science* 23 (1993): 129–73.

Olesko, Kathryn M. 'The Measuring of Precision: The Exact Sensibility in Early Nineteenth-Century Germany'. In M. Norton Wise (ed.), *The Values of Precision*, 103–34. Princeton, NJ: Princeton University Press, 1995.

Olesko, Kathryn M. 'Precision, Tolerance, and Consensus: Local Cultures in German and British Resistance Standards'. In Jed Z. Buchwald (ed.), *Scientific Credibility and Technical Standards in 19th and Early 20th Century Germany and Britain*, 117–56. Archimedes 1. Dordrecht: Kluwer, 1996.

Oreskes, Naomi, and John Krige (eds). *Science and Technology in the Global Cold War*. Cambridge, MA: MIT Press, 2014.

Osterberg, Harold. 'A Multiple Interferometer for Analyzing the Vibrations of a Quartz Plate'. *Physical Review* 43 (1933): 819–29.

Osterberg, Harold. 'The Temperature Coefficients of Shear and Longitudinal Modes of Vibration'. *Review of Scientific Instruments* 7 (1936): 339–41.

Pancaldi, Giuliano. 'Vito Volterra: Cosmopolitan Ideals and Nationality in the Italian Scientific Community between the Belle Époque and the First World War'. *Minerva* 31 (1993): 21–37.

Pang, Alex Soojung-Kim. 'Edward Bowles and Radio Engineering at MIT 1920–1940'. *Historical Studies in the Physical and Biological Sciences* 20 (1990): 313–37.

Pantalony, David. *Altered Sensations: Rudolph Koenig's Acoustical Workshop in Nineteenth-Century Paris*. Dordrecht: Springer, 2009.

Patterson, Samuel. 'Kurt Heegner—Biographical Notes'. *Mathematisches Forschungsinstitut Oberwolfach Report* No. 24 (2008): 1354–6.

Perrier, Albert. 'Zur Temperaturabhängigkeit der Piezoelektrizität'. *Zeitschrift für Physik* 58 (1929): 805–10.

Petržílka, Václav. 'Turmalinresonatoren bei kurzen und ultrakurzen Wellen'. *Annalen der Physik* 15 (1932): 72–88.

Petržílka, Václav, and A. Žáček. 'Radial and Torsional Vibrations of Annular Quartz Plates'. *Philosophical Magazine* 25 (1938).

Pierce, George W. 'Piezoelectric Crystal Resonators and Crystal Oscillators Applied to the Precision Calibration of Wavemeters'. *Proceedings of the American Academy of Arts and Sciences* 59, no. 4 (October 1923): 81–106.

Pierce, George W. 'Electrical System'. US patent 2,133,642 (A), filed 18 April 1930, and issued 18 October 1938.

J. C. Poggendorffs. *Biographisch-literarisches Handwörterbuch für Mathematik, Astronomie, Physik, Chemie und verwandte Wissenschaftsgebiete*. Leipzig [etc.]: Verlag Chemie [etc.], 1925–70.

Pol, Balthasar van der. 'Het gebruik van piëzo-electrische kwarts-kristallen in de draadlooze telegrafie en telefonie'. In *Gedenkboek ter herinnering aan het tienjarig bestaan van de Nederlandsche vereeniging voor ra-diotelegrafie, 1916–1926*, 293–6. Zutphen: Nauta, 1926.

Pol, Balthasar van der. 'On "Relaxation-Oscillations" '. *Philosophical Magazine* 2 (1926): 978–92.

Pol, Balthasar van der. 'Ober "Relaxationsschwingungen"'. *Jahrbuch der drahtlosen Telegraphie und Telephonie, Zeitschrift für Hochfrequenztechnik* 28 (1926): 178–84.

Pol, Balthasar van der. 'Über "Relaxationsschwingungen" II'. *Jahrbuch der drahtlosen Telegraphie* 29 (1927): 114–18.

Pol, Balthasar van der, and J. van der Mark. 'Frequency Demultiplication'. *Nature* 120, no. 3019 (September 1927): 363–4.

Pol, Balthasar van der. 'The Nonlinear Theory of Electric Oscillations'. *IRE Proceedings* 22 (1934): 1051–86.

Potts, David B. *Wesleyan University, 1910–1970: Academic Ambition and Middle-Class America*. Wesleyan University Press, 2015.

Powell, J. H., and J. H. T. Roberts. 'On the Frequency of Vibration of Circular Diaphragms'. *Proceedings of the Physical Society of London* 35 (1923): 170–82.

Rasmussen, Nicolas. *Picture Control: The Electron Microscope and the Transformation of Biology in America, 1940–1960*. Stanford CA: Stanford University Press, 1997.

Ratcliffe, J. A. 'William Henry Eccles. 1875–1966'. *Biographical Memoirs of Fellows of the Royal Society* 17 (1971): 195–214.

Rayleigh, J. W., and Strutt Baron. *The Theory of Sound*, 2nd edn, Vol. 1. New York: Dover, 1945.

Reich, Leonard S. 'Irving Langmuir and the Pursuit of Science and Technology in the Corporate Environment'. *Technology and Culture* 24 (1983): 199–221.

Reich, Leonard S. *The Making of American Industrial Research: Science and Business at GE and Bell, 1876–1926*. Cambridge, UK: Cambridge University Press, 1985.

Richardson, Lewis Fry. 'Apparatus for Warning a Ship at Sea of its Nearness to Large Objects Wholly or Partly under Water'. GB191211125 (A), filed 10 May 1912, and issued 27 March 1913.

Richardson, Lewis Fry. 'Apparatus for Warning a Ship of its Approach to Large Objects in a Fog'. GB191209423 (A), filed 20 April 1912, and issued 6 March 1913.

Riordan, Michael, and Lillian Hoddeson. *Crystal Fire: The Invention of the Transistor and the Birth of the Information Age*. New York: Norton, 1997.

Roblin, Gérard. 'L'Institut d'Optique a 75 ans'. *Opto* 118 (1996): 13–24.

Russo, Arturo. 'Fundamental Research at Bell Laboratories: The Discovery of Electron Diffraction'. *Historical Studies in the Physical Sciences* 12 (1981): 117–60.

Rutherford, Ernest. *Radio-Activity*. Cambridge Physical Series. Cambridge, UK: Cambridge University Press, 1904.

Rutherford, Ernest. *Radioactive Substances and Their Radiations*. Cambridge, UK: Cambridge University Press, 1913.

Rutherford, Ernest. 'Henry Gwyn Jeffreys Moseley'. *Nature* 96 (1915): 33–4.

Saunders, Frederick A., and Frederick V. Hunt. 'George Washington Pierce'. *Biographical Memoirs. National Academy of Sciences* 33 (1959): 351–80.

Schaefer, Clemens, and Ludwig Bergmann. 'Über neue Beugungserscheinungen an schwingenden Kristallen'. *Die Naturwissenschaften* 22 (1934): 865–90.

Schaffer, Simon. 'Making Up Discovery'. In Margaret A. Boden (ed.), *Dimensions of Creativity*, 13–51. Cambridge, MA: MIT Press, 1994.

Schaffer, Simon. 'Rayleigh and the Establishment of Electrical Standards'. *European Journal of Physics* 15 (1994): 277–85.

Schaffer, Simon. 'Accurate Measurement Is an English Science'. In M. Norton Wise (ed.), *The Values of Precision*, 135–72. Princeton, NJ: Princeton University Press, 1995.

Schaffer, Simon. 'Metrology, Metrication, and Victorian Values'. In Bernard Lightman (ed.), *Victorian Science in Context*, 438–74. Chicago: University of Chicago Press, 1997.

Schallenberg, Richard H. *Bottled Energy: Electrical Engineering and the Evolution of Chemical Energy Storage*. Philadelphia: American Philosophical Society, 1982.

Schatzberg, Eric. *Technology: Critical History of a Concept*. Chicago: University of Chicago Press, 2018.

Schilowsky, Konstantin. 'Verfahren und Vorrichtung zum Nachweis unterirdischer Erzlager oder von Grundwasser mittels elektrischer Schwingungen'. DE322040 (C), filed 16 November 1913, and issued 19 June 1920.

Schimmler, Jörg. *Alexander Behm (1880–1952): Erfinder des Echolots—Eine Biographie*. Norderstedt: Books on Demand, 2013.

Schirrmacher, Arne. 'Die Physik im Großen Krieg: Warum wissen wir so wenig über den Einfluss des Ersten Weltkriegs auf die Forschung, technische Anwendungen und Karrieren in der Physik?' *Physik Journal* 13, no. 7 (2014): 43–8.

Schirrmacher, Arne. 'Sounds and Repercussions of War: Mobilization, Invention and Conversion of First World War Science in Britain, France and Germany'. *History and Technology* 32 (2016): 269–92.

Schneider, Werner. 'Alexander Behm und 100 Jahre Echolotpatente'. *Hydrographische Nachrichten*, no. 96 (Oktober 1996): 11–14.

Schrödinger, Erwin. 'Studien über Kinetik der Dielektrika, den Schmelzpunkt, Pyro- und Piezoelektrizität'. *Sitzungsberichte der kaiserlichen Akademie der Wissenschaften in Wien. Mathematischnaturwissenschaftliche Klasse (IIa)* 121 (1912): 1937–72.

Schroeder-Gudehus, Brigitte. 'Challenge to Transnational Loyalties: International Scientific Organizations after the First World War'. *Science Studies* 3 (1973): 93–118.

Schubert, H. 'Industrielaboratorien für Wissenschaftstransfer. Aufbau und Entwicklung der Siemensforschung bis zum Ende des Zweiten Weltkrieges anhand von Beispielen aus der Halbleiterforschung'. *Centaurus* 30 (1987): 245–92.

Schulwas-Sorokina, R. D. 'Is It Possible to Determine the Piezoelectric Constant at High Temperature by the Statical Method?' *Physical Review* 34 (1929): 1448–50.

Seidl, Franziska. 'Elektrische Leitfähigkeit von belasteten Piezoquarzen'. *Zeitschrift für Physik* 75 (1932): 488–503.

Servos, John W. *Physical Chemistry from Ostwald to Pauling: The Making of a Science in America*. Princeton, NJ: Princeton University Press, 1996.

Shapin, Steven. *The Scientific Life: A Moral History of a Late Modern Vocation*. Chicago: University of Chicago Press, 2008.

Shea, T. E., and C. E. Lane. 'Telephone Transmission Networks Types and Problems of Design'. *Transactions of The American Institute of Electrical Engineers* 48 (1929): 1031–44.

Shinn, Terry. *Research-Technology and Cultural Change: Instrumentation, Genericity, Transversality*. Oxford: The Bardwell Press, 2008.

Slotten, Hugh Richard. *Radio and Television Regulation: Broadcast Technology in the United States, 1920–1960*. Baltimore: Johns Hopkins University Press, 2000.

Smits, F. M. (ed.). *A History of Engineering and Science in the Bell System: Electronics Technology (1925–1975)*. A History of Engineering and Science in the Bell System 6. Indianapolis: AT& T Bell Laboratories, 1985.

Sokoloff, Sergei J. 'Schwingungen piezoelektrischer Quarzstäbe, hervorgerufen durch ein ungleichförmig verteiltes Feld'. *Zeitschrift für Physik* 50 (1928): 385–94.

Spencer, Percival William. 'Delay Device for Use in Transmission of Oscillations'. US patent 2,263,902 (A), filed 8 February 1938, and issued 25 November 1941.

Stephens, Carlene, and Maggie Dennis. 'Engineering Time: Inventing the Electronic Wristwatch'. *British Journal for the History of Science* 33 (2000): 477–97.

Stephens, Carlene E. 'Reinventing Accuracy: The First Quartz Clock of 1927'. In Johannes Graf (ed.), *Die Quarzrevolution: 75 Jahre Quarzuhr in Deutschland 1932–2007*, 12–23. Furtwanger Beiträge Zur Uhrengeschichte; N.F., Bd. 2. Furtwangen: Dt. Uhrenmuseum, 2008.

Stevenson, Randall. 'Mechanical and Electrical Clocks'. In Derek Howse (ed.), *Greenwich Time and the Discovery of the Longitude*. Oxford: Oxford University Press, 1980.

Straubel, Harald. 'Fundamental Crystal Control for Ultra-High Frequencies'. *QST* (April 1932), 10–13.

Sullivan, D. B. 'Time and Frequency Measurement at NIST: The First 100 Years'. In *Frequency Control Symposium and PDA Exhibition, 2001. Proceedings of the 2001 IEEE International*, 4–17, 2001.

Süsskind, Charles. 'Pierce, George Washington'. In *Dictionary of Scientific Biography*, 10: 604–5. Detroit: Charles Scribner's Sons, 1975.

Tebo, J. 'Lloyd Espenschied—Radio Imagineer'. *Communications Society* 10, no. 4 (September 1973): 3–6.

Thiessen, Arthur E. *A History of the General Radio Company, 1915–1965*. Concord, MA: General Radio Company, 1965.

Thrower, Keith R. *History of the British Radio Valve to 1940*. Hants: MMA International, 1992.

Topham, W. R. 'Warren Marrison—Pioneer of the Quartz Revolution'. *National Association of Watch and Clock Collectors Bulletin* 31 (April 1989): 126–34.

Tyne, Gerald F. G. *Saga of the Vacuum Tube*. Prompt Publications, 1977.

US National Bureau of Standards. *Standards Yearbook*. Washington, DC: US Governmentt Printing Office, 1927.

Valasek, Joseph. 'Piezo-Electric and Allied Phenomena in Rochelle Salt'. *Physical Review* 17 (1921): 475–81.

Vallauri, Giancarlo. 'Confronti fra misure di frequenza per mezzo di piezorisuonatori'. *L'Elettrotecnica* 14 (1927): 445–52.

Vallauri, Giancarlo. 'Confronti fra misure di frequenza per mezzo di piezorisuonatori'. *L'Elettrotecnica* 14 (1927): 682–4.

Van Dyke, K. S. 'The Electric Network of a Piezo-Electric Resonator'. *Physical Review* 25 (1925): 895.

Van Dyke, K. S. 'The Piezo-Electric Resonator and Its Equivalent Network'. *IRE Proceedings* 16 (1928): 742–64.

Varcoe, Ian. 'Scientists, Government and Organised Research in Great Britain 1914–16: The Early History of the DSIR'. *Minerva* 8 (1970): 192–216.

Vecchiacchi, Francesco. 'Applicazione all'oscillografo catodico della demoltiplicazione statica di frequenza'. *L'Elettrotecnica* 15 (1928): 805–14.

Vigoureux, Paul. *Quartz Resonators and Oscillators*. London: H.M. Stationery Office, 1931.

Vincenti, Walter G. *What Engineers Know and How They Know It: Analytical Studies from Aeronautical History*. Baltimore: John Hopkins University Press, 1990.

Vizgin, Vladimir, and Viktor Frenkel. 'Vsevolod Frederiks, Pioneer of Relativism and Liquid Crystal Physics'. In *Einstein Studies in Russia*, 149–80. Boston: Birkhäuser, 2002.

Voigt, Woldemar. 'Allgemeine Theorie der piëzo- und pyroelectrischen Erscheinungen an Krystallen'. *Göttingen Abhandlungen* 36 (1890): 1–99.

Voigt, Woldemar. *Lehrbuch Der Kristallphysik*. Leipzig: B. G. Teubner, 1910.

Vries, Marc J. de. *80 Years of Research at the Philips Natuurkundig Laboratorium (1914–1994): The Role of the Nat. Lab. at Philips*. Amsterdam: Amsterdam University Press, 2005.

Williams, Rosalind H. 'Lewis Mumford's Technics and Civilization'. *Technology and Culture* 43 (2002): 139–49.

Wills, Ian. 'Instrumentalizing Failure: Edison's Invention of the Carbon Microphone'. *Annals of Science* 64 (2007): 383–409.

Wilson, David. *Rutherford, Simple Genius*. Cambridge, MA: MIT Press, 1983.

Wise, M. Norton (ed.). *The Values of Precision*. Princeton, NJ: Princeton University Press, 1995.

Wittje, Roland. *The Age of Electroacoustics: Transforming Science and Sound*. Cambridge, MA: MIT Press, 2016.

Wood, A. B. 'Stephan Butterworth, OBE: An Appreciation'. *Journal of the Royal Naval Scientific Service* 1 (1945): 96–8.

Wood, A. B. 'From Board of Invention and Research to Royal Navy Scientific Service: Reminiscences of Underwater-Sound Research, 1915—1917'. *Sound: Its Uses and Control* 1, no. 3 (May 1962): 8–17.

Wood, A. B. 'Admiralty Experimental Station, Parkeston Quay (Harwich)1917 to 1919'. *Journal of the Royal Naval Scientific Service* 20, no. 20 (A. B. Wood, O.B.E., D.Sc. Memorial number) (July 1965): 27–41.

Wood, A. B. 'Admiralty Experimental Station, Shandon (Gareloch) Dunbartonshire'. *Journal of the Royal Naval Scientific Service* 20, no. 20 (A. B. Wood, O.B.E., D.Sc. Memorial number) (July 1965): 42–78.

Yeang, Chen-Pang. 'Characterizing Radio Channels: The Science and Technology of Propagation and Interference, 1900–1935'. PhD, Massachusetts Institute of Technology, 2004.

Yeang, Chen-Pang. *Probing the Sky with Radio Waves: From Wireless Technology to the Development of Atmospheric Science*. Chicago: University of Chicago Press, 2013.

Ziehm, Günther H. 'Kiel—Ein Frühes Zentrum Des Wasserschalls.' *Deutsche Hydrographische Zeitschrift, Ergänzungsheft Reihe B*, 29 (1988): 42.

Zimmerman, David. 'Paul Langevin and the Discovery of Active Sonar or Asdic'. *Northern Mariner* 12 (2002): 39–52.

Zimmerman, David. '"A More Creditable Way": The Discovery of Active Sonar, the Langevin–Chilowsky Patent Dispute and the Royal Commission on Awards to Inventors'. *War in History* 25 (2018): 48–68.

INDEX

For technical reasons, indexed terms that span two pages (e.g., 52–53) may, on occasion, appear on only one of those pages.

Tables are indicated by an italic *t* following the page/paragraph number.